D0079393

TRUST IN NUMBERS

TRUST IN NUMBERS

THE PURSUIT OF OBJECTIVITY IN SCIENCE AND PUBLIC LIFE

Theodore M. Porter

PRINCETON UNIVERSITY PRESS PRINCETON, NEW JERSEY

Published by Princeton University Press, 41 William Street,
Princeton, New Jersey 08540
In the United Kingdom: Princeton University Press,
Chichester, West Sussex

Library of Congress Cataloging-in-Publication Data

Porter, Theodore, 1953–
Trust in numbers : the pursuit of objectivity in science and public life /
Theodore M. Porter.
p. cm.
Includes bibliographical references and index.
ISBN 0-691-03776-0
1. Science—Social aspects. 2. Objectivity. I. Title.
Q175.5.P67 1995
306.4′5—dc20 94-21440

This book has been composed in Galliard

Princeton University Press books are printed
on acid-free paper and meet the guidelines
for permanence and durability of the Committee
on Production Guidelines for Book Longevity
of the Council on Library Resources

1 3 5 7 9 10 8 6 4 2

Contents

Preface

SCIENCE is commonly regarded these days with a mixture of admiration and fear. Until very recently, though, English-language historians of science were more likely to resent its pretensions than to fear its power. Here resentment grew out of reverence. Karl Popper and Alexandre Koyré, who gave form to brilliant traditions in the philosophy and history of science beginning especially in the 1950s, agreed that science was about ideas and theories. Koyré gave priority to thought experiments over the work of hands and instruments, and wondered, famously, if Galileo had ever performed any experiments at all. Popper allowed that experimentation could falsify theories, but held that the real work was done when the theory was adequately articulated. Experimenters had no more than to carry out what the theory dictated. Both praised science as a model of intellectual and philosophical achievement. Neither provided any reason for thinking that science could have much to do with technology. Still less could the hierarchical imagination of the historian or philosopher of science conceive that social science was authentically powerful.

This problem of the relations of science to technology inspired nothing like the heated (and, it now seems, empty and incoherent) controversy over the relative merits of "externalist" and "internalist" explanations of scientific change. Rather than arguing, much of the profession took for granted that science had the loosest connections with the practical world of engineering, production, and administration. In retrospect, I can see that my graduate training provided ample opportunity to form a more judicious view. My teachers learned earlier than I did to appreciate the limitations of seeing the scientific enterprise mainly as a pursuit of theory. Still, I think I was not unusual among historians of science of my generation in thinking that the widespread linking of science and technology or of science and administrative expertise involved something fundamentally spurious, that these supposed connections brought undeserved credit to each enterprise by making science seem more practical and its "applications" more intellectual than either really is.

A critique of this nature underlay my original formulation of this project. I planned to examine the history of neoclassical economics, the most mathematical of social science disciplines—indeed, possibly the most mathematical of all disciplines. Economics values most highly this supremely abstract mathematics, yet somehow economists sustain the image of a discipline capable of telling businesses and governments how to manage their affairs more effectively. I expected to show through an analysis of the relations of economics to policy that academic economics was a kind of sport, empty of implications for economic practice.

That is not the book I have written. It didn't take long to realize that neoclassical economics has had many critics who were better informed than I was likely to become. I found also that the economics discipline involves a greater variety of tools, aims, and practices than I had appreciated, and while I still think there is need for a more profound consideration of the relations between economic mathematics and the practices that support forecasting and policy advice, I am not the one to undertake it. In any case, my earlier suspicion that mathematics and policy were almost independent worked badly as a way of formulating a historical project. Its validity was even more damaging than its shortcomings. If, indeed, neoclassical mathematics is irrelevant to the economic world, my history of the relations between economics and policy would turn into the history of nothing at all.

So I have taken here a different tack. The interpenetration of science and technology, I now concede, is unmistakable, especially in the current century. That of social knowledge and social policy is only slightly less so. How are we to account for the prestige and power of quantitative methods in the modern world? The usual answer, given by apologists and critics alike, is that quantification became a desideratum of social and economic investigation as a result of its successes in the study of nature. I am not content with this answer. It is not quite empty, but it begs some crucial questions. Why should the kind of success achieved in the study of stars, molecules, or cells have come to seem an attractive model for research on human societies? And, indeed, how should we understand the near ubiquity of quantification in the sciences of nature? I intend this book to display the advantages of pointing the arrow of explanation in the opposite direction. When we begin to comprehend the overwhelming appeal of quantification in business, government, and social research, we will also have learned something new about its role in physical chemistry and ecology.

My approach here is to regard numbers, graphs, and formulas first of all as strategies of communication. They are intimately bound up with forms of community, and hence also with the social identity of the researchers. To argue this way does not imply that they have no validity in relation to the objects they describe, or that science could do just as well without them. The first assertion is plainly wrong, while the latter is absurd or meaningless. Yet only a very small proportion of the numbers and quantitative expressions loose in the world today make any pretense of embodying laws of nature, or even of providing complete and accurate descriptions of the external world. They are printed to convey results in a familiar, standardized form, or to explain how a piece of work was done in a way that can be understood far away. They conveniently summarize a multitude of complex events and transactions. Vernacular languages are also available for communication. What is special about the language of quantity?

My summary answer to this crucial question is that quantification is a technology of distance. The language of mathematics is highly structured and rule-bound. It exacts a severe discipline from its users, a discipline that is very nearly uniform over most of the globe. That discipline did not come automatically, and to some degree it is the aspiration to a severe discipline, especially in education, that has given shape to modern mathematics.[1] Also, the rigor and uniformity of quantitative technique often nearly disappear in relatively private or informal settings. In public and scientific uses, though, mathematics (even more, perhaps, than law) has long been almost synonymous with rigor and universality. Since the rules for collecting and manipulating numbers are widely shared, they can easily be transported across oceans and continents and used to coordinate activities or settle disputes. Perhaps most crucially, reliance on numbers and quantitative manipulation minimizes the need for intimate knowledge and personal trust. Quantification is well suited for communication that goes beyond the boundaries of locality and community. A highly disciplined discourse helps to produce knowledge independent of the particular people who make it.

This last phrase points to my working definition of objectivity. It is, from the philosophical standpoint, a weak definition. It implies nothing about truth to nature. It has more to do with the exclusion of judgment, the struggle against subjectivity. This impersonality has long been taken to be one of the hallmarks of science. My work broadly supports that identification and tends to the view that this, more than anything else, accounts for the authority of scientific pronouncements in contemporary political life. Once again, though, I am reluctant to make science the unmoved mover in this drive for objectivity. In science, as in political and administrative affairs, objectivity names a set of strategies for dealing with distance and distrust. If the laboratory, like the old-regime village, is the site of personal knowledge, the discipline, like the centralized state, depends on a more public form of knowing and communicating. Quantification is preeminent among the means by which science has been constructed as a global network rather than merely a collection of local research communities.

Some of the best and most fashionable recent work in science studies has aimed to understand science as a thoroughly local phenomenon. The genre of microhistory, which has enjoyed brilliant success in cultural history, has become influential also in the history of science. I have learned a great deal from this work, and I hope I have adequately appreciated its virtues. It provides a superb point of departure for studies of science, precisely because it renders the universality of scientific knowledge problematical. But it does not simply negate it. Science has, after all, been remarkably successful at pressing universal claims and gaining international acceptance. Explaining this achievement, and unpacking its implications, ought to be central problems of the history of science. The account I give here is mainly cultural and, broadly, political. I suggest that the problems of organization and communication faced by science are analogous to those of the modern political order. This is not meant to imply that science is not constrained in important ways by the properties of natural objects, nor even that the forms of language and practice I discuss are independent of those properties. I do not claim that quantification is nothing but a political solution to a political problem. But that is surely one of the things that it is, and our understanding of it is poor indeed if we do not relate it to the forms of community in which it flourishes.

The argument, as I have presented it so far, is as much sociological or even philosophical as historical. Since I am unlicensed in both the former domains, I tremble at the thought of writing a book that is not securely historical. The flow of topics and arguments in the book, however, is hard to reconcile with narrative or analytical history. Indeed, the book does not conform well to any established genre of scholarly writing. But there is, I like to think, some method to this madness. I should perhaps explain at the outset the pressures and strategies that have given shape to this study.

I began, as I have already explained, with the intention of studying the modern history of social quantification in relation to academic disciplines. Soon I found myself paying more attention to professions and bureaucracies. This research, much of it in primary sources, is presented in chapters 3 and 5–7, and is used in support of various arguments elsewhere. It is the heart of the book. These chapters attest to my allegiance to the standards of my own discipline, which requires general explanations to prove themselves in analytic narratives that respect the cultural richness of real historical situations. The other chapters are more general, even theoretical, and draw heavily on other scholarship. They appear here partly as conclusions from my properly historical material, but the more empirical chapters are not at all innocent of the perspective they present. On the contrary, I found that I needed to think through the issues with which they grapple before I was able to write the narrative sections.

As it appears here, the book is divided into three parts and nine chapters. The first part is about how numbers are made valid—that is, how they are standardized over wide areas. Chapter 1 is concerned with aspects of the natural sciences, chapter 2 with the social. Chapter 3 is about their relation, and argues that this practical quantifying activity has been at least as central to the identity and ethos of modern science as any aspiration to formulate broad theoretical truths. Chapter 4 discusses the forms of political order that permit or encourage quantification. It examines some of the moral and political issues raised by this drive to create rigorous quantitative rules in domains previously occupied by a more informal style of judgment.

The second part presents some notable attempts at social and economic quantification in an explicitly political and bureaucratic context. I argue that the transition from expert judgment to explicit decision criteria did not grow out of the attempts of powerful insiders to make better decisions, but rather emerged as a strategy of impersonality in response to their exposure to pressures from outside. Chapter 5 treats nineteenth-century British actuaries, who were able to resist these pressures, and twentieth-century American accountants, who were not. Chapters 6 and 7 support a similar but subtler contrast involving the use of the economic analysis of costs and benefits by nineteenth-century French engineers and twentieth-century American ones. While, as I urge in part 1, numbers and systems of quantification can be very powerful, the drive to supplant personal judgment by quantitative rules reflects weakness and vulnerability. I interpret it as a response to conditions of distrust attending the absence of a secure and autonomous community.

Part 3 undertakes to apply the perspectives developed for professions and bureaucracies in part 2 back to the academic disciplines. Chapter 8 assesses the bearing of bureaucratic cultures on science, then shows how inferential statistics became standard in medicine and psychology as a response to internal disciplinary weakness and external regulatory pressures. Finally, chapter 9 examines the moral economy of scientific communities. I argue there that the seemingly relentless push for objectivity and impersonality in science is not quite universal, and must be understood partly as an adaptation to institutional disunity and permeable disciplinary boundaries.

I make no pretense to having written a general history of quantification. I include very little before 1830, and almost nothing from outside of western Europe and North America. The geographical limitations are perhaps less forgivable than the temporal ones, and the history of colonialism, of international organizations, and of centrally planned economies all provide extremely rich materials for the history of quantification. I discuss frequently the best-established academic disciplines, but treat none of them in depth, preferring to concentrate on the role of quantification in applied fields such as accounting, insurance, official statistics, and cost-benefit analysis. Even within these constraints, I have been anything but exhaustive. Each of the topics just mentioned could form the subject matter for an entire historical subfield. So could many others that I have not discussed at all. Perhaps the highest ambition I can reasonably entertain for this book is that some of them will. If so, it may be possible in some decades to survey the field systematically. My main reason for discussing a range of topics and countries rather than writing a monograph on one is to suggest something of the potential richness of the field. This strategy presupposes another of my central goals: to convince readers that the history of quantitative objectivity is after all a potential subject of inquiry, and not simply a miscellany.

The last thing I would want, though, would be for this topic to become a new, autonomous specialty. One of the really heartening developments in history of science in the last decade or so has been the breakdown of its isolation. It brings me no small satisfaction that history of statistics has been noticed and increasingly is being studied in academic units devoted to literature, philosophy, sociology, psychology, law, social history, and various of the natural sciences, as well as in history of science and statistics itself. I am even more hopeful for the history of quantification as it bears on the cultural study of objectivity. Indeed, there is already a considerable literature, most of it very recent, that relates directly to the questions I ask in this book. So far there is nothing like a single discussion, but rather a variety of local conversations, largely isolated by discipline. I think the barriers are breaking down, and hope that this book will help to level a few sections of the wall(s). I have drawn freely and extensively on several bodies of scholarly literature, mainly because they are indispensable to my argument, but also in the hope that those who have contributed to or learned to appreciate one of them will find themselves unexpectedly in an integrated neighborhood—and like it.

Acknowledgments

SINCE so much of this book is a synthesis of other people's work, the text and notes themselves must stand in for a proper expression of my obligations. Nor is there space to acknowledge individually all those friends and antagonists who asked provocative questions or made helpful comments in response to earlier presentations and publications of ideas that have found their way into this book. I thank Ayval Ramati for expert research assistance, and David Hoyt for help in preparing the manuscript. I have benefited from comments on the entire text by Lorraine Daston, Ayval Ramati, Margaret Schabas, Mary Terrall, and Norton Wise, and on specific chapters by Lenard Berlanstein, Charles Gillispie, and Martin Reuss.

This research has been supported most generously during its excessively long gestation by a number of foundations and other sources of research funds: the Earhart Foundation, the Sesquicentennial Fund and summer faculty fellowship fund of the University of Virginia, the Thomas Jefferson Memorial Foundation, the Academic Senate of the University of California, Los Angeles, the National Endowment for the Humanities, the John Simon Guggenheim Memorial Foundation, and National Science Foundation grant DIR 90–21707. For access to archival materials I gratefully acknowledge the Archives Nationales, the Bibliothèque Nationale and the Bibliothèque de l'Ecole Nationale des Ponts et Chaussées in Paris, and the Bibliothèque de l'Ecole Polytechnique in Lozère, France; the National Archives in Washington, D.C., in Suitland, Maryland, in Laguna Niguel, California, and in San Bruno, California; the Water Resources Library at the University of California, Berkeley; and the Office of History, Army Corps of Engineers, Fort Belvoir, Virginia.

Finally, it is a pleasure to acknowledge several debts of a more personal nature. Diane Campbell and I spent a decade trying to find two academic jobs in the same place. This became rather more desperate for me when I reluctantly left my position at the University of Virginia to follow her to a new job in biology at the University of California at Irvine. During all these years I drew heavily on friends and colleagues for support and encouragement, and more tangibly for letters and telephone calls. I will always be grateful. Astonishingly, it ended well—a job offer in 1991 provided a fine solution (plus or minus sixty miles) to this problem of academic geography. Finally I want to thank my parents, Clinton and Shirley Porter, my wife, Diane Campbell, and my son, David Campbell Porter, for their love and patience.

University of California, Los Angeles
March 1994

TRUST IN NUMBERS

INTRODUCTION

Cultures of Objectivity

"Whatever logic is good enough to tell me is worth WRITING DOWN," said the Tortoise. "So enter it in your book please."
(*Lewis Carroll*, "What the Tortoise Said to Achilles," *Mind*, 1895)

"OBJECTIVITY" arouses the passions as few other words can. Its presence is evidently required for basic justice, honest government, and true knowledge. But an excess of it crushes individual subjects, demeans minority cultures, devalues artistic creativity, and discredits genuine democratic political participation. Notwithstanding such criticism, its resonance is overwhelmingly positive. Attacks are rarely directed at true objectivity, but rather at pretenders who use it to mask their own dishonesty, or perhaps the falseness and injustice of a whole culture. Most often it is not closely defined, but simply invoked to praise or blame. In the United States, scientists, engineers, and judges are generally presumed to be objective. Politicians, lawyers, and salesmen are not.

There remains the delicate question of what these attributions of objectivity mean. It is not merely an all-purpose honorific, for it applies more readily to the despised bureaucrat than to the indispensable entrepreneur. It has, however, several distinct senses, which tend to reinforce the positive associations of the term and at the same time to obscure it. Its etymology suggests an acquaintance with objects. Paradoxically, to us, until the eighteenth century these were usually objects of consciousness rather than physical things; real entities existing outside of us were called subjects. But in current philosophical usage, objectivity is very nearly synonymous with realism, while "subjective" refers to ideas and beliefs that exist only in the mind. When philosophers speak of the objectivity of science, they generally mean its ability to know things as they really are.[1]

An earlier generation, the positivists, considered such claims merely metaphysical, and hence meaningless. But they did not disdain using the term. There are other ways of construing the objectivity of science. The most influential has defined it by an ability to reach consensus. Normally it suffices if that consensus holds within a specialist disciplinary community. We might, with Allan Megill, call this "disciplinary objectivity," by contrast to the "absolute objectivity" of the preceding paragraph. This form of objectivity is not self-subsistent. Its acceptability to those outside a discipline depends on certain presumptions, which are rarely articulated except under severe challenge. Specialists who claim objectivity should provide some evidence of their expertise. They should comport themselves appropriately. They should appear reasonably disinterested, or at least should not expect to speak authoritatively where their own individual or professional interests are at stake. We trust physicists to tell us about phase transitions in supercooled helium, but we are more skeptical if they appear as paid expert witnesses in court, or when they tell of the great economic advantages that will attend the construction of a superconducting supercollider.

Still, physicists control a large territory on which they are not called upon by outsiders to justify their conclusions. Disciplinary objectivity is made conspicuous mainly by its absence. Where a consensus of experts is hard to reach, or where it does not satisfy outsiders, mechanical objectivity comes into its own. Mechanical objectivity has been a favorite of positivist philosophers, and it has a powerful appeal to the wider public. It implies personal restraint. It means following the rules. Rules are a check on subjectivity: they should make it impossible for personal biases or preferences to affect the outcome of an investigation. Following rules may or may not be a good strategy for seeking truth. But it is a poor rhetorician who dwells on the difference. Better to speak grandly of a rigorous method, enforced by disciplinary peers, canceling the biases of the knower and leading ineluctably to valid conclusions.

The tension between the disciplinary and the mechanical senses of objectivity is a central concern of this book. But these two senses will not be discussed only on the terrain of science, and so it is important to consider also the meanings of objectivity in explicitly moral and political discourse. In most contexts, objectivity means fairness and impartiality. Someone who "isn't objective" has allowed prejudice or self-interest to distort a judgment. The credibility of courts depends on an ability to elude such charges. They do so in large part by placing disputants in a highly controlled situation and authorizing independent judges and jurors to resolve the facts and apply the law. The objectivity of jurors means little more than their presumed disinterestedness, since by definition they lack special expertise. Judges too are expected to be impartial, though they should also be trained professionals. Their expertise must include an ability to follow the rules—mechanical objectivity—but there is no avoiding the judicious exercise of discretion.

Two of the three meanings discussed in Kent Greenawalt's *Law and Objectivity* pertain directly to objectivity as fairness. "Legal determinacy" refers to the ability of any lawyer or other intelligent person to reach the same conclusions about what the law means. It does not require that existing law be morally defensible, but only that different judges will apply the law to most cases in the same way. So defined, this kind of objectivity is not the preserve of disciplinary insiders, though it may be that only those who have immersed themselves in the culture of law can attain this consistency of judgment. Greenawalt observes, next, that treating people impersonally according to "objective standards" is central to what we call the rule of law. This generally entails a rigid schedule of punishments for various criminal acts, and a minimum of opportunity for discretionary adjustments based on subjective inferences about character and intentions. Both these senses of objectivity imply that rules should rule, that professional as well as personal judgment should be held in check. They point to the alliance of objectivity as an ideal of knowing and objectivity as a moral value.[2]

It is important to understand that mechanical objectivity can never be purely mechanical. Greenawalt offers as an example the simple instruction, spoken by a manager as a subordinate enters her office: "Please shut the door." It requires some experience of the world, and perhaps also of the office in question, to know which door, and when; to judge whether to mention first some reasons why it should remain open; and also to understand that if the company president suddenly appears at the door, the directive should be put aside. Rarely does any of this need to be spelled out, at least within one culture. Similar questions, including some much harder ones, will arise in filing papers, keeping accounts, taking a census, or preparing a graph. Especially in law, philosophy, and finance, where clever people make a business of exploiting ambiguities, much of what would otherwise go without saying ends up having to be said.

Mathematical and quantitative reasoning are especially valued under these circumstances. They provide no panacea. Mapping the mathematics onto the world is always difficult and problematical. Critics of quantification in the natural sciences as well as in social and humanistic fields have often felt that reliance on numbers simply evades the deep and important issues. Even where this is so, an objective method may be esteemed more highly than a profound one. Any domain of quantified knowledge, like any domain of experimental knowledge, is in a sense artificial. But reality is constructed from artifice. By now, a vast array of quantitative methods is available to scientists, scholars, managers, and bureaucrats. These have become extraordinarily flexible, so that almost any issue can be formulated in this language. Once put in place, they permit reasoning to become more uniform, and in this sense more rigorous. Even at their weakest point—the contact between numbers and the world—methods of measurement and counting are often either highly rule-bound or officially sanctioned. Rival measures are thereby placed at a great disadvantage. The methods of processing and analyzing numerical information are now well developed and sometimes almost completely explicit. Once the numbers are in hand, results can often be generated by mechanical methods. Nowadays this is usually done by computer.[3]

The growing role of quantitative expertise in the making of public decisions is a development well known to scholars. Yet we have no satisfactory histories of it. This is due mainly to a failure to integrate two rival views of the development of quantitative methods, and of expertise generally. One narrative treats their history as the progressive accumulation of truer, or at least more powerful, methods. The other reduces them to ideology, to be explained mainly in terms of social structures of domination, though with due regard to the often nefarious aims of their individual purveyors. These are the arguments of partisans, who for the moment have forgotten the value of nuance. But it is not merely moderation that is called for. Expertise, much more even than science, is not understandable as simply the result of solitary thinking and experimenting, or even of the dynamics of a disciplinary community. It is a relation between professionals—often academic scientists or social scientists—and public officials. Their appreciation for expertise, in turn, reflects their relationship to a still wider public. To understand the circumstances under which quantitative objectivity has come into demand, we need to look not only at the intellectual formation of experts, but even more importantly at the social basis of authority.

We now have a few studies that have taken this insight as their point of departure. One argument, particularly influential among American historians, holds that the social science of the 1890s and 1900s arose from a new sense of interdependence among Americans, and ultimately from the social and economic processes that produced that interdependence.[4] There is doubtless something to this, even if a world economy did not abruptly form in the late nineteenth century. But the form of expertise that arose in specific response to this sense of interdependence is not the most important kind, and it is not at all characteristic of public uses of social science. It amounts, in Thomas Haskell's account, to a philosophical understanding of human interdependence, providing the consolation of explanation to a bewildered public. In fact there were a variety of rival forms of explanation of the industrialized social world, not all of them consoling, and most coming from preachers or labor organizers rather than professors. Academic social scientists have had only the most modest success in forming public opinion. The principal audience for their expertise is a bureaucratic one, usually with the acquiescence of elected officials.[5] The public culture licenses academic specialists not to issue general pronouncements, but to assemble very specific findings.

To be sure, this is not the only kind of expertise. There is a kind of wisdom that comes from long experience, which often is passed on from parent to child or master to disciple. In modern times, personal experience and contact with a master have increasingly been supplemented or replaced by formal instruction at a university or other educational institution. There the ineffable skill of the craft or guild is, so far as possible, made formal and explicit, and thus the secrets of the trade are deemphasized. To citizens of large-scale democratic societies, this is more acceptable because it is more open and less personal. Nevertheless, expert knowledge is almost by definition possessed by only a few, and no such art is ever reduced to a handful of rules that can be looked up and mastered by anyone with a textbook. Thus the intuition or judgment of specialists continues to command a degree of respect, even if the doctor, for example, cannot explain exactly why the problem must be in the liver. Still, both physicians and patients have learned not to be satisfied with an opinion based on little more than intuition. Better to apply an instrument, to take a culture, to produce some specific evidence.

In public even more than in private affairs, expertise has more and more become inseparable from objectivity. Indeed, to recur to the previous example, it is in part because the relation of physician to patient is no longer a private one—due to the threat that it might be opened up in a courtroom—that instruments have become central to almost every aspect of medical practice. In public affairs, reliance on nothing more than seasoned judgment seems undemocratic, unless that judgment comes from a distinguished commission that can be interpreted as giving representation to the various interests. Ideally, expertise should be mechanized and objectified. It should be grounded in specific techniques sanctioned by a body of specialists. Then mere judgment, with all its gaps and idiosyncrasies, seems almost to disappear.

This ideal of mechanical objectivity, knowledge based completely on explicit rules, is never fully attainable. Even with regard to purely scientific matters, the importance of tacit knowledge is now widely recognized.[6] In efforts to solve problems posed from outside the scientific community, informed intuition is all the more crucial. The public rhetoric of scientific expertise, however, studiously ignores this aspect of science. Objectivity derives not mainly from the wisdom acquired through a long career, but from the application of sanctioned methods, or perhaps the mythical, unitary "scientific method," to presumably neutral facts. There should be no room for the biases of the researcher to corrupt the results. It is, of course, possible for investigators or officials to be impartial as a result of their inherent fairmindedness, or perhaps their utter indifference to the outcome, but how can we know? In a political culture that idealizes the rule of law, it seems bad policy to rely on mere judgment, however seasoned.

This is why a faith in objectivity tends to be associated with political democracy, or at least with systems in which bureaucratic actors are highly vulnerable to outsiders.[7] The capacity to yield predictions or policy recommendations that seem to be vindicated by subsequent experience doubtless counts in favor of a method or procedure, but quantitative estimates sometimes are given considerable weight even when nobody defends their validity with real conviction.[8] The appeal of numbers is especially compelling to bureaucratic officials who lack the mandate of a popular election, or divine right. Arbitrariness and bias are the most usual grounds upon which such officials are criticized. A decision made by the numbers (or by explicit rules of some other sort) has at least the appearance of being fair and impersonal. Scientific objectivity thus provides an answer to a moral demand for impartiality and fairness. Quantification is a way of making decisions without seeming to decide. Objectivity lends authority to officials who have very little of their own.

Part I

POWER IN NUMBERS

Now it must here be understood that ink is the great missive
weapon, in all battles of the learned, which, conveyed
through a sort of engine, called a quill, infinite Numbers of
these are darted at the enemy, by the valiant on each
side, with equal skill and violence, as if it were an
engagement of porcupines.
*(Jonathan Swift, "The Battle . . . between
the Ancient and Modern Books," 1710)*

CHAPTER ONE

A World of Artifice

I thought it was the task of the natural sciences to discover the facts of nature, not to create them.
(Erwin Chargaff, 1963)

MAKING KNOWLEDGE IMPERSONAL

The credibility of numbers, or indeed of knowledge in any form, is a social and moral problem. This has not yet been adequately appreciated. Since the 1970s, debates about objectivity between philosophical and sociological camps have been polarized mainly over the question of realism. The claim that science is socially constructed has too often been read as an attack on its validity or truth. I consider this a mistake, as well as a diversion from more important issues. Perhaps there is something to be accomplished by arguing whether science can get at the real nature of things. But the answer can scarcely be peculiar to science, unless we are to suppose that systematic research is incapable in principle of identifying real entities, even though we can do so as if by instinct in our everyday lives. I find this and the opposite doctrine equally implausible. This book does not presuppose and will not defend any position on the much-vexed philosophical issue of realism.

If a declaration of faith is called for at the outset, I would say that interested human actors make science, but they cannot make it however they choose. They are constrained, though not absolutely, by what can be seen in nature or can be made to happen in the laboratory. Experimental interventions, guided but not dominated by theoretical claims, have often been remarkably effective. There remain subtle questions about what should count as truth. I am content to invoke Ian Hacking's modest but elegant formulation, "It is no metaphysics that makes the word 'true' so handy, but wit, whose soul is brevity."[1] Let us suppose for the sake of argument that scientific investigation is able to yield true knowledge about objects and processes in the world. It must nonetheless do so through social processes. There is no other way.

To accept this point is only to fix the terms for discussing a problem, not to solve one. Through what specific social processes is scientific knowledge made? How wide a circle of inquirers and judges is involved in the process of deciding what is true? The standard view has long held that in mature sciences, the truth is worked out or negotiated by a community of disciplinary specialists whose institutions are strong enough to screen out social ideologies and political demands. I will try to show toward the end of the book that the effectiveness of this segregation has been exaggerated—that the sciences have been compelled to redefine their proper domain in order to monopolize it, and that much of what passes for scientific method is a contrivance of weak communities, partly in response to the vulnerability of science to pressures from outside. But for the moment it is enough to think about processes of constructing knowledge that are internal to disciplines.

According to the individualist form of rhetoric about science, still much used for certain purposes, discoveries are made in laboratories. They are the product of inspired patience, of skilled hands and an inquiring but unbiased mind. Moreover, they speak for themselves, or at least they speak too powerfully and too insistently for prejudiced humans to silence them. It would be wrong to suppose that such beliefs are not sincerely held, yet almost nobody thinks they can provide a basis for action in public contexts. Any scientist who announces a so-called discovery at a press conference without first permitting expert reviewers to examine his or her claims is automatically castigated as a publicity seeker. The norms of scientific communication presuppose that nature does not speak unambiguously, and that knowledge isn't knowledge unless it has been authorized by disciplinary specialists. A scientific truth has little standing until it becomes a collective product. What happens in somebody's laboratory is only one stage in its construction.

In recent times, peer review has achieved an almost mythical status as a mark of scientific respectability.[2] It rivals statistical inference as the preeminent mechanism for certifying a finding as impersonal and, in that important sense, objective. It is by no means sufficient in itself to establish the validity and importance of a claim, however. Indeed, it is a mistake to speak as if the validity of truth claims were the principal outcome of experimental researches. Experimental success is reflected in the instruments and methods as well as the factual assumptions of other laboratories. Day-to-day science is at least as much about the transmission of skills and practices as about the establishment of theoretical doctrines.[3] Experimental truth claims depend above all on the ability of researchers in other laboratories to produce results sufficiently similar, and to be convinced that the similarity is indeed sufficient.

Just how this transmission of skills, practices, and beliefs takes place is among the crucial issues in contemporary studies of science. Significantly, the problem has arisen in the context of the new interest in laboratories and experiments. Already in the 1950s, Michael Polanyi argued that science involved a crucial element of "tacit knowledge," knowledge that could not be articulated or reduced to rules. In practice, this meant that books and journal articles must necessarily be inadequate vehicles for the communication of such knowledge, since what matters most cannot be conveyed by words. Following his reasoning, the crucial institution for the transmission of science is an apprenticeship undertaken by a student with a master scientist.[4]

To argue this way is to diminish the importance of the published paper or textbook, to locate knowledge first of all in the laboratory and not the library. It is to doubt the universality of science, to confine it to particular spaces. In principle, of course, the barriers around those spaces are easily breached. Nature, we suppose, is uniform: another researcher carrying out the same procedures, even on another continent or in another century, should obtain the same results. Such a principle, though, counts for little unless it can be instantiated in practices. In practice, replication is anything but easy. This insight has been developed most fully by Harry Collins, who considers that independent replication is effectively impossible. Those who try to build their own copy of a new instrument or experimental setup, on the basis of printed information alone, normally fail. Detailed reports and private communications make it easier to reproduce an experiment, but also compromise any claims to independence. The usual way of learning to use a new instrument or technique is to experience it directly. This, argues Collins in a case study that is now widely regarded as paradigmatic, is the only way that the TEA laser was ever reproduced.[5] He may exaggerate the point, but this is a phenomenon that practicing scientists have long understood. Ernest Lawrence warned in the 1930s, for example, that it would be foolhardy to attempt to build a cyclotron without sending someone to work with one in his Berkeley laboratory. "It is rather ticklish in operation," he explained, "and a certain amount of experience is necessary to get it to work properly."[6]

This line of argument may have important implications for our understanding of claims to scientific truth. If experimental setups are really so ticklish, and the phenomena so difficult to produce reliably; if experimental findings are almost never independently replicated, but instead are always reproduced using instruments that have been calibrated against the original: then experimental regularities should perhaps be interpreted in terms of human skill rather than of stable underlying entities and the operation of general laws of nature. Or if these alternatives are not incompatible, then at least the problem of transporting skills beyond the confines of a single laboratory must be seen as a critical one. Without such communication there could be nothing like objectivity, since every laboratory would have its own science. To recur again to Polanyi's language, science would be nothing but "personal knowledge."

Polanyi himself didn't think that it was: "Whenever connoisseurship is found operating within science or technology we may assume that it persists only because it has not been possible to replace it by a measurable grading. For a measurement has the advantage of greater objectivity, as shown by the fact that measurements give consistent results in the hands of observers all over the world."[7] Here, though, he attributed to the very nature of measurement what had in fact been accomplished within certain domains through heroic efforts. The construction of measurement systems that could claim general validity was not simply a matter of patience and care, but equally of organization and discipline. Administrative achievements of this kind lie at the heart of most experimental and observational knowledge. Mathematics and logic were less intractable from this standpoint.

Theoretical reasoning is of course not beyond criticism. It is, for example, vulnerable to the charge that it has been spun out by a fevered brain, and bears no relation to any actual world. On the other hand, it adapts very nicely to the printed page, which in retrospect seems its natural medium. Thus it can be communicated far more easily than anything depending on special experience. And rigorous deduction can almost compel assent. In the extreme case of pure mathematics, those who accept, even as useful fictions, the axioms, should be led ineluctably to the conclusions. To be sure, mathematized theory in science is rarely so pellucid or so rigorous that its significance and bearing can be grasped immediately by distant readers. Appreciation of this kind of science, too, is easier for those who share an intellectual community with the author. As Polanyi observed, even inference according to formulas remains an art: "There exist rules which are useful only within the operation of our personal knowing." Collins argues similarly about mathematical deduction and artificial intelligence.[8] Still, distance is much less of an obstacle for purely theoretical sciences than for sciences of experience, and the problem of reproduction is correspondingly attenuated. Little wonder that the term "science," meaning demonstrated knowledge, was applied to logic, theology, and astronomy long before there were communities of experimental researchers.[9]

In the seventeenth century, experimentation was still associated with practices like alchemy, with all its connotations of mystery and secrecy.[10] How was this private knowledge transformed into fit material for a culture of objectivity? The historical literature has only just begun to deal with this question. Sociologists have taken it more seriously. At least two lines of response are being developed. One focuses on how experimental results, which can normally be witnessed by only a few people, came to be accepted as truthful by nearly everyone. This was above all a triumph of rhetoric—of what I call here technologies of trust—and also of discipline. Parts 1 and 3 of this book are centrally concerned with these issues, though not mainly with respect to laboratories.

The other broad explanation for the objectification of experiment emphasizes the spread of laboratory practices. Independent replication may be rare, but the reproduction of methods is not. By the eighteenth century, experimental knowledge had to a large degree come to be defined in terms of potential reproducibility. Seventeenth-century experimental philosophers, such as Robert Boyle, exhibited a great fondness for the odd happening, whose intractability was taken as testimony to the advantage of experience over vain theorizing. But singular events provided a poor basis for making communities of researchers, since those who were not present could do little with them but hope they had been faithfully reported. Lorraine Daston instances Charles Dufay, a French researcher of the 1720s and 1730s, to epitomize a different experimental ideal. Whereas Boyle was famously prolix, Dufay was austere, informing his readers only of what was essential for producing an effect. And he considered that the effect should not be reported until it had been brought under good experimental control.[11] Such practices enhanced the lawlikeness of nature, since well-behaved laboratory phenomena would thereafter have a more secure ontological status than mere events. They also promoted a spirit of public knowledge, at least within the specialist community, since close laboratory control offered the best chance for reproducing work at other sites.

Still, the obstacles to the replication even of what seem to us the most basic of experiments, like Newton's separation of colors using a prism, could be formidable.[12] Personal contact, often involving extended visits to other laboratories, was and remains invaluable for the sharing of methods and results. Boyle's contemporaries seized every opportunity to view his air pump in operation, and to witness the results he claimed to have produced.[13] In our own time the spread of instruments and techniques through direct contact has been institutionalized in a variety of ways. Most involve brief or lengthy visits. Those who want to master a new instrument or technique travel to a laboratory where it is already working if they are young, or import a graduate student or postdoctoral researcher from such a laboratory if they are well established. Knowledge, then, does not diffuse uniformly outward from the place of its discovery. It travels along networks to new nodes, and what appears as universal validity is in practice a triumph of social cloning.[14]

In the early life of a new technique, when it is still on the cutting edge, personal contact will most often be crucial for its spread to other laboratories. Indeed, this may be just what "cutting edge" means in experimental science. But experiments that succeed, again perhaps by definition, will not long remain in the domain of intricate craft skill and personal apprenticeship. The air pump may again be taken as emblematic. Boyle required the most prodigious efforts of glass-blowing and the most adept handlers of leather and sealing wax, as well as a large personal fortune, to build a pump that worked some of the time. Already in Boyle's day, though, there were shops specializing in scientific instruments, and they soon added air pumps to their repertoire. Any air pumps that were incapable of producing the experimental phenomena associated with a vacuum would be sold at first to unhappy customers, and then not at all. As the pumps were improved and standardized, the phenomena became more easily reproducible.[15] In recent times, such technologies have proliferated. Not only have instruments been standardized; nature has too. Chemists buy purified reagents from catalogs—and they would be quite helpless if they had to extract them from the soil. Cancer researchers depend on patented strains of mice and would not know how to interpret results derived from ordinary field mice.

The growth of science has to a large degree involved the replacement of nature by human technologies. Ian Hacking has made this insight into the basis for an important general book on philosophy of science. Experiments succeed, he observes, when they permit the reliable manipulation of objects. At least some of these objects, such as lasers, may never exist outside the laboratory. Most or all cannot be found in anything like a pure form, except when they are created by human interventions. But as these artificial or purified objects come to be more reliably manipulated, they begin to be incorporated into other experiments, and perhaps also into processes outside the laboratory. This is perhaps the most crucial sense in which laboratories are self-vindicating.[16]

Bruno Latour argues that science is now inseparable from technology, and uses the term "technoscience" to symbolize their merger. Both, he suggests, aim to construct black boxes, artificial entities that are treated as units and that nobody is able to take apart. The black boxes of the scientist may be laws or causal claims as well as material technologies, but these depend on instruments and reagents for their production, just as instruments cannot be built, operated, or interpreted without the benefit of scientific knowledge. Our interventions have become too powerful for us to talk usefully any more about science in terms of learning what happens in nature, independent of human activity. Every scientific claim succeeds by mobilizing a network of allies: reagents, microbes, instruments, citations, and people. If the network is strong, a new fact is created. It is an artifact, but it is nonetheless real, for it can be enlisted in the networks that support new facts. The progress of experimental science is the increasing ability to make and use new things, and at the same time to transform the world that science purports to describe. Latour also affirms Elie Zahar's argument that the success of mathematics in scientific theory "is not a miracle but the result of an arduous process of mutual adjustment."[17]

This adjustment even extends beyond theories and experiments to the scientists themselves. The "self-vindicating laboratory" depends also on an appropriate selection of people, and the exclusion of those who refuse to accept its discipline. For example in psychology, as Liam Hudson explains, "tough" experimentalists disdain humanists, though they prefer not to admit it.

> If cornered, they point to the unfortunate fact that, among psychologists, it is the weaker students who specialize in the more humane branches: those with lower seconds, young ladies with an interest in people. It follows, the tough point out with evident regret, that standards are lower in the more humane fields. The argument is a tricky one to combat, especially as its prophecies are self-fulfilling. As teachers and examiners, the tough-minded are in a position to give their own assumptions weight. With minds as open as any can be, they design courses and set papers that favour candidates whose style of intelligence suits them to experimental research. They thus operate a self-perpetuating social system.

The corresponding argument seems at first less credible for the most prestigious natural sciences, but this is only because they have no clinical or humanistic branches. Or rather, those branches have been expelled from the domain of science, and are found now only within such genres as nature writing, poetry, and environmental activism. But social selection, including a gendered dimension in physics and biology at least as strong as in psychology, provides an important part of the explanation for the distinctive character of modern science as a form of knowledge and practice.[18]

QUANTIFICATION AND POSITIVISM

Numbers, too, create new things and transform the meanings of old ones. This is especially significant in the human sciences, as the next chapter will undertake to show. But measurement activities were central in forming some of the most basic ideas of the physical sciences as well. Less than three centuries ago temperature was a medical concept, useful for describing the atmosphere in much the way that temperament was invoked to characterize the human body. Experimental physicists created a narrower and more operational concept of temperature. They did it with very little input from theory; the idea that heat is motion and that temperature is a measure of mean molecular energies was not developed until the late nineteenth century. The standard view in the late eighteenth century was that heat might be motion or it might be a substance, and that measurement could go forward in either case. The mercury thermometer at least rose when things became hotter, and fell when they were cooled. Liquids of different temperatures could be mixed to learn about mean degrees of heat. Promiscuous measurement, informed by a few simple analogies, gave birth to quantitative concepts such as "heat capacity" and "latent heat." The phenomena, it seemed, could be described with as much precision as in mechanics.[19]

It should be observed that this infatuation with measuring led to the neutralization of concepts as well as their creation. Temperature had less human meaning after the experimental physicists laid hold of it. Diderot, in his more romantic moods, complained of the alienation from nature implied by mathematics. In the 1830s, the Hegelian natural philosopher Georg Friedrich Pohl compared Georg Simon Ohm's mathematical treatment of the electrical circuit to a travel book that ignored a charming landscape and its inhabitants in favor of recording precisely the times of arrival and departure of trains.[20]

The late-eighteenth-century quantifiers of experimental natural philosophy were quite prepared to sacrifice rich concepts in order to promote rigor and clarity. This, indeed, had been advocated explicitly in the influential philosophy of Etienne Bonnot de Condillac. Condillac was a nominalist. He saw no reason to hanker after an understanding of the true nature of things, nor even to suppose that things have true natures. In a world without fixed types, humans are free to impose on nature whatever order best serves their purposes. Condillac admired rigorous classification. He also favored thoroughgoing quantification. He considered algebra the model language, since it permitted reasoning from known to unknown quantities. This did not mean finding mathematical laws of natural philosophy, but rather, as Charles Gillispie remarks, balancing the accounts.[21] Measurement and even mathematization were often favored as evasions of theory: it was not necessary to choose between substance and motion theories of heat, or to find the correct force law pertaining to capillary action. Lavoisier and Laplace, for example, offered the quantitative results of experiments using their ice calorimeter as data upon which researchers of diverse theoretical persuasions could readily agree.[22]

Max Horkheimer and Theodor Adorno complained in their *Dialectic of Enlightenment* that positivist science replaces "the concept with the formula, and causation by rule and probability."[23] Of course mathematics has not always been allied to the positivist retreat from causal understanding that so vexed the Frankfurt critics. In practice, as Nancy Cartwright argues, it is impossible even to set up a statistical analysis without assuming some explanatory structure.[24] In theoretical writings, currents of mathematical realism, tending sometimes to geometrical or numerological mysticism, have run through science since Pythagoras. But the view of mathematics as mere description has been no less influential. This provided much of the rationale for keeping mathematical astronomy in its place, as against the higher, causal disciplines of (Aristotelian) physics and theology, in Renaissance universities. The Catholic Church attempted in the same way to neutralize Galileo's Copernicanism. Often, scientists have adopted this rhetoric to protect themselves. Newton, unable to find a satisfactory mechanism for the forces he posited, inveighed against mere hypotheses like Descartes' ether. Quantifiers occupied with measurement rather than the formulation of mathematical laws have often found the language of descriptionism especially appealing.

At first glance it seems a humble, self-effacing language, and there is no doubt that it could serve that function. John Heilbron, who has written most incisively on descriptionism as a cultural phenomenon, attributes its popularity among physicists toward the end of the nineteenth century to their need not to offend the higher powers in lands still dominated by the traditional estates of aristocracy and church.[25] But Uriah Heep was humble too. Metaphysical modesty brought compensating advantages. Positivist philosophers and working scientists have not been shy about seizing them.

Not least among these advantages was the compatibility of positivism with the pursuit of control over nature. Something of this was already presupposed by the low status of Renaissance mathematicians, who were considered to be technicians and tradesmen rather than genuine seekers of truth.[26] In more modern times this hierarchy has been flattened, or even reversed, and experimental domination has become itself an accepted form of knowledge. In the natural sciences, Ernst Mach's positivism was especially influential among experimenters. Biologists such as Jacques Loeb and a host of admirers treated "nature" the way B. F. Skinner treated the mind. It was at best unknowable, and perhaps merely a metaphysical conceit. If the rat runs through the maze, or the experimental trial yields consistent results, we know all that we can.[27]

Rigorous certainty was another virtue identified with a mode of science that did not long for deep understanding. Partly in response to the profusion of representations of electricity in the later nineteenth century, many physicists retreated to purely mathematical descriptions of the phenomena. Perhaps the most influential were Gustav Kirchhoff and

Heinrich Hertz, each of whom wrote general treatises in almost purely mathematical form. They aimed to provide rigorous descriptions of observable phenomena, descriptions that permitted deduction without introducing any causal hypotheses. Hertz, for example, constructed his mechanics without invoking force, which seemed a doubtful entity. Forces could be quite adequately replaced in the equations by accelerations. By giving up a pretended acquaintance with causes and mechanisms, he hoped, physics could gain an almost timeless validity.

Descriptionism, or perhaps we should say positivism, had a third and perhaps still more important advantage. Since it presupposed nothing about the real causes operating, it was very nearly neutral as to subject matter. Not by accident has positivism become almost a synonym for scientism. Auguste Comte, its founder, wanted to characterize science in a way that would apply as well to sociology as to astronomy, without in any way reducing one to the other. More than a century later, the Vienna Circle positivists left as their last testament the revealingly named *Encyclopedia of Unified Science.* Toward the turn of the century, Ernst Mach and his allies argued repeatedly that a philosophy of science could not be valid if it applied only to physics. For him, positivism weakened the hold of materialism, clearing the way for a psychophysics that unified physics and psychology by joining mind and matter.[28]

The resonance of the positivist mania for quantification with vast social ambitions for science is exemplified best of all by the career of Karl Pearson. From the early 1890s until his death more than forty years later, Pearson harnessed his prodigious talents to the development of a statistical method and its application to biological and social questions. He was practically the founder of mathematical statistics, and he believed firmly that it provided the proper discipline to reasoning in almost every area of human activity. This included government and administration, which for too long had been in the hands of scientifically illiterate gentlemen and aristocrats.

Pearson, though English, acquired a lasting affinity for German culture during his student years. His positivism, like Mach's, arose from antimaterialism. His was a world not of real objects, but of perceptions. The proper goal of science was to put them in order. Nature in itself had no definitive form. It did not follow, however, that what we call knowledge is arbitrary or merely personal. Nature, or rather our understanding of it, was to be ordered by method. This could apply as well to the social and biological realms as to the physical. "The field of science is unlimited; its material is endless; every group of natural phenomena, every stage of past or present development is material for science. *The unity of all science consists alone in its method, not in its material.*" That method consisted in "the careful and laborious classification of facts, in the comparison of their relationships and sequences, and finally in the discovery by the disciplined imagination of a brief statement or *formula*, which in a few words resumes a wide range of facts. Such a formula . . . is termed a scientific law."[29]

Nature was not quite passive in the face of scientific investigation. Though Pearson doubted the usefulness of talking about an independently existing world, he did invoke "normal" perceptive faculties to explain how the sciences could achieve consensus. Such faculties, he understood, were given by nature—that is, by natural selection. Nature also presented phenomena to perception. But we could never gain access to entities or causes. It might be reasonable to speak of force, for example, but only as "a convenient measure of motion, not its cause." Atoms and molecules are "conceptions" that may usefully "reduce the complexity of our description of phenomena." Their status was roughly the same as that of "geometrical conceptions," such as the circle, which is no more than a limit of perceptual experience. Their validity was in every case defined by their usefulness, which might even vary from one situation to another. For this reason, Pearson saw nothing objectionable in the use of apparently contradictory expressions by different disciplines.[30]

What he liked best, though, was not modeling, but an austerely quantitative description and analysis. Here there were no inconsistencies among disciplines, but a coherent set of concepts that could be applied universally. Preeminent among these were the tools of statistics, mental constructs that could readily be mapped onto the world. Nowhere do we find perfect lawlikeness, he stressed. Everywhere we find correlations. That is, even in mechanics there is always some unexplained variation. This should cause us no distress. The possibility of science depends only in the most general way on the nature of the phenomena being investigated. A correlation, after all, is not a deep truth about the world, but a convenient way of summarizing experience. Pearson's conception of science was more a social than a natural philosophy. The key to science he found not in the world, but in an ordered method of investigation. For Pearson, scientific knowledge depended on a correct approach, and this meant, first of all, the taming of human subjectivity.[31]

STANDARDIZING MEASURES

One may object that Pearson's philosophy has more to do with administering the world than understanding it. But the bureaucratic imposition of uniform standards and measures has been indispensable for the metamorphosis of local skills into generally valid scientific knowledge. Science as we know it depends on the administration of nature, a stunning social achievement. Pearson captured brilliantly the spirit behind much quantifying activity, whether bureaucratic or scientific. His philosophy applies especially well to the campaigns to standardize measures. We may take the rectangular land survey of the United States as exemplary. The surveyors could not quite ignore the curvature of the earth, but that was the only concession they made to nature. Watersheds and mountains were no obstacle to the imposition of a uniform grid over the land.[32]

This does not mean that quantification is inherently opposed to nature. The uniform grid and its equivalents are not the only form that quantified knowledge can take. The land surveyors were quite capable of charting the positions of rivers and using contour lines to depict landforms in detail. A land surface can be described quantitatively in an infinite variety of ways. But a square grid has usually been preferred by central governments on account of its greater simplicity. A highly organized labor force was required to produce one, but once in place it permitted land claims to be registered and enforced from hundreds of miles away, with a bare minimum of judgment or local knowledge.

Social measurement, as Otis Dudley Duncan has observed, is rarely simply imposed from outside. Instead, quantification is implicit "in the social process itself, before any social scientist intrudes."[33] Natural measurement, in contrast, is apparently imposed from outside. Yet it too may appropriately be regarded as implicit in a social process, the social process of exploiting and investigating nature. This, certainly, went on long before any people we would recognize as natural scientists began to intrude. Yet there is something fundamentally misleading in posing the issue this way. Of course there was measurement, but of what kind? Scientists, both social and natural, fundamentally altered these social processes. What they brought was a kind of objectivity—measurement that aspired to independence from local customs and local knowledge. In this they were allied to the centralizing state and to large-scale economic institutions. Almost the same problem of separating knowledge from its local context is faced in the political, economic, and scientific spheres.

It would be hard to say whether keeping time means social or natural measurement. Until a few centuries ago, social time was suffused with the natural. Time by sundials was divided first of all into day and night. Each part lasted twelve hours. The boundary between them was marked off by the rising and setting of the sun. In terms of the homogeneous time now in effect, daytime hours lasted longer in summer than in winter. This was entirely appropriate, since the working day also lasted longer in summer than in winter. The identification of time with natural cycles was even more pronounced for calendrical than for diurnal time. To every thing there was a season: planting, flooding, weeding, mowing, grazing, sending the animals up to the mountain pastures. For nomadic peoples the seasonal cycles were still more elaborate: a time to go to the woods to hunt deer, to the meadows to pick berries, to the rivers to fish for spawning salmon, to the estuaries to catch migrating birds. The positions of the sun and the stars, or a tabulation of days, helped in identifying these times, but there were other, biological signs to temper the inflexibility of the heavens.[34]

The demand for a more rigid and predictable calendar was created by administrative needs of church and state, for whom there was a time to pay taxes, a time to report for military service, and a time to observe Lent or celebrate Easter. Clock time, too, acquired religious significance, and the punctual observance of matins in monasteries was among the first incentives for living by the clock.[35] Industrialized work relations had a more pervasive influence, and ever since the beginnings of industrialization the clock has been among the principal agencies of discipline in factories, schools, and offices. Its growing sovereignty necessarily came at the expense of natural, diurnal rhythms of light and darkness, warmth and cold. It was, in short, part of an artificial regime, the technological, economic, and social conquest of time. By the late nineteenth century, with the spread of rail networks, it even began to seem desirable to impose uniform hours on wide swathes of land running from north to south. A bit later, against strong opposition from farmers and others still residually committed to natural cycles, governments first declared that time should be moved forward every spring and set back every fall.[36]

Similar considerations apply to measures of length, weight, and volume. These are physical measures, but they are social measures as well, and like most social measures they long predated any concern with science. It is scarcely possible to imagine an economy of markets and trading without prices and measures, and hence without extensive quantification. Since many of the units were anthropomorphic in origin, we can identify a move away from nature in the gradual shift toward arbitrary units. But it matters only a little whether a measuring system is based on the foot and the pound or the meter and the kilogram. The really important shift was toward standardization and interconvertibility. The culture of quantification has changed radically in the last three centuries, and this has involved the intrusion of scientists as well as bureaucrats.[37]

In our own time, measurement means nothing if not precision and objectivity. Our ideal exchange is an impersonal one. Consumers rarely lay eyes on the owner or maker of the items they purchase; traders and brokers may never even see the goods in which they deal. An important element of personal trust goes into some of these transactions, but they depend even more on faith in impersonal technological and regulatory mechanisms to assure that scales give good weight and that boxes are honestly labeled. Volume measures, which are more difficult to control than weights, have become almost obsolete except for liquids. Few of us have ever imagined that there could be disagreement about what constitutes a pound of butter or a hectare of land. Scientific laboratories accept without comment or special scrutiny instrumental readings in nanoseconds, milligrams, and angstroms.

In old-regime societies, by contrast, measurement was always a matter for negotiation. Not quite everything was negotiable. Witold Kula remarks that town halls in eighteenth-century Europe were likely to display a bushel vessel, valid for that region. If anybody questioned the accuracy of any particular bushel, its contents could be poured into the official one to see if they were equal. But this was by no means the end of the matter. Everybody knew that grain could be packed more densely by pouring it from a greater height, and for certain purposes the method of filling might be specified in contracts or by law. Most crucially, there was the matter of the heap on top of the bushel vessel. Even flattened bushels would contain variable amounts depending on whether the strickle was applied with or without pressure. There was always room for power, negotiation, and fraud in determining the size of the heap.

This system of discretionary measures could work rather well in the right circumstances. Grain had a just price, and the flexibility of measures provided room to keep the system functioning. For example, since wheat was esteemed more highly than oats, it would generally be exchanged in flattened measures, whereas oats were sold in heaped ones. A suitable heap might be negotiated for wheat as well if it were dirty, chaffy, or musty. The practice of merchants, who preserved the just price by buying in heaped bushels and selling at the same rate in flattened ones, was indispensable to their livelihood. Kula mentions that Polish land measures often varied by soil quality, so that a unit of land would represent more or less equal productive value. This unit was often defined as the territory upon which a certain quantity of seed could properly be sown. If a dispute arose, it would be resolved by calling in "the most honest and experienced sower, who could be trusted to be right to within a gallon."[38] Without such honest mediators, the system could scarcely function. But in a regime of trust, these discretionary measures could be far more useful than some result of indiscriminate objectivity produced by a surveyor.

We should not suppose that we have here a happy *Gemeinschaft*, in which trust was universal and abuse impossible. Measures could be an important source of dispute and resentment as well, especially in transactions between unequals. Kula observes that this discretionary system of measurement was intimately tied to a regime based on social privilege rather than uniform law. Noble seigneurs almost always received their rents and feudal dues in heaped bushels. The more enterprising ones would periodically introduce a new bushel vessel. Even if it had the same interior volume as its predecessor, it might be made lower and flatter, so that it could support a larger heap. Peasants did not fail to notice these changes, or perhaps on occasion to imagine them, but they lacked the social power to complain effectively. When, during the early stages of the French Revolution, they were given the opportunity to compose *cahiers de doléance*, measures were among the most frequently mentioned grievances. The local bushel, they said, had been growing ever larger, to the profit of seigneurs. It was time to declare a single, true bushel, valid for the whole of France.

Kula concludes that in the preindustrial world, the qualitative was always dominant over the quantitative. The regime of discretion and negotiation clearly favored local interests over central powers, as was universally recognized. The privileging of judgment over objectivity in measures was only the tip of the iceberg. Every region, sometimes every village, had its own measures. Kula notes that in old Silesia, "newly-enfranchised towns would determine their own bushels as a symbol of liberty and sovereignty."[39] Indeed it was more than a symbol, since it complicated administration and tax collection by higher authorities. Even the government of a relatively centralized state like France faced innumerable jurisdictions with their own measures. Moreover, there were different units for different materials or substances. Silk would be exchanged in different measures from linen, and milk from wine. None of the measures were decimalized. Neither was coinage. The arithmetic could be so complicated that even local merchants would be pressed to the limit of their skills working with the rule of three. Converting from the units of one region to another generally required the assistance of masters of reckoning, thereby supporting most of the mathematicians in early modern Europe.[40] This was at least an inconvenience, if not an obstacle, to the growth of large-scale trading networks, and the expansion of capitalism was one important source of the impetus to unify and simplify measures.

The other, of course, was the state—sometimes collaborating with large industrial or commercial interests, and sometimes acting for its own reasons. Standard measures and uniform classifications were at least as useful for centralized governmental activity as for large-scale commerce and manufactures. English measures had achieved a fair degree of standardization before the eighteenth century, but the French Revolution was the signal event for the creation of uniform measures on the Continent. Kula, who links metrological to juridical equality, observes that political revolutions brought the metric system to Russia and China as well. Precise, uniform measures helped to move the economy away from an order based on privilege into the domain of law. They also enhanced administrative control over matters of taxation and economic development. At the same time, an impressive display of state power was required to enact the new system in the first place. In France, it took more than forty years. Since nobody knew what liters and kilograms were, the state had to begin by expressing them in local units. The first scheme cooked up by the authorities was to gather up all the local measures and send them to Paris to be converted into metric equivalents. This would truly have made Paris a center of calculation. But it was quite unworkable.

It was especially difficult because of resistance from the provinces. Liters and kilograms were not what French peasants had prayed for as they drew up their *cahiers de doléance.* For the metric system was not designed for peasants. It did not bring back the true bushel, but discarded the bushel in favor of a system of wholly unfamiliar quantities and names, most of them drawn from an alien dead language. The institutionalization of the metric system involved special difficulties because of the aspiration to universalism that helped to give it form. This universalism was consistent with the ideology of the revolution, and more particularly with the ideology of empire. It was also nicely consonant with the ideals of scientists, who after all designed it. The new units were given Greek names, just as Lavoisier and his collaborators made up Greek names for the new elements of chemistry.

More impressively, the designers of the metric system aspired to a wholly cosmopolitan frame of reference for their measures. The really egregious instance of this is the meter, which was defined as one 10,000th part of the distance from the pole to the equator. This, said the committee of scientists that first proposed it, was a natural unit, independent of every nation. It seems to exemplify a typically scientific aspiration to perfect objectivity, like Max Planck's admiration for constants of nature that are wholly separate from every human trait and interest, and hence must be equally valid even for nonhumans.[41] This definition of the meter, though, was also a response to a more local political uncertainty. Most French scientists preferred a unit defined as the length of a pendulum that beats out seconds. But there was a distinct possibility that time also would be decimalized, and it appeared unwise to define the meter in terms of a thing so fleeting as the second.[42]

The extreme unworldliness of the earth-based meter was not essential for constructing a rationalistic system of measures. But the collaboration of science with the state in the definition of the metric system reflects a certain commonness of interest. Each, in its way, aspired to the rule of law. The validity of law was not supposed to depend on intimate knowledge or personal contact, but should be effective over great distances and enforceable by strangers. Not surprisingly, the involvement of scientists in the setting of standards has become even more crucial since the 1790s. In some ways, the high point of this activity came in the late nineteenth century in the setting of electrical standards, which involved research scientists of the very highest rank.[43] A new phase in this relationship was inaugurated by the creation in 1871 of the first real bureau of standards, the Physikalische-Technische Reichsanstalt in Berlin, with Hermann von Helmholtz as its founding director.[44] There has been little evidence of disharmony between the interests of science and those of the state or of large industries. Peter Lundgreen remarks: "The alliance of scientific neutrality and public authority brings about a very persuasive tool for settling or at least diminishing conflicts." He quotes Ulysses Grant, who appealed unsuccessfully to Congress in 1877 for governmental testing of materials: "These experiments cannot be properly conducted by private firms, not only on account of the expense, but because the results must rest upon the authority of disinterested persons. . . ."[45] Bureaus of standards normally involve the collaboration of science, government, and industry.

Public bureaus are not the only place where measurement procedures are established and coordinated. Trade groups perform the same function for particular industries. Scientists have often been able to achieve uniformity without calling on a centralized government agency. But it necessarily involved active intervention. As Latour argues, all measures "construct a commensurability that did not exist before their own calibration." The drawing of weather maps giving air pressure data exemplifies the difficulties. By the end of the nineteenth century there was already a network of observatories covering most of Europe. The instrumental readings could be assembled almost immediately by means of telegraph. In principle, everybody was measuring the same quantities. But instruments and practices remained discrepant, and it was enormously difficult to coordinate them. For years, as the Norwegian Vilhelm Bjerknes complained, the failure of coordination appeared on most weather maps in the form of a wholly artifactual cyclone over Strasbourg. Evidently the Strasbourg observatory produced systematically lower pressure readings than most others. Coordinating the observatories was as great an achievement as defining a theoretical framework by which to analyze their output.[46]

Still, it pales before some of the tasks faced by modern public bureaus of standards. Their job is to provide officials at every level of government with specifications and tolerances for all kinds of measures. These have some value for pure scientific research, but their main purpose lies at the intersection of science and regulation. An especially important one these days relates to the control of air, water, and ground pollution. In order to regulate potentially harmful substances, there must be prescribed ways to measure them. J. S. Hunter writes of the United States National Bureau of Standards: "We have now reached the stage where there is a federally mandated method for measuring almost every physical, chemical, or biological phenomenon."[47] The reason for mandates, of course, is not mainly to protect against fraud in science, though having an officially sanctioned measurement protocol will often be useful for scientists. It is to prevent economic agents, such as polluters, from choosing a method of measurement in order to present themselves in the most favorable light. It has been officially estimated that all this measuring absorbs about 6 percent of the gross national product of the United States. Hunter laments that nearly all the measures remain deeply inadequate despite all these resources and all these specifications. For regulatory purposes even more than for scientific ones, the measures have no value unless they are reasonably standardized. It has proved overwhelmingly difficult to get farms, laboratories, factories, and retailers to report the quantities of the myriad substances they discharge in the same form following the same measurement protocol.

To measure for public purposes is rarely so simple as to apply a meter stick casually to an object. Hunter speaks grandly but appropriately of "measurement systems." In the case of waste discharges, he proposes, an adequate measurement system must include criteria for (1) choice of samples; (2) manipulation and preservation of samples; (3) control of analytical reagents; (4) methods of measurement, including the calibration of instruments; (5) custody of samples; (6) methods of recording, manipulating, and recording data; (7) training of personnel; and (8) control of interlaboratory bias. Adequate measurement, clearly, means disciplining people as well as standardizing instruments and processes. Until this has been achieved, measurements will be unreliable. So long as inconsistencies remain, the discharges measured cannot be effectively quantified, no matter how many numbers have been gathered. Indeed, specifying them is not enough; the specifications must be put into effect at millions of diverse locations, by calibrating millions of instruments and millions of people to the same standard.

Even if all this could be accomplished, one still might have doubts. Hunter does not worry openly about whether we know the true amount of a given substance discharged. The more pressing and practical problem is to assure that everyone is measuring and reporting their discharges the same way. Then at least we can reasonably talk of adequate quantification. Then it is possible to combine and manipulate data—for example, to add together all reports along a given river as a measure of the total emission of some substance into it. Accommodating variation in measurement practices is almost impossible. If an eccentric but conscientious manufacturer were to invest extra resources and hire a particularly resourceful chemist to perform the analysis with great care using the newest research methods, this would be viewed by the regulators as a vexing source of interlaboratory bias and potentially of fraud—not a welcome improvement in accuracy. There is a strong incentive to prefer precise and standardizable measures to highly accurate ones. For most purposes, accuracy is meaningless if the same operations and measurements cannot be performed at other sites. This is especially true, and especially urgent, where the results of research are to be put to work outside the scientific community.

BIOLOGICAL STANDARDIZATION

In no other field must high-level research results be put to work at so many sites as in medicine. The relationship of research to practice has become important mainly in the last century. It was made possible in part by subjecting physicians to an intensely academic training in the relevant sciences before they could be licensed to practice. This would accomplish very little, though, if clinicians did not have access to diagnostic tests and images, producing information identical in form to that in the research laboratories. Therapeutics is no less dependent on the standardization of drugs. Many thousands of pharmacists working mainly with plant-based substances could not possibly provide uniform medicines. Even the big pharmaceutical companies of the late nineteenth century found that drugs were highly variable in different batches. Around 1900, the principal role of science in the pharmaceutical industry was not the development of new medicines, but testing and standardizing.[48]

The most important methods of standardization were chemical. The isolation of active ingredients permitted the synthesis of drugs, which removed or greatly lessened the problem of natural variability. A significant class of medicines, though, resisted chemical isolation. These provided the subject matter, early in the present century, for a new and emphatically international discipline of "biological standardization." The basic idea here was to test drugs suspected of high natural variability on animals, and measure their effects. Dosages could then be modified depending on whether the lot in question proved relatively strong or weak.

The centralizing implications of this project were resisted by pharmacists as a threat to their autonomy. Their job description, after all, included the performance of chemical tests on drugs, and biological assays seemed not so very complicated in principle. In 1910, two Americans explained a method for testing digitalis "so simple that it may be mastered by the retail pharmacist, and conducted with the apparatus which he has at hand." This meant no fancy physiological measurements. The "progressive pharmacist" need only test each harvest of leaves by determining the minimal fatal dose per kilogram of cat. This should be called the "cat unit." Cats are easy to use, the authors explained, and their deaths "do not affect the sentimental portion of the community to the same extent that the employment of dogs does." They display also an "extraordinary uniformity" of response.

Or so it seemed at first. A footnote, perhaps added in page proofs, warned of cats recently found to tolerate 50 percent extra, so that the reliability of the method would now require "a somewhat larger number of observations."[49] The sentimental portion of the community may not have been delighted by this. And there were other problems. Digitalis extracted from foxglove was found to have several active components. Doctors resisted the simplification of the drug, preferring the ineffable advantage of a union of constituents. It seemed that potential test animals were sensitive to different active ingredients. Already the "frog unit" had fallen into disrepute because frogs tolerated the drug differently in summer and winter, and because they were often killed by its effect on their nerves rather than their heart. By 1931 there were more than seven hundred papers on the quantitative testing of digitalis, involving a variety of animals. Joshua H. Burn, one of the leaders of the field, remarked in 1930 that biological assay "remains a subject for amusement or despair, rather than for satisfaction or self-respect. We have cat units, rabbit units, rat units, mouse units, dog units, and, latest addition of all, pigeon units. The field of tame laboratory animals having been nearly exhausted, it remains for the bolder spirits to discover methods in which a lion or elephant unit may be described."[50]

These disagreements, often tinged by national pride, about the categories of laboratory guinea pig (Paul Ehrlich's favorite sacrificial animal) may not have much inconvenienced the progressive pharmacist testing digitalis. The evidence of variability within species, and the consequent need to test drugs on many animals was a more serious problem. In practice, biological standardization was one of the forces leading to the consolidation of a pharmaceutical industry, and to a redefinition of the art of the pharmacist. Large companies had the resources to hire scientific personnel to conduct the necessary tests.[51] Still, researchers and governments aspired to something better than conventional units varying by manufacturer, even if these could be presumed reliable. Scientists worked to defeat the variability of nature by breeding well-standardized laboratory animals. But this was unlikely to succeed when they couldn't even agree on the best species to use in testing a drug. The most promising course of action was to form a set of standards, like the platinum meter, against which all drugs of each type would be tested. This entailed monumental feats of organization, and eventually required the collaboration of national governments and international organizations.

Diphtheria antitoxin provided the exemplar. Paul Ehrlich, working in the last years of the nineteenth century, found that while diphtheria toxin was unstable, the antitoxin could be maintained in a dry state. He compared other samples of antitoxin with the standard one by testing both in identical systems against toxins from a single source. His prestige as the discoverer sufficed to make his antitoxin the standard against which others should be compared. Ehrlich maintained the standard by sending out samples of his antitoxin to researchers who wanted it.

During the First World War, the German materials were no longer available, and one of the satellite samples, in Washington, D.C., became for a time the international standard. In 1921, the League of Nations convened a conference to compare this with Ehrlich's standard and to learn whether it had varied. Satisfied that it had not, the conference fixed it as "the international unit for diphtheria antitoxin." The next year, 1922, it established one for tetanus antitoxin. Many others followed, including digitalis, whose standard was constituted as an average out of a mixture of leaves from different places. The League set up a Permanent Commission on Biological Standardisation in 1924. It gave custody of serum standards to the State Serum Institute in Copenhagen, and of all others to the National Institute of Medical Research in London.[52]

The standardization of insulin provides a good illustration of the system in action. The Toronto researchers who discovered it initially defined a unit as the dose required to produce a certain degree of hypoglycemia in rabbits weighing two kilograms. But, as was pointed out by the leading British researcher on biological standardization, Henry H. Dale, such a unit could not "maintain the requisite uniformity when determined in different institutions in a number of different countries, on animals kept under different conditions." So an international conference decreed that preparing insulin in a dry and stable form was the best way of "defining and stabilising the unit." "The standard preparation would then serve as a convenient currency, by means of which the unit could be transmitted to every country concerned." Indeed, they sent one-tenth of a gram to "some responsible organisation in each country," or at least each country that was deemed to have a responsible organization. The scientists of every nation could then conduct their own comparisons as they thought best. The official conference publication nevertheless included articles describing in detail the two existing methods: measuring blood sugar levels in rabbits, and inducing convulsions in mice.[53]

The League of Nations, and later the World Health Organization of the United Nations, developed an elaborate system for maintaining and diffusing the standards. A. A. Miles explained in 1951 how this worked. Most standards were dried, sealed, surrounded with an inert gas such as nitrogen, and kept in the dark at −10°C. From time to time they would be brought out of the vault and compared with samples closer to the scene of action. Unfortunately, the standards slowly wore out, and the really difficult job was confirming their stability. Animal response could not be the official standard, since "the animals themselves cannot be specified precisely." It remained, Miles explained, the "hidden standard." "[W]ithin one laboratory, where the workers are familiar with their animal stock and its breeding and feeding, and are continuously performing a certain type of assay, their combined experience of the standard, though it is largely incommunicable, constitutes a valuable check on its potency."[54]

"The adoption of stable standards of this kind brings the estimation of biological properties into the same position as the measurement of length and weight," explained J. H. Burn in his handbook of biological standardization.[55] He conceded, though, that the problems were greater. Indeed, heroic efforts were required to extend the benefits of standardization to technology, regulation, medicine, and society itself. Important as standardization has been to well-established sciences without close ties to applications, they can accomplish much without it. The organization of science into disciplinary and subdisciplinary communities promotes extensive sharing of personal knowledge. Also, the self-interest of scientists is less likely to provide an incentive to deception, so rules and standards need not be defined so rigorously. In the anonymous and multifarious world of medicine, industry, agriculture, and regulation, informal working methods are almost impossible to harmonize. Unambiguous rules, supported by regular surveillance, are correspondingly more important.

Still, these are differences of degree, not of kind. Whatever validity scientific laws and measures may claim with respect to the external world, this has never been enough to make them operationally valid across boundaries of culture, language, and experience. What we call the uniformity of nature is in practice a triumph of human organization—of regulation, education, manufacturing, and method. Numbers, too, had to be made valid, but they have also proven indispensable in advancing this project. Karl Pearson was neither the first nor the last to worship quantification, which he regarded as integral to scientific method. Its appeal has been the appeal of impersonality, discipline, and rules. Out of such materials, science has fashioned a world.

CHAPTER TWO

How Social Numbers Are Made Valid

Mathematics . . . is a machine that . . . can think for us; we derive as much advantage from its service as from machines in industry that work for us.
(Jules Dupuit, 1844)

DISCIPLINE AND VALIDITY

The Latin root of *validity* means "power." Power must be exercised in a variety of ways to make measurements and tallies valid. Nobody seriously doubts that phosphorus, say, exists in some real quantity in any given discharge of waste water. But it requires a massive exercise of social power to establish valid measures of such discharges. This involves not only a disciplined labor force, but also good public relations. If manufacturers or environmentalists think the measurement process is unreliable or, worse, biased, it may well break down. If the most accurate methods are too expensive, inferior ones may become standard. To use the best methods in some particular case will then raise suspicions, or at least will present problems of interpretation in relation to sites that use the conventional methods. None of these uncertainties depend on any doubts about the facts of the matter. More than one solution is possible because more than one measurement regime is possible, and this means that there is a range of potentially valid measures.

An example from public statistics reveals what is at stake. In principle, the population of a country is a relatively unproblematical number. But it is not fully determined by the distribution of bodies over a landscape. First a decision must be reached about how to count tourists, legal and illegal aliens, military personnel, and persons with more than one residence or multiple citizenship. Even after these issues are resolved, population numbers will depend on the methods specified for getting them. In the United States, there have been lively controversies about whether to incorporate the Census Bureau's own estimate of its undercount into the official numbers. Since the undercount is assumed to affect particularly the inner-city homeless, these estimates are anything but politically neutral. For the 1990 census, the secretary of commerce decided not to use them, on the ground or pretext that such adjustments can never be sufficiently objective.

But of course the enumeration itself is only made objective by specifying in detail what efforts will be made to locate and tally people who reside at new addresses, or who can never be found at home, or who have no fixed residence. Any method that works systematically to the disadvantage of specific jurisdictions or racial and ethnic categories is certain to be contested, since the apportionment of political power and of federal revenues depends on the numbers. The census bureau is so vulnerable to outside criticism that it cannot rely on professional judgment in defiance of politics. Population measures have so far proved too sensitive for ad hoc corrections to gain the acceptance that could make them valid.[1]

Equally crucial in determining modes of quantification are the forms of expertise and power relations within a work force. The differences between public opinion polls and academic surveys of attitude are instructive. Both strategies of inquiry were worked out mainly in interwar United States. Opinion polls enforced a strict discipline on employees and respondents. Having learned that logically equivalent forms of the same question produce quite different distributions of responses, pollsters used rigid standardization to minimize this source of variation. Their employees were instructed to recite each question with exactly the same wording and in a specified order to all subjects, who were required to choose one of a small number of packaged statements as the best expression of their opinions. In contrast, academic studies of "attitude" generally encouraged the interviewer to rephrase questions and to vary their order, and allowed subjects to respond in their own words. The researchers hoped in this way to make certain that the question was correctly understood and that the response was a genuine expression of beliefs or feelings.

This reflected a different conception of the subject matter: the academics were not content to collect what they took to be superficial expressions of opinion. It naturally required some probing to get at the deeper level of commitments and beliefs that would permit the researchers to give explanations of behavior. These divergent interview styles were also closely tied to different forms of social organization. The academic researchers performed much of the work themselves, or used the labor of graduate students who could be trained to exercise their discretion in prescribed ways. Opinion polling, in contrast, involved numerous, large-scale studies conducted by poorly paid assistants, such as housewives, who were not initiated into the arcana of the craft. Their judgment was not to be relied upon, and the relatively rigid and objective form of the multiple-choice questionnaire was for that reason de rigueur.[2] Strict rules are almost indispensable unless those gathering the numbers are themselves very well socialized in the craft. As Jacques Bertillon remarked in 1903, in relation to the extraordinary problem of gathering international statistics on causes of death, it is always better in cases of difficulty to have clear standards rather than to depend on judgment. "Whatever solution is adopted, it is preferable that this solution be uniform." The point was expressed still more epigramatically in 1978 by two researchers on the coding of death certificates: "Comparable statistics cannot be obtained if everyone does what he or she thinks is correct."[3]

In more extreme cases, the available forms of social organization might determine whether you could count at all. A complete census of a large population requires sophisticated bureaucratic structures, which few states possessed before the nineteenth century. The French relied on a form of sampling and probabilistic calculation during the eighteenth century to estimate their population.[4] The first four censuses in Britain, from 1801 to 1831, were conducted through the Anglican Church. A particularly interesting and ambitious attempt at a census, discussed in an admirable book by Marie-Noëlle Bourguet, was carried out in France in the year 9 of the Republic (1800–1801). This was a time of relatively benign politics when the incessant wars of the revolutionary period were at least in remission. The Bureau de Statistique, operating to a large degree on its own initiative, was dominated by men who conceived the project in terms of promoting liberal government. They hoped that by gathering up and disseminating great masses of information about all the regions of France, they could promote national unity and an informed citizenry. They also wanted to know whether France was flourishing under republican government. They sent out questionnaires to the prefects in each département asking for a wealth of information, most of it quantitative. They wanted to know the population, of course, but they also requested detailed information about the economy. What was the land area, how much of it was arable, and how much in vineyards, orchards, meadows, and forests? They asked about domestic animals: how many cows, goats, and sheep were there in the region, and of what breeds, and how much milk, wool, leather, and meat did they produce? They wanted the population divided up by occupation, property holdings, and wealth, though certainly not according to the distinctions of status that had prevailed before the revolution.

The prefects, newly installed and badly overworked, were baffled and overwhelmed by these demands. They had been told to fill out a table that was several pages long, and they commanded nothing like the bureaucracy that would have been necessary to do so. So they looked for assistance to local scholars and notables, worthy citizens whose families had been in the area for a long time and who prided themselves on their intimate sense of the traditions, customs, and produce of their regions. The fruit of their investigations, in those départements where it yielded any result at all, was a collection of monographs, full of helpful information about the character of the landscape and its people, their dress, habits, customs and festivals, produce and manufactures. The scholars were not ideologically opposed to numbers, and where information could be obtained about births, marriages, or exports, the reports might pass this on. But these elite volunteers were unlikely to travel from household to household asking dozens of probing questions about the inhabitants and their wealth and production. Even if they had wanted to, there were not enough such scholars to survey more than a small fraction of the population. And even if the information could somehow have been collected, neither the prefects nor the Bureau de Statistique itself commanded the resources to digest it.

We can see in the relations between the statisticians and the local notables a collision of cultures. The Bureau de Statistique wanted a kind of information that only a large and disciplined bureaucracy could have provided. The authors of the reports were *savants* and *érudits* who nourished a quite different ideal of knowing. They were not to be converted into automated agents of other people's investigations. A third culture entered forcefully a few years later, in the form of the emperor and general, Napoleon Bonaparte. The liberal aims of the statisticians meant nothing to him. He wanted specific, focused information for purposes of conscription, requisitions, taxes, and wartime management of the economy. The Bureau de Statistique was unable to supply what he demanded, and eventually in 1811 he shut it down.

These administrative and political difficulties point to a more general obstacle faced by French statisticians in 1800. France was not yet capable of being reduced to statistics. Lack of centralization and bureaucratic administration made it impossible to discipline a labor force, but it also meant that many aspects of the French nation could not be described in statistical form. Revolutionary France remained, in important ways, an old-regime society. Of course the population could be counted, though in a highly stratified society it was unclear to most people that anything very useful was accomplished by tallying up such a diverse lot of beings. The task of classifying people was particularly thorny. It was hard to keep the ranks and orders that the revolution had officially abolished out of the reports. And the Bureau de Statistique quickly learned that no single set of categories could be adequate to the whole of France. J.A.C. de Chaptal, recognizing this, sent out a circular inviting local authorities to introduce new categories into the tables where necessary. Bourguet points out, though, that this was a damaging concession, for it meant "recognizing the existence of a diverse, local reality, irreducible to the categories of a national accounting."[5] The trades were often fluid, and in any case they varied from region to region. Hierarchies of labor, professions, and administration were both unsettled and diverse. The local scholars, it seems, were right in preferring a verbal, descriptive statistics to a uniform, rigidly quantitative one. Complex and sensitive to regional differences, their work was for this very reason poorly adapted to the demands of a centralized administration. An adequate statistics for bureaucratic purposes had to await the remaking of the country.

RULES AND INTERVENTIONS

A few decades later, Balzac considered that France had been reconstructed according to the requirements of the statisticians. "Society isolates everyone, the better to dominate them, divides everything to weaken it. It reigns over the units, over numerical figures piled up like grains of wheat in a heap."[6] Since this move toward individualism was not merely the result of "society," but also of the growing administrative power of the state, the statistical enterprise was, to a degree, self-vindicating. Indeed, the concept of society was itself in part a statistical construct. The regularities of crime and suicide announced in early investigations of "moral statistics" could evidently not be attributed to the individual. So they became properties instead of "society," and from 1830 until the end of the century they were widely considered to be the best evidence for its real existence.[7]

The creative power of statistics is not limited to such global entities as society. Every category has the potential to become a new thing. The tables for marriage revealed that each year a small number of men in their twenties married septuagenarian women. Here was a phenomenon that could be investigated. The curious statistician could compare the rates in different countries, or according to religious faith and inheritance laws, in order to understand this aspect of social life. A more commonplace statistical entity, to us, is a crime rate. There were, of course, crimes before the statisticians occupied this territory, but it may be doubted whether there were crime rates. Similarly, people sometimes found themselves or people they met to be out of work before this had become a statistical phenomenon. The invention of crime rates in the 1830s and of unemployment rates around 1900 hinted at a different sort of phenomenon, a condition of society involving collective responsibility rather than an unfortunate or reprehensible condition of individual persons.[8]

Ian Hacking provides a vivid example of the creation of a statistical entity. In 1825, John Finlaison testified before a select committee of the House of Commons that while mortality was subject to a known law of nature, sickness was not. Such a state of affairs was unacceptable to the government, especially because many thousands of friendly societies of workers had undertaken to insure their subscribers against the consequences of illness. The select committee was concerned that they might soon be bankrupt. By April, the committee had browbeat Finlaison into admitting the possibility of laws of sickness. The committee report then misleadingly summarized his testimony as confirming that sickness "may be reduced to an almost certain law." An 1852 commentary, taking the committee's summary as valid, wondered why these laws of sickness hadn't been calculated, given the abundant materials contained in the quinquennial returns of friendly societies. To this the council of the newly formed (English) Institute of Actuaries responded by denying the validity of laws in this whole domain. "The notion that there is a 'fixed' rate of mortality and a 'fixed' rate of sickness is evidently untenable. There is reason to believe that these rates differ in every [insurance] association, not widely perhaps, but characteristically."[9]

This variability, the actuaries considered, explained why an insurance company needed expert, professional management by men such as themselves. It did not mean, however, that the companies were left at the mercy of nature and the habits of their subscribers. The companies could take care of themselves by arranging matters so that sickness in any given organization would remain within the bounds of its own set of laws. One William Sanders explained in 1849 to another of the many parliamentary select committees on friendly societies how he kept the Birmingham General Provident and Benevolent Institution solvent. Tables giving rates of sickness were important, he told them, but the crucial element was strict rules, to define the bounds of appropriate sickness. The testimony proceeded as follows:

> T. H. SUTTON SOTHERON [of the committee]: A mere calculation of good tables would not be sufficient to secure the society; you must have good rules as well?
>
> SANDERS: So far from that, I would rather trust a society with moderate tables and good rules, than a high one with bad rules.
>
> SIR H. HALFORD [of the committee]: The stringency of the rules consists in the smallness of the payments?
>
> SANDERS: Of course, it consists in the limitations we place upon insurance; we do not allow our members to insure such an amount in sickness, as, looking at their circumstances and income, would prove a temptation to fraud.

HALFORD: You do not refer to any strict supervision as to the reality of sickness?

SANDERS: That is inquired into, of course; we pay nothing but upon a surgeon's certificate. In addition to that, the parties are visited by ordinary members, and those visits are weekly reported to the secretary.

HALFORD: Of course you interdict their work during sickness?

SANDERS: Our rules on that point are more stringent than most.[10]

Sickness, in short, could not be reliably quantified until it was mapped out and subdivided. This policing of sickness has become all the more important in recent times. Otherwise the public treasury would be drained by epidemics of impermissible maladies, and, following the logic of the new Ricardianism, all surplus value would pass ineluctably into the hands of physicians.

Life insurance was somehow less vulnerable to malingering, and the prospect of reliable quantification without intervention was correspondingly favorable. For some purposes, such as monitoring the health of an entire national population, general life tables were considered suitable. These typically assumed a birth cohort of 10,000 of each sex, and provided the number who could be expected on average to remain living every year up to age 100. The regularities were of course subject to fluctuation caused by cholera or potato blight. Life insurance companies, though, considered this the least of their problems. A society that admitted all applicants would soon have a membership made up overwhelmingly of the sick and dying, which would be fatal to the company as well as its membership. Even if there were general "laws of mortality," a matter of controversy among the actuaries, they provided no adequate basis for the institution of life insurance. Nineteenth-century actuaries recognized that their work required creating a domain of artificial order. This they aimed to accomplish mainly through the skillful selection of lives.

Modern insurance historians support the view of Victorian actuaries on the importance of this selection. Clive Trebilcock explains that the Pelican was unprofitable throughout the nineteenth century because it "simply was not proficient at selecting which lives to insure."[11] It seems they insured too many dissolute aristocrats, while other companies enlisted the sober, middle classes. The key importance of proper selection was universally recognized. The Anglo-Bengalee Disinterested Loan and Life Assurance Company, created by Charles Dickens in his novel of 1843–44, *Martin Chuzzlewit*, advertised to Dickens's readers its irresponsibility by admitting lives indiscriminately. Dr. Jobling, the company doctor, received a commission on every policy issued.[12]

The selection of lives presented a difficult problem of trust and surveillance. A sound company would take care that medical as well as financial expertise was represented on its board. The customary practice among life insurance companies in the early decades of the industry was to require a personal appearance of every applicant before the assembled directors. There an inspection would take place, and a decision would be reached about whether this was indeed a "select" life. But sometimes an inspection was grossly inconvenient, especially if the applicant lived far from London. Charles Babbage reported in his study of insurance institutions in 1826 that most companies were willing to dispense with this visit for a certain percent. How much this ought to be, he added disapprovingly, had never been calculated.[13]

Clearly the companies would in any event require some information about the lives they were considering. The most convenient source of advice was their agents in other cities who had solicited the business in the first place. But the agents might have no medical expertise, and in any case it was dangerous to rely on the discretion of persons working on commission. Trebilcock shows that in the case of fire insurance, at least, the poor judgment or cupidity of some agents caused Phoenix Assurance to suffer huge early losses in St. Thomas and then Liverpool.[14] The Pelican appointed a medical adviser to its Board in 1828 and tried to keep checks on the quality and credentials of its doctors. But their attention to medical matters was only fitful. The large number of canceled policies attests to the frequency of mistakes. The Board was generally more interested in investments than in actuarial or medical work. Perhaps this was why it suffered such high mortality.[15] Royal Exchange Assurance was more successful with its actuaries and medical examiners, and hence also with its life insurance business. It appointed a medical adviser fourteen years later than the Pelican, in 1842. It did not require a medical certificate of applicants for insurance until 1838. We should probably read this not as a sign of indifference, but rather of intense personal interest, a reluctance to delegate to others these crucial decisions about the quality of lives.[16]

Four actuaries called before a Parliamentary Select Committee on Joint Stock Companies in 1843 described the identification of quality lives in some detail. First, the candidate was asked if he had suffered "certain named diseases." He was to supply a reference to his "medical attendant, and to some private friend who is acquainted with his habits of life and general state of health." Letters of inquiry were then sent to the friend as well as the doctor, and the candidate himself was required to appear "before either the directors at the insurance office, or some medical officer they may appoint, or both." The select committee chairman, Richard Lalor Sheil, was not convinced that an appearance before the board could achieve any useful end. "I consider it very useful," replied Charles Ansell. "But the main reliance is placed upon the medical report, is it not?" he was asked. "I am not prepared to say that; indeed, I know cases in which the directors are bold enough to differ *in toto* from the medical officer, and accept lives which their medical man rejected, and sometimes the contrary." Another actuary, Griffith Davies, interjected that the directors almost never accept a life the medical officer has rejected, but often reject applicants the medical officer has approved. "There is another advantage," continued Ansell, "which is sometimes derived from men of the world seeing the lives which are proposed for assurance; and that is, that men's habits are frequently indicated by their appearance; and it leads often to inquiries as to the parties' habits of life," such as use of spirits.[17] Life insurance was not for the loose or disreputable.

Since the companies were not yet very large or bureaucratic by midcentury, actuaries too were involved in selection of lives, and occasional bits of advice on this matter were printed in the journal of the institute of actuaries, the *Assurance Magazine.* In 1859–60 it published a collection of medical maxims for identifying bad lives. "The practised eye of the medical examiner will at once detect the advanced drunkard in the characteristic bloated countenance," and reject his application. An attack, however slight of apoplexy "renders a life quite ineligible," and no respectable company would seriously consider "a gouty person who is a free liver and of sedentary habits."[18]

MAKING THINGS

Official statistical categories occupy contested terrain. The numbers they contain are threatened by misunderstanding as well as self-interest. Statisticians confront a problem of replication very much analogous to that faced in the measurement of effluent concentrations. Thousands of agents must be trained to arrange an unruly humanity into conformable categories. Craft skills are developed in each office, as employees discuss with each other the appropriate occupational classification for a retired dentist managing vacation rentals or a budding novelist who for the moment is waiting tables. Alain Desrosières and Laurent Thévenot of the French national statistical office, INSEE, discuss the problems of coding, and report that even in this exemplary statistical agency, a repeat interview will assign an employee to a different occupational category from what was reported initially in up to 20 percent of cases.[19]

On occasion the uncertainties go deeper, and the categories themselves are challenged. Racial and ethnic categorization inspires great passion, and is always highly contentious in the United States. Activists and bureaucrats have managed to create the category, "Hispanics," out of

Americans of Mexican, Cuban, Puerto Rican, Iberian, and Central and South American descent, though it was by no means universally supported among the people it labels.[20] In Germany, the United States, and France one finds three rather different forms of humanity pertaining to what in English are called professionals. Desrosières and Thévenot discuss the political and administrative ambitions that gave rise to them. All three reflect a shift away from categorization by sector, which would put doctors with nurses and auto executives with assembly-line workers, and toward a stricter observance of hierarchy. In each case there is also a more local story. The German category *Angestellte*, a name for salaried employees outside the public sector, was invented at the time of Bismarck's social insurance laws so that these respectable types would not be classed with wage workers, nor represented by socialist unions. The American "professional" arose early in the twentieth century to distinguish men of knowledge committed to an ideal of service from business managers. French statisticians formed the *cadre* as part of economic planning in the 1930s and 1940s.

The dependence of categorization on particular circumstances would seem to imply that the categories are highly contingent, and hence weak. Once put in place, though, they can be impressively resilient. Legions of statistical employees collect and process numbers on the presumption that the categories are valid. Newspapers and public officials wanting to discuss the numerical characteristics of a population have very limited ability to rework the numbers into different ones. They thus become black boxes, scarcely vulnerable to challenge except in a limited way by insiders. Having become official, then, they become increasingly real.

Desrosières offers a striking illustration. In 1930 nobody in France talked of *cadres*, or even knew what they were. The germ of the concept is to be found in a movement of middle-class solidarity, in opposition to plutocrats and the working classes. The term *cadre* was first applied to these engineers and managers under Vichy. In postwar *planification* it became a category in the official statistics. This required a close definition so that its members could be counted, and soon attached to it a legion of numerical characteristics. Now one can read in French newspapers about what the *cadres* think on the issues of the day, or how they dress and what they read.[21] Increasingly, the statistical categories form the basis for individual and collective identity. Thévenot makes stories like these central to the formation of social classes, which, he argues, are inseparable from the instruments of social statistics that contribute to their articulation.[22] National identity, too, may be formed in part through the articulation of public statistics—or a conspicuous lack of statistical uniformity, as in Italy, may threaten it.[23] Public statistics are able to describe social reality partly because they help to define it.

In the industrialized West, as in the centrally planned economies formed in the name of Marxist socialism, quantification has been part of a strategy of intervention, not merely of description. The novelist Alexander Zinoviev characterized the Soviet case nicely, and with only a little sarcasm:

> Any hopes that one can make scientific discoveries in the sphere of predicting the future are without foundations. First of all, in the Soviet Union predictions about the future are the prerogative of the highest party authorities, and so scientific small fry are simply not allowed to make any discoveries in this area. Secondly, the Party authorities don't predict the future, they plan it. It is in principle impossible to predict the future, but it can be planned. After all, in some measure history is the attempt to correspond to a plan. Here it's like the five-year plans: they are always fulfilled as a guide to action, but never as predictions.[24]

Theodor Adorno made a related point regarding the relation of quantification to capitalism in the culture industry. As an American refugee, by one of the odder quirks of fate in intellectual history, he became associated with a study of radio headed by another German-language emigré, the archquantifier Paul Lazarsfeld. Adorno reminisced: "When I was confronted with the demand to 'measure culture,' I reflected that culture might be precisely that condition that excludes a mentality capable of measuring it." But, he determined, this need not rule out the quantitative study of mass entertainment. "It is a justification of quantitative methods that the products of the culture industry, second-hand popular culture, are themselves planned from a virtually statistical point of view. Quantitative analysis measures them by their own standard."[25]

As with the methods of natural science, the quantitative technologies used to investigate social and economic life work best if the world they aim to describe can be remade in their image. If psychological tests predict school grades, this is in part because quite similar tests are used in schools to evaluate students. If they correlate with success in business, this owes something to the culture of quantitative puzzle-solving imported from business schools. Zinoviev's remark about Soviet economic plans applies with few changes to bureaucratic business corporations in the West: quantification is simultaneously a means of planning and of prediction. Accounting systems and production processes are mutually dependent. Cost accounting, for example, was impossible until manufactured products, as well as machinery and the workers, were highly standardized. At the same time, sophisticated accounts were indispensable to the creation of economies of mass production. A world of craft production and barter would have little use for the quantifier's tools, and would be impervious to them.

It has been urged that accounts have less to do with representing conditions than with guiding behavior in large firms. This is undoubtedly so, though there is more than a hint here of a false dichotomy. Numbers that have no credibility as truth claims will be less effective also at projecting power and coordinating activity. But the imperative mood tends to define the indicative. Adequate description counts for little if the numbers are not also reasonably standardized. Only in this way does calculation establish norms and guidelines by which actors can be judged and can judge themselves. Business corporations began early to evaluate laborers by quantity of production, which had the dual advantage of being easy to measure and unambiguously related to the profitability of the firm. One of the crucial goals of accounting was to apply such objective evaluation to ever higher levels of responsibility, and hence to manage large, multicentered firms so far as possible according to clear and open standards. It was, as G. C. Harrison observed in 1930, much easier to accomplish this for "the five dollars a day man" than for top executives. But already such corporations as Du Pont and General Motors were judging their operating divisions using a standard index of profitability, return on investment or ROI.[26]

Any such measures necessarily involve a loss of information. In some cases, as with accounting, the credibility of the bottom line may be such that this loss seems largely irrelevant. But such an attitude presupposes that the bottom line is determined unambiguously by the activities it summarizes. It never is. When business managers are judged by the accounts, they learn to optimize the accounts, perhaps through such artifices as putting off needed maintenance and other long-term costs.[27] Nonfinancial measures may be even looser. A congressional mandate permits the United States Forest Service to cut no more lumber than is renewed by annual growth. Since that law was put into effect, growth rates have been greatly enhanced, at least in the Forest Service accounts, by new herbicides, pesticides, and tree varieties. Through such doubtful forecasts, it drew the teeth from the law.[28]

Given the ways that measures can be undermined through self-interested manipulations, we may doubt that they correspond to anything in the world. But a plausible measure backed by sufficient institutional support can nevertheless become real. Accounting measures like return on investment are exemplary. As Peter Miller and Ted O'Leary point out, this one does not function merely as a piece of information passed along to the top levels of management to keep them informed. Neither is it a servant of coercive power, enabling a centralized administration to make decisions over the heads of middle management. To the extent that it has become real, it provides the basis for a crucial kind of self-discipline, harnessing the interests of managerial employees to those of the firm. Successful firms depend on vigorous decentralized activity. Numbers alone never provide enough information to make detailed decisions about the operation of a company. Their highest purpose is to instill an ethic. Measures of profitability—measures of achievement in general—succeed to the degree they become, in Nikolas Rose's phrase, "technologies of the soul." They provide legitimacy for administrative actions, in large part because they provide standards against which people judge themselves. Grades in school, scores on standardized examinations, and the bottom line on an accounting sheet cannot work effectively unless their validity, or at least reasonableness, is accepted by the people whose accomplishments or worth they purport to measure. When it is, the measures succeed by giving direction to the very activities that are being measured. In this way individuals are made governable; they display what Foucault called governmentality. Numbers create and can be compared with norms, which are among the gentlest and yet most pervasive forms of power in modern democracies.[29]

INFORMATION

This creative activity of making things is a precondition also for much of what we know as information. Some form of knowing, of course, is presupposed by virtually all human activities, and no society could function without the sharing of this knowledge. In this sense, the modern term "information society" is quite meaningless, for a village of peasant farmers could no more get by without information than can the head office of a large business firm. But only a little attention to nuance is required to see that much has changed. One, much noticed by the gurus of information, is that the census tables reveal a huge increase in the number and variety of people who live mainly from the accumulation and exchange of knowledge, and whose hands remain white and soft. Another is the explosion of printed, factual material, so that basic literacy and numeracy have become essential to function in the industrial (or post-industrial) world.

This explosion of knowledge is in important ways less impressive than we are often urged to believe. Knowing does not in general depend on print, and if early modern farmers, carpenters, butchers, and smiths had been as industrious about describing their work as they were while doing it they could have filled volumes, just as our researchers do now. But theirs was an order based on more private ways of sharing skills and exchanging goods. Children of peasants acquired the subtle skills of agricultural life from their parents. Tradesmen learned their crafts in a long apprenticeship that combined technical with moral instruction. Outsiders had no need to know any of this, and indeed to share the skills indiscriminately would tend to undermine that insistence on quality and self-regulation by which the life of the guilds was ordered.[30]

Public affairs, too, were kept largely private until at least the late eighteenth century. This did not require elaborate mechanisms to preserve secrecy, though public as well as private institutions often had good reasons to maintain secrets.[31] It reflected, rather, the weakness of institutions promoting public knowledge. Political and business information alike was spread mainly through networks of personal acquaintances. Indeed, political and business connections were often inseparable, and neither could be readily distinguished from friendship. Eighteenth-century Americans treated private letters as public business, and a letter might be opened and read several times as it made its way along a chain of acquaintances from sender to recipient. Family was central to much information exchange, and letters within elite families often mixed family and public news. Those who lacked the connections to learn of political affairs informally were assumed to have no real need to know. Elites viewed local newspapers as an extension of personal knowledge. Only newspapers from abroad were experienced as something like pure information. Even printed material often bore a personal stamp, and someone arriving with a newspaper or proclamation from afar would be expected to interpret and explain its contents.[32]

How could it have been otherwise? What reason was there to put faith in an anonymous document? Impersonal information was very hard to come by. As Bourguet's study shows, even the French bureaucracy in 1800 was unable to create much of it. Scientific reports depended for their credibility on the social standing of the author and of witnesses, who often were named and identified in print. Lack of trust was compounded by problems of comparability, the result of diverse institutions and unstandardized commodities and measures. In the information society, information means first of all communication with people who are unknown to one another, and who thus have no personal basis for shared understanding. Such information was of little importance as recently as the eighteenth century. Since most news was privately circulated, good sources of information were synonymous with power. This remains true, in a way, but much of what had to be learned privately two centuries ago has since been replaced by formalized, printed knowledge. This was promoted by the vast expansion of newspaper publishing beginning in the late eighteenth century, associated with what R. R. Palmer called the "age of the democratic revolution," and with Jürgen Habermas's "public sphere."[33] But routine reliance on published factual information presupposed a shared discipline specifying how it was to be generated and interpreted. In most cases it required also the administrative creation of new things.

The work of the Chicago Board of Trade, discussed in a book by William Cronon, provides an outstanding example. Standard practice in the grain trade before the railroads came through was for farmers to load their wheat in bushel sacks and send it down river by boat. A miller or wholesaler downstream would offer a price for the wheat based on a close examination of a sample. Under such circumstances it is difficult to talk of "the price of wheat," or of information at all. The Midwest appeared flat and uniform, but the produce of each farm was unique. It might be possible to say that good quality wheat was bringing a certain price, but a merchant would be unwise to buy any unless he or a trusted deputy were on the scene, running his fingers through the grains. Such personal inspection continued all the way down the line until at last it reached consumers as flour or bread.

By the 1850s, though, markets were becoming more centralized. The Chicago Board of Trade, founded in 1848 as a voluntary organization of businessmen, began almost immediately to impose some uniformity on this highly variegated world. It first redefined the bushel in terms of weight. Bushel sacks were fine for riverboats, but inconvenient for grain elevators. An even greater problem for the elevators was quality. It was inconvenient to keep each farmer's grain in a separate compartment. Beginning in 1856, the Board of Trade undertook to define uniform categories of wheat. Their initial efforts nearly led to disaster. When farmers discovered that they would receive about the same price for excellent clean wheat as for dirty, damp, or sprouted wheat, they began to complain bitterly. They also began mixing their wheat with dirt and chaff, or at least taking little care to keep it clean. Soon the price of Chicago wheat in the markets of New York fell five to eight cents below that of Milwaukee. The new system proved itself adequate to generate impersonal information in the form of a uniform price, but to the immense disadvantage of local farmers and traders.

In 1857 the Board introduced grading of wheat on the basis of quality. To this end it appointed a city grain inspector, to keep watch over the grading operations at the various elevators. But grading by the elevator operators, an interested party, proved unsatisfactory. In 1860 the chief inspector was ordered to train his own assistants, thus forming a little bureaucracy. For a set fee, these inspectors would certify the grade of any shipment of grain to be traded on the Chicago Exchange. To do this they had to be given the right to enter the elevators and inspect the grain personally. Every lot was placed in one of four grades, from club class to rejected. The elevator operators had only to keep the four grades and three main varieties separate.

But of course they did not, for quality is continuous, and the categories were discrete. They soon learned that they could increase their profits by mixing all grain down to the lower threshold of the grade. This did not long remain a secret. Soon farmers began complaining that this mixing was diverting to shady operators revenue that rightly belonged to them. They won the sympathy of newspapers and elected officials who threatened to intervene in the grain trade. Controlling the politics was as crucial as grading the wheat for the standardization of grain, and the Board of Trade joined the farmers in support of laws against mixing wheat of different grades.

In the end, bureaucrats and traders managed to create what had never existed on farms, much less in nature: uniform categories of produce. Thereafter, wheat could be bought and sold on the Chicago Exchange by traders who had never seen it and never would—who couldn't distinguish wheat from oats. They could even buy and sell futures, commodities that didn't yet exist. Thus a net of regulatory activity created a space for information, in the modern sense. A successful trader of wheat no longer had to spend his time at farms, ports, and rail terminals judging the quality of each farmer's produce. By 1860 the knowledge needed to trade wheat had been separated from the wheat and the chaff. It now consisted of price data and production data, which were to be found in printed documents produced minute by minute. Of course the need for personal contacts and private sources did not disappear. Increasingly, though, even rumors originated where the action was—not on the farms, but on the floor of the exchange.[34]

CHAPTER THREE

Economic Measurement and the Values of Science

The social engineer . . . conceives as the scientific basis of politics something like a *social technology*.
(Karl Popper, 1962)

QUANTIFICATION AS A SOCIAL TECHNOLOGY

Textbook science is predominantly about theory. This is especially true of physics, the currently reigning queen of the sciences, which beginners and other outsiders sometimes confuse with mathematics. This class of outsiders includes most social scientists who have thought at all about the achievements of natural science and their implications for human studies. When the issue is posed in such abstract terms, even experimenters will often say that their business is to test theory. I discussed in chapter 1 some of the reasons for believing that experiment has a life of its own, a life of instrumental practices. But of course it is also a life of literary practices, of analyzing, writing, and arguing. Quantification plays a role in modern experimental life scarcely less central than that of mathematics in physical theory. One of its purposes is to serve as a bridge between the material culture of the laboratory and the predictions derived from formal theory. This is often taken as the decisive role of experimental quantification in the practice of science. It is not. Researchers on topics that lack mathematical theory are often equally assiduous in reporting methods as well as results in quantitative form, and filtering out findings that cannot be so expressed.

Quantification is a social technology. Whereas modern mathematical ideals have their roots in ancient geometry, which emphasized demonstration and was largely separate from the domain of number, arithmetic and algebra were born as practical arts. They were associated with activities of merchants, the keeping of accounts. This remained true in the sixteenth century, and to a degree even in the nineteenth. In science, too, quantitative measurements and manipulations of numbers go back to ancient times, but their place was distinctly subordinate to mathematical demonstration. In the Renaissance, such activities made up much of mathematical astronomy, which was considered useful for predicting planetary positions and determining the date of Easter. To this end the positions of stars and planets were carefully measured. Until Kepler, few worried much about fitting the measurements to a physical theory. The life of measurement was not a life utterly apart, but it did not exist simply for the sake of theory.

Even at the end of the eighteenth century, when the experimental sciences were won over to an ethic of measurement, that life remained as closely allied to the practical world of commerce and administration as to exact theory. The chemical balance came to chemistry from mine assaying, with the encouragement of state mining bureaucracies. For Lavoisier it was the conclusive test of experimental proficiency, but even then it had almost nothing to do with the testing of theories. Another fine example is the use of the barometer to measure elevations. Pascal realized in 1648, according to a qualitative theory, that the mercury should fall when the barometer was carried to higher elevations, and it evidently did. Eighteenth-century military engineers needed a good deal more precision if the barometer was to be of use for drawing topographical maps of mountainous regions, and this was the principal incentive to exactitude in barometric hypsometry.[1]

In many fields, including barometry, there soon were mathematical theories to test. Tests of theories have sometimes provided an important inducement for increasing the degree of precision in measurement. A notable early case of this was the dispute between Newtonians and Cartesians over whether the earth was a flattened or elongated sphere. Significantly, as Mary Terrall shows, the latter claim was not a consequence of Cartesian theory, but an early finding of French mapmakers that was subsequently contested on Newtonian grounds. The famous midcentury expeditions to measure the earth's curvature in Lapland and Peru thus had theoretical reasons to seek greater precision and reliability, but precision was already important enough to cartography for the issue to have arisen independently.[2] And in any case, the use of exact measurement to decide between theories is not at all routine. For about two centuries, quantitative precision has been understood as central to experimental science, even where measurements cannot be related to any mathematical theory. The quest for precision has been sustained in science for reasons having more to do with moral economy than theoretical rigor. Precision has been valued as a sign of diligence, skill, and impersonality. Quantification has also been a crucial agency for managing people and nature.

This practical imperative is part of what I call the "accounting ideal." To use such a term in relation to science may seem an act of *lèse majesté*, though it should be inoffensive enough to those who can live without an absolute monarch. Accounting is manifestly a mundane activity, and alerts us to the craft dimension of quantification. It is a way of organizing commercial and bureaucratic life, and calls attention to the analogous role of measurement in giving shape to experimental investigation in science. We must be wary of dismissing it as routine and unoriginal. The reputation of accounts and statistics for grayness helps to maintain their authority. Considered as a social phenomenon, accounting is much more powerful and problematical than scholars and journalists generally realize.

The moral dimension of accounting, as the exemplar of inoffensive impersonality and objectivity, is defined in chapter 4 and historicized in chapter 5. Here I aim to call attention to its efficacy in administration. Accounts and statistics, broadly speaking, are the lines connecting the world to what Latour calls "centers of calculation."[3] Inevitably, the goal of managing phenomena depends also on convincing an audience. When the French state, or any other, decided to begin providing accident insurance to industrial workers, it needed statistics for budgeting purposes. When it charged taxes to towns in proportion to census results, controversy about population was inevitable, and with it a demand for the stamp of objectivity to certify the figures.[4] Scientists have been keenly aware of these aspects of quantification. With rather few exceptions, they have been reluctant to engage with theory, including mathematical theory, that could not be incorporated somehow into a world of experimental control and measurement. It is easy enough to support this with pronouncements by distinguished scientists, and I mention a few later. The appropriateness of the accounting metaphor, though, is most graphically revealed in the approaches taken by natural scientists to economic questions. That is the main topic of this chapter.

BARREN THEORY

William Whewell, like most of the scientists and engineers considered in this chapter, looked to statistics as an alternative or at least an indispensable supplement to abstract theory in economics. The leading advocate of a statistical and historical economics in England during the 1830s and 1840s was Richard Jones. Whewell was his close friend and frequent correspondent, and at his death became his literary executor. Both were among the earliest members of the Statistical Society of London. Whewell looked to Jones to perform those empirical economic investigations that he favored but did not care to undertake himself. He was by no means too proud to perform the hard work of gathering and analyzing facts. But he supported Jones mainly in a different way: by writing mathematical theory. This may seem an improbable alliance: why should the great enemy of deduction in economics have tried to mathematize it? To destroy his enemies, of course. Whewell looked to mathematics to impose discipline on theoretical political economy, and to block its indiscriminate application.

Political economy was not Whewell's major scientific concern. He was a polymath—a leading scientific organizer; master of Trinity College, Cambridge, and thinker and writer on educational subjects; an astronomer physicist, geologist, and mineralogist. He devoted much of his scientific effort to "tidology," the science of tidal movement, involving the collection of enormous amounts of quantitative data, which he hoped could be brought into accord with mathematical predictions. He is best known now as the author of a three-volume *History of the Inductive Sciences*, followed by two more on the *Philosophy of the Inductive Sciences* and a last, *On the Philosophy of Discovery.*

Whewell's philosophical outlook is the obvious place to begin in seeking to understand his critical approach to political economy.[5] We find, first, that political economy is not a topic of Whewell's history or philosophy. This was, after all, history teaching by example, and its author found nothing in political economy that could fit it to be a model for other scientific investigations. Rather, political economists had much to learn from the more successful disciplines, meaning the natural sciences. Whewell criticized Ricardian economics not because he thought the model of natural science inappropriate for political economy, but because political economists had departed too far from the historical pattern of successful scientific investigation.

That pattern involved, first of all, induction. Whewell considered himself a devoted follower of Francis Bacon, and he argued repeatedly that science should proceed by induction to successively broader generalizations. The temptation must be resisted to leap from a few casually observed facts to vast, all-embracing principles, and proceed thereafter by the easy path of deduction. This last is what he thought David Ricardo had done. To join mathematics to Ricardian political economy would be to "make nonsense of it." If the political economists "will not understand common sense because their heads are full of extravagant theory, they will be trampled down and passed over."[6]

Verbal reasoning, he argued, is too slippery. It does not require that the premises be made clear, and it permits auxiliary hypotheses to slip in unnoticed. It provides no clear checks against errors of reasoning. It is too imprecise for its results to be tested against those uncompromising judges, experiment and observation. Mathematical economics could overcome these defects. The result, of course, might often be to show that we are not yet able to succeed at deductive reasoning, that our premises are not sufficiently in accord with the world. But this, too, is valuable knowledge. Exact results, even if faulty, are to be preferred to imprecise, sweeping conclusions, to "the statements which we perpetually receive from the economists, of that which must necessarily be but yet is not, and to general 'truths,' to which each particular case is an exception."[7]

Given all this, it is hard to be surprised at Whewell's conclusions. Ricardo had allowed dubious tacit assumptions to creep into his argument. Once exposed and made explicit, Ricardo's qualitative findings could be judged against historical and empirical work of men such as Jones. Whewell seemed not to anticipate their total vindication. He claimed also to find mistakes in Ricardo's abstract verbal reasoning. Ricardo erred, for example, in his inference of the effect on rent and profits of growing English prosperity, and of the sector upon which taxes of various descriptions would ultimately fall. Not that Whewell believed the mathematician could reach decisive, exact conclusions on these points. His purposes were more critical than constructive: to show "of what kind and how many are the data on which the exact solution of such problems may depend."[8] Mathematics should not supplant empirical investigation, but clear the ground for it by revealing the weakness of verbal deductions.

This use of mathematics to show the inconclusiveness of existing theory was not uncommon in the nineteenth century. Another British scientist working with similar aims was Fleeming Jenkin. Jenkin was a close friend of William Thomson, James Clerk Maxwell, and Peter Guthrie Tait, and himself professor of engineering at the University of Edinburgh. He structured his economics after the physics of heat engines.[9] His papers of 1868 and 1870 used graphical rather than analytic mathematics, and his purposes were at least partly constructive. Yet he was inspired in large part by a distaste for one of the main conclusions of classical political economy, the so-called wages-fund doctrine. This held that a limited sum of money is available for wages at any given time, and that since trade unions can do nothing to expand it, they cannot improve the conditions of workers. Jenkin objected that this doctrine is meaningless so long as we do not know how the fund is determined. "No economist has hitherto stated the law of demand and supply so as to allow this calculation to be made."[10] To work out the interaction of causes required, if not an abstract mathematical formulation, at least generalizable quantitative techniques. He pronounced the solution indeterminate without a considerable improvement in the empirical data.

He proceeded by seeking the equilibrium between supply and demand. These are, of course, functions of price—or, in the particular problem here addressed, of the wage rate. The shape of these curves is not given timelessly by nature, but depends, as Jenkin put it, on states of mind—of the capitalist, and of the workers. "The laws of prices are as immutable as the laws of mechanics, but to assume that the rate of wages is not under man's control would be as absurd as to suppose that men cannot improve the construction of machinery." Hence so-called laws of demand and supply "afford little help, or no help, in determining what the price of any object will be in the long run."[11] The structure of the market matters: unorganized laborers are like goods to be unloaded in a bankruptcy sale. Hence organization into trade unions most certainly can improve the worker's lot. How much? In a subsequent paper, Jenkin suggested empirical measurement of supply and demand schedules to resolve the effects of taxation, and the same methods would apply to wages.[12] But given the mental component that he emphasized so heavily in the determination of wage rates, prediction here might well be beyond the political economist's art.

We may be tempted to regard this empirical attitude as characteristically British, especially in the time of Whewell and of Charles Babbage, whose economic writings emphasized accounts, statistics, and machinery.[13] In fact it was never stronger than in imperial Germany, where historical economics won a complete victory over classical theory. The German historical school was a statistical school. A few of its members, most notably Wilhelm Lexis and Georg Friedrich Knapp, used higher mathematics, though generally as tools of criticism. They aimed to refute "atomistic" individualism and deny the possibility of "natural laws" of society.

It is curious but revealing that, in the *Methodenstreit* between historicist followers of Gustav Schmoller and the deductivist Austrian school of Carl Menger, quantification was plainly on the side of history. Though antideductive, it provided from another standpoint a middle way between the verbal theories of Menger and the new mathematical marginalist theory that Lexis criticized as excessively abstract. Deductive theory, he charged, can show no more than tendencies. Its propositions do not give a "reliable predetermination of actual events, and cannot by themselves decide the measures to be taken in pursuit of goals in economics."[14] For the historical school, the goals of economics were first of all practical and administrative ones. Its members aimed above all at social reforms, to improve the lives of workers. Effective state intervention in economic affairs, they believed, depended on expertise that had proved itself by its empirical adequacy. This of course was more easily said than achieved. But given the choice, they preferred descriptive accounts and statistics to formal, deductive theory. The same outlook was typical of most natural scientists who wrote on economic matters.

THE ECONOMICS OF ENGINEERS AND PHYSICISTS

Engineers are often required by their profession to practice economics. Physicists, at least as researchers, generally are not. But the line between physics and engineering has not always been very sharp. The gap was kept narrow through most of the nineteenth century as a result of the great importance in physics and engineering first of heat engines, and then of electricity. Especially in the early part of the century, relations between thermodynamic and economic ideas could be very close. Each made use of concepts from the other. By no means was economics simply parasitic on physics; economic and physical ideas grew up together, sharing a common context. An economic point of view, the idea of balancing energy accounts through transformations and exchanges, formed the central metaphor of thermodynamics. That view did not come mainly from the likes of Ricardo or Jean Baptiste Say. The economic mentality at issue here was associated more closely with accounting than with high theory. This economic conception itself already integrated a labor theory of value with a set of analogies involving engines.[15]

This form of economics was perhaps best developed in Great Britain. There, as Norton Wise has shown, work, meaning energy, became the basis for an alternative economics. The economics of energy was ideally suited to become an economics of measurement, for it permitted the productivity of labor to be assessed against an absolute standard. It made the labor of machines, animals, and men commensurable. The champions of energy economics were not generally hostile to free trade, laissez-faire, or the other leading doctrines of classical political economy. Neither, though, were they content with an economic science that was mainly theoretical. Here was a form of economic reasoning, and more crucially a system of economic practice, that would permit scientists to judge the productivity of machines and labor, and to improve them. In this economics, the statistics of factories, of workers, and of production meant something. Quantification could aid administration, could guide the improving activities of engineers and reformers.

In Britain, the most important early champion of the new French physics of work was Whewell, author of an 1841 textbook on the *Mechanics of Engineering*. He wanted to raise engineering above mere craftsmanship, to introduce physical theory in alliance with physical measurement. His book made the foot-pound the common unit of laboring force. In this case, machines could be compared with humans and animals, and their advantages understood in familiar terms. James Thomson, brother of the famous physicist William and himself a distinguished engineer, gave a typical calculation in 1852. His pump, he ascertained, could lift water at the rate of 22,700 foot-pounds per minute. A man can lift only 1,700 foot-pounds per minute, and that only for eight hours in a day. Hence the pump did the work of forty men. Physical work, as Wise remarks, was here literally labor value.[16]

Even more crucially, this formulation permitted a clear distinction between useful work and waste, and indeed gave a quantitative expression of efficiency. This was invaluable to the industrial engineer, for calculation could then be used to determine an optimal mix of machine and human labor. William Thomson showed how energetic and monetary calculations could be combined to reach an optimum in telegraphy. Having determined how to calculate the retardation of signals in a wire, it became "an economical problem, easily solved . . . to determine the dimensions of wire and covering which, with stated prices of copper, gutta-percha, and iron, will give a stated rapidity of action with the smallest initial expense." At about the same time, James Thomson calculated to determine whether it was energetically advantageous to boil urine as fertilizer, thereby producing an increase in food for human workers, or to employ the coal fire directly for productive work.[17]

With this we begin to discover the benefits of energetic calculations for friends of the poor and working classes, especially those philanthropists hailing from the Gradgrind school. R. D. Thomson, of the Glasgow Philosophical Society, looked forward to the day "when the light of science will enable the guardians of the poor to manage our poverty-stricken fellow men by precise and definite rules."[18] To this end, the Glaswegians were pleased to make use of a tabular presentation of the nutritive value of various food items: beans, peas, wheat, rye, oats, cabbage, and turnips. R. D. Thomson determined a ratio of nutritive value to cost for various types of bread, with the aim of minimizing the cost of supplying energy to human labor power. For him it was rather like measuring the energy content of coal, or the efficiency of machines. Lewis Gordon, the first professor of engineering in a British university, shared this perspective. Thorough energy accounts would enable the engineer to design and run factories with a maximum of efficiency.

The economics of energy was not inconsistent with the more customary medium of economic quantification, money. The crucial feature here is the pursuit of measurement—of quantification in standard, comparable units. This was a form of economics patterned after physics that aimed less at theoretical elegance than at practical management and efficiency. The contrast with the mathematical economics developed by William Stanley Jevons and Léon Walras two decades later could scarcely be more vivid. The economics of quantified energy, unlike that of mathematized utility, won the interest and even enthusiasm of contemporary physicists.

This was true also in France, where in fact the fruitful confrontation of physics with engineering and economics first took place. Members of the Académie des Sciences had been required to assist in technological and economic decisions already under the Old Regime. Many were involved also in quantitative demographic or economic studies, such as Lavoisier's attempt to draw up a national account for the French nation at the time of the Revolution.[19] The study of energy and work was closely associated with the culture of the Ecole Polytechnique, the first institution in the world to make mathematics and science central to the engineering curriculum. Soon after its founding in 1795, a Polytechnique education became prerequisite for entry into two distinguished state engineering corps, the Corp des Mines and the Corps des Ponts et Chaussées ("bridges and highways," but also canals, harbors, and railroads). The mathematics taught to these engineers was often very abstract, and its role in their formation was anything but straightforward. Many have charged that it was better adapted to educate mathematicians than engineers, even that it had more to do with credentialing than with practice. Whatever its deep significance, it guaranteed that polytechnicians were adept at the manipulation of numbers and of formulas. At this modest level of abstraction, at least, French engineers put their mathematical knowledge to work.

A notable instance of this was the study of engines. When the Napoleonic wars ended in 1815, the French found themselves decades behind the British in the technology of steam engines, which became an important topic of scientific as well as engineering inquiry.[20] French engineers were not content to approach engines as a problem of craft skill and technical ingenuity. C.L.M.H. Navier, G. G. de Coriolis, J. V. Poncelet, and Charles Dupin believed in the unity of engineering and science, and they sought an adequate scientific vocabulary for talking about the effectiveness of engines. An adequate vocabulary, naturally, presupposed the possibility of measurement. They introduced in this context the crucial physical notion of work, the action of a force through a distance, most easily measured as the product of weight and the height to which it was raised. Like their British followers, they meant this also to be a measure of labor power, of work in the colloquial and economic sense.[21]

Measurement of work, and of other quantities, was central to the French tradition of engineering economics. "Engineers do economics while others talk about it,"[22] proclaimed one twentieth-century French polytechnician. The Ecole Polytechnique and the Ecole des Ponts et Chaussées had long recognized that the business of the engineer required a familiarity with economic ideas. There were enduring doubts about whether the writings of those who called themselves political economists were capable of supplying what the engineers needed. Classical economics, some charged, was too impractical, too qualitative, too dogmatic. More typically, the engineers approved liberal economics as dogma only.[23] They cultivated their own practical economic tradition, which borrowed only a little from Say, Joseph Garnier, and other classical French economists.

Both the longstanding concern of French engineers with economic matters and their suspicion that the economists did not have quite what they needed are evident from the decision of the Council of the Ecole Polytechnique in 1819 to institute a new course called *Arithmétique sociale*. It declared:

> When we consider the development taking place every day in French industry, and the necessary relations of this industry with the government established by the charter, it is clear that the execution of public works will tend in many cases to be handled by a system of concessions and of private enterprise. Hence our engineers must hereafter be able to regulate and direct these developments. They must be able to evaluate the utility or inconvenience, whether local or general, of each enterprise; they must consequently have true and precise knowledge of the elements of such investments. They must, that is, be informed of the general interests of industry and agriculture, of the nature and effects of currencies, of loans, of insurance, of company assets, of amortization; in a word, of all that can help them appreciate the probable benefits and costs of all these enterprises: such is the collection of subjects that should be treated in this program.[24]

The council went on to argue that in the current world, public tranquility could be assured only when the superior classes are able to justify their wealth and power with virtue and knowledge. The study of social arithmetic was designed to promote such qualities in the French elite.

The course was indeed set up. It was taught not by an economist, but by the physicist François Arago, until Félix Savary took it over from him in 1830. Arago seems in retrospect a natural choice, since he was active politically as well as scientifically. But he taught a rather uninspired compilation of topics centered on mathematical probability, few of which bore directly on the needs of engineers and administrators. Emmanuel Grison observes that the course was created during the time of Laplace's effort to shift the curriculum toward pure mathematics.[25] But Laplace did not extirpate the economic perspective from French engineering. The real threat to economy came from an ethic of monumentality. This was less typical in the nineteenth century than in the eighteenth, but French state engineers showed a persistent preference for permanent structures over inexpensive ones.[26] Still, economy was part of the standard practice of engineers, and required no special instruction.

This is clear from a number of papers on engineering topics published by the Corps des Ponts et Chaussées in its *Annales.* Efficiency could never be ignored by engineers. Even in the planning of public works that concern came frequently to the fore. Navier, whose commitment to inexpensive construction in general may be doubted, stressed the need to incorporate economic considerations in defining the best route for a railroad or canal. For this purpose, physical parameters such as mechanical efficiency had to be made commensurable with costs of construction, maintenance, and loading and unloading. The engineer would then seek to minimize the mean cost of transporting a ton of merchandise one kilometer. Navier's paper on this topic gives him some claim to be a pioneer of modern accounting. The involvement of this distinguished physicist and leader of the Corps des Ponts in economic and accounting matters shows how seriously such subjects were taken by French engineers.[27] The problem was particularly pressing, though correspondingly difficult to quantify, when a choice had to be made about what cities should first have railway lines, or how much to invest in railroads and how much in canals.[28] But it arose also in the most mundane details of civil engineering. The choice of materials in a road, or the decision about steepness of grades and sharpness of curves on a railroad were economic problems, as was recognized in any number of papers by state engineers on the construction of routes.[29]

Jules Dupuit, the only French engineer of the nineteenth century whose economic writings have won him a lasting reputation, began his economic career writing on engineering problems that he confronted as chief engineer in Châlons sur Marne. He won two gold medals from the Corps des Ponts in 1842 for engineering papers: one on the force needed to draw wagons over highways as a function of type of wagon and load; the other on minimizing road maintenance costs.[30] The two were related; Dupuit argued successfully for lifting restrictions on weight and wheel width because the greater economy of transport overbalanced the increased costs of road maintenance. More generally, he showed how to raise this discussion out of the mire of day-to-day necessity. Dupuit proposed to bring "mathematical rigor" to this subject by evaluating the expenses of regular maintenance. This meant restoring to the road precisely what is worn away, thereby preventing the considerable and expensive damage caused by ruts. Formulated this way, road maintenance became a quantitative problem. The wear on roads, the rate at which the surface is ground into dust, should be a linear function of the traffic, and could be measured as a volume of rock per kilometer of road. It was then easy to calculate the expense of maintenance for any given material, and to reduce this to a minimum by choosing a surface suitable for the level of traffic.

Dupuit's solution to the problem of road maintenance was an economic one, though he had to begin with physical measurements before translating into money terms. He concluded with a broader economic perspective, noting that almost twenty times more money is spent by the traffic on roads than for maintenance on them. If by increasing maintenance 20 percent we could reduce these costs by 10 percent, "society" would receive a return of more than eight to one. Similarly, to build a bridge that reduces by one kilometer the daily journey of five hundred colliers is worth 36,500 francs per year, an excellent investment if building and maintaining the bridge cost only 10,000 francs per year. "In vain will one attempt to struggle against the irresistible power of these figures."[31]

THE PRICING OF PUBLIC WORKS

The question of tolls was another unavoidable economic problem faced by railroad engineers. No single standard ever won general assent, though a considerable literature was devoted to it. The usual approach, introduced by Navier, took this to be a problem of distributive justice, and allocated expenses in proportion to use. In an 1844 paper for the *Annales des Ponts et Chaussées*, Adolphe Jullien worked to define a homogeneous unit of rail travel. He did so by defining conversion factors between passengers and freight, and then by constructing a *convoi moyen*, which consisted of 6.25 passenger cars, 1.7 baggage cars, 0.29 post cars, and 0.03 horse cars. This made a total of 118.61 passenger equivalents. The mean expense per train is 1.4877 francs per kilometer, so the cost per unit of traffic is 0.01254 francs. Jullien then, somewhat arbitrarily, doubled this to take account of administration and interest on capital. Here was a just price for rail traffic.[32]

But it was not an adequate basis for setting rates, urged Alphonse Belpaire, an engineer with the Belgian Ponts et Chaussées. Jullien's promiscuous use of mean values, he argued, mixes together such a miscellany of causes and results that we cannot uncover the influence of any of them. "What can be the use of such an amalgam?"[33] An allocation of costs to causes is pointless if it does not enable us to predict the expenses for any kind of train. He thought it crucial that costs are not linear with volume. We want to know how much the cost goes down as the volume increases, so we can decide if rates can be reduced. This requires an allocation of costs to their particular causes, hence an *analyse minutieuse*.

This he provided, in a 600-page book about the operations of the Belgian rail system in 1844. He undertook the formidable task of identifying the causes of variable costs, and distributing fixed costs uniformly over some appropriate unit, such as cars, passengers, or passenger trips. He did not look for a single, grand mean, but tried to compute separately for each line, or at least each category of lines. He did not insist too strenuously on the mathematical rigor of his calculations. He recognized, for example, that his numbers were highly dependent on the particular circumstances of the various lines. "If the observer is one of those men committed to exact and absolute ideas, who admits no approximations and rejects everything that lacks rigorous mathematical exactitude, he will have no use for this calculation, and the question will rest eternally at the same point, at least until a less scrupulous spirit takes it up."[34]

An alternative, and evidently not a less scrupulous one, was adumbrated in another of Navier's papers, first published in 1830. Navier aimed there not to allocate costs, but to measure benefits, and to show how works could be operated to maximize that benefit. To build a canal, he observed, costs about 700,000 francs per league. This can be converted to an annual interest charge of 35,000 francs (at 5 percent). Maintenance and administration add 10,000 francs per league per year. Now, the difference between the cost of transport on canals and that on roads for a ton of merchandise is 0.87 francs per league. It is easy to calculate, then, that the canal becomes a worthy investment if 52,000 tons (that is, 45,000 francs divided by 0.87 francs per ton) are transported on it each year. The problem is that if a toll of 0.87 francs per ton were charged for the use of the canal, much of the traffic would return to the roads on account of the slowness of canal travel. The obvious conclusion was that the revenue for building and operating canals should not be extracted from users. The British, infatuated with private enterprise, refuse to provide the necessary subsidies, but the French state could. Its administration displays "experience, superior enlightenment, power, wealth, credit, and dedication."[35]

It may be noted that Navier did not include economy in this list of virtues. He generally preferred a solid structure, embodying the latest advances of science, to one that was merely cheap. Navier was known to reject privately proposed bridge projects on behalf of the Corps because the principles underlying their design could not be formulated mathematically.[36] But mathematics did not always triumph. At the very time he was defending by calculation the benefits of public works, he was at the center of a scandal involving a bridge at the Invalides in Paris. Navier wanted a monumental structure, and also one that would display the superiority of the refined mathematical calculations of state engineers over the mere empiricism of untutored builders. Suspension bridges, a new technology, permitted mathematization in a way that traditional structures had so far resisted. His design, too expensive to be erected by entrepreneurs, was opposed by them. What was much worse, the anchorages of his bridge ruptured after construction was nearly complete. Solid anchorages depended on an intimate knowledge of ground types, the one aspect of suspension bridges that had not been colonized by mathematics. Navier's bridge was torn down and the materials used to construct three cheaper, privately built structures. This unhappy story is doubtless an aberration. Navier's disdain for the need to make a bridge pay, however, was not. Neither was his fondness for mathematical analysis.[37]

His ideal of quantitative public management became rather common in the Corps des Ponts. The most theoretical of the economic writings in this tradition were published in the 1840s by Dupuit. The concept of diminishing marginal utility, which was perhaps implicit in the writings of some predecessors, was explicit and fundamental in his. The benefit of rail travel is not constant for all users, but is identical to what they are willing to pay. Some individuals will pay an extremely high price for the convenience and speed of a railroad journey; others might use the railroads only if they are free. The only coherent way to represent the value of a good or service is as a demand schedule. At very high prices, demand will approach zero. At low prices, it may be very great.

Dupuit's form of economic accounting became influential in the Corps des Ponts beginning in the 1870s, but it was at first received with a mixture of opposition and incomprehension. His comparatively low measure of the utility of public works was suspect among Ponts engineers. Still worse was his argument that really useful works could pay for themselves, provided charges were allocated not in proportion to expenses, but to the various utilities of transport. Passengers and shippers who benefit most from rail transport should pay most. In this way, the increase in public utility brought by a new rail line could be turned into revenue without discouraging any shipment that can at least pay the variable costs of transport. This economic strategy, he pointed out, is equally applicable to state and to private industry, and leaves no special reason for public ownership of rail lines or canals.[38] Dupuit was a militant laissez-faire liberal. He backed his convictions with mathematics. "Custom treats [politics] as a moral science: time, we are convinced, will make it an exact one, borrowing its methods of reasoning from analysis and geometry, to give its demonstrations a precision they now lack." "Those who prefer to apply their economic doctrines with moderation are like geometers who see admirable flexibility in the view that the sum of angles in a triangle is sometimes a little more, sometimes a little less, than two right angles." Moreover, he held the certainty of mathematical political economy to be definitive for policy. The proper role of the lawmaker, he explained, is "to consecrate those facts demonstrated by political economy."[39]

Dupuit's liberalism cut rather too close to home for many engineers of the Corps des Ponts. He was criticized by the engineer Louis Bordas for confusing utility with mere prices. Bordas also challenged Dupuit from the standpoint of practice. These schedules of demand as a function of price, he held, are at best purely hypothetical curves, and can never be known. "How can we build a theory on so variable a foundation, one that depends entirely on the taste and the fortune of every consumer?"[40] Dupuit acknowledged that some trial and error (*tâtonnements*) would be necessary. But even if "a rigorous solution is impossible for practical reasons, this science can at least provide means to approximate it." He added that political economists, like geometers, "have all the more reason to apply rigorous principles to the elements of this science because the available data are relatively incomplete or uncertain."[41] A few decades later, as chapter 6 shows, Dupuit's arguments were in fact translated into strategies of quantification, suitable for managing public works.

One more general line of approach to the problem of tolls within the tradition of French public engineering was developed in the 1880s by Emile Cheysson. Cheysson's career exemplifies better than any other the union of administration, reform, economics, and statistics that was available to engineers of the Ponts et Chaussées. After graduating from the Ecole Polytechnique and the Ecole des Ponts, Cheysson worked during the 1860s as a railway engineer, and then in the early 1870s for the iron works at Le Creusot. In 1877 he rejoined the French bureaucracy with an assignment in statistics and the general economy of public works. Later he directed the preparation of a new topographic survey of France, and he soon became known for his elegant statistical charts and maps. He worked with statistics as a patron and reformer, and not merely as a statistician. In the mid-1860s he had become associated with Frédéric Le Play (himself a product of the Ecole des Mines), and thereafter he was deeply committed to the ideals of social reform championed by Le Play's group.

Cheysson did not want to see the value of statistical study diminished by an excessive devotion to mathematical rigor. As one of the judges in an 1886 prize competition on mean values, he was so disappointed at receiving only one entry—and that one merely mathematical—that he wrote an essay of his own for the commission report. The prize topic had been his idea, part of his campaign to develop a general method of statistics. He considered that engineers needed to understand statistics to promote the skillful management of workers. He wanted to use numbers to divert economics from its abstractions, emphasizing instead the "study of the conditions that produce the well-being, the peace and the life of the greatest number." This would promote contentment as well as efficiency. Sanford Elwitt suggests that this engineering ideology became the basis for a hegemonic social liberalism of the fin-de-siècle. To use his pun, Cheysson built a bridge between Le Playist reformers and republican social liberals.[42] But Cheysson's social engineering seems rather less of a departure from old-fashioned employer paternalism than Elwitt implies. The objectification of workers must remain incomplete when decisions about how to treat them have not been reduced to formulas. Employee relations remained under Cheysson a matter of the good judgment of patrons, informed by statistics but not determined by them.

A reverence for good sense and sound judgment, as opposed to mechanical calculation, is also to be found in Cheysson's views on political economy. Like so many engineers, he took physics as his model. Economics, as usual, suffered by comparison. It lacked, he said, a common unit: the value of money is too changeable, and utility is impossible to measure. Unlike many others, he did not pursue energy as an alternative.[43] Instead, he conceded that economics can make no pretense of being an exact science. This remark was directed against certain pretenders, such as the marginal utility theorists. "Despite ingenious attempts, the rigorous procedures of algebra have proven sterile in application to this order of phenomena, for the equations are incapable of embracing all the facts."[44]

Still, Cheysson did develop ideas tending to automatic decision criteria. His outstanding contribution to the mechanization of judgment was an article on the geometry of statistics, first published in an engineering journal in 1887. This was written in defense of specialized commercial education, and against the view that there is no school but that of practice to prepare a good businessman or industrial manager. All the skills of the engineer in improving efficiency and reducing costs will come to nought if bad decisions are made about products, materials, markets, and prices. Such, he argued, was the situation then prevailing in France. Geometrical statistics was put forward as a remedy. Unlike political economy, it was not a mere abstraction, "speculative analysis," but a quantitative tool developed to solve practical problems in public and private affairs. It would permit the maker of decisions to avoid blind groping toward a best price or optimal tax rate, and instead to calculate directly a valid solution.

Cheysson defended the use of graphical methods for solving optimization problems, though he conceded that analysis could attain the same results. Analysis required fancy mathematics, while lacking the intuitive appeal of that *langue universelle*, graphical statistics. Suppose we want to determine how much to charge for railway travel on some line or network. We must plot two curves: one, like Dupuit's, for demand, and one for costs, each as a function of charge per kilometer. These curves may be hard to measure, he conceded, but they really exist. Once they have been drawn, it is easy to plot a curve of net revenue, and to locate its peak. This, from the standpoint of the railway company, is the quantity to be maximized. It can, he claimed, be a rigorous solution. In some cases extrapolation may be required, but only if the optimum rate is outside the range that the railway companies have tried. Such was the case for the Austrian Nordbahn, whose *zone experimentale* he found to be far above even the profitability optimum.

It was natural to apply such analysis first of all to railways, where rates were closely regulated. Cheysson argued, however, that his methods had great generality. His curves could be used to find optimum wages, and for that reason should not be ignored by friends of workers. They could guide investment decisions, or the choice of sources from which to purchase materials, or even tax rates and tariffs. He recognized one serious limitation of his method: it could not reconcile discrepant aims. The best price from the standpoint of the producer is not the same as that for the consumer, nor will the treasury and the taxpayer easily agree. For that reason, other engineers sought a basis for calculating a just price as well as a revenue-maximizing one. Cheysson left such considerations to the good judgment of the responsible parties. But to solve the problem even from one standpoint seemed to him a great advance, and in some cases, as with the Austrian Nordbahn, it pointed to changes favorable both to consumers and to the company.[45]

WALRAS CONFRONTS THE POLYTECHNICIANS

"Economics!" exclaimed Divisia in his celebration of French engineer-economists.

> How far are we from its resonant controversies that go round and round through the decades or the centuries, from its clever and subtle dissections, the games of mandarins, from its previsions that are just the opposite of reality one time in two, from its experiments that really aren't and that lack even the value of a lesson in facts. Economics! After all, is it anything more than a job well done, as all our engineers must know how to do?[46]

The object of Divisia's scorn was the economics profession, and the methods of neoclassical economics. The nineteenth-century French have as strong a claim as the British to the invention of this theoretical economics. Was mathematical economics really so remote from the practical quantifying urge that prevailed among French engineers?

Philip Mirowski argues for what would at first appear to be the opposite view, namely that economics began to become mathematical late in the nineteenth century as a result of a concerted effort to copy the physicists and engineers. He adds, however, that they failed, that the mathematical analogies upon which they fastened were impossible to defend.[47] The arguments of economic critics like Whewell and Cheysson tend perhaps to support his imputation of failure. Not all engineer-economists, however, rejected classical political economy. Navier and his admirers stood behind the definitions and conceptual framework, if not the politics, of Say. Dupuit was more critical, and those who followed him seem to have felt little need to look outside the French engineering tradition of economic calculation. Contemporary economists generally returned the compliment. As François Etner points out, these engineers were in the business of solving problems by calculating utilities, not explaining the mechanisms of the economy. Their work did in fact lead often to general formulas, but for reasons less economic than administrative.[48]

The career of Léon Walras, the great French nineteenth-century protagonist of mathematical economics, highlights the differences between the calculating engineers and the economic school that would seem to be closest to them. Like A. A. Cournot before him, Walras was almost entirely unsuccessful in winning support or even interest from among what he regarded as the liberal ideologues who dominated political economy in France. He spent his entire career in exile, as he viewed it, at the University of Lausanne, Switzerland. Recent studies of Cournot and Walras, noting their almost complete isolation from the French legal and literary school of political economy, have linked them instead to engineering and scientific traditions—to "a scientific ideology that enshrined the example of classical mechanics, and . . . an institution—the Ecole Polytechnique—where a problem crystallized: the 'application' of mathematics."[49] Indeed, these economists did draw on its mathematical culture, but the history of their relations with the practical quantifiers is one of perpetual misunderstandings rooted in incompatible aims.

The Ecole Polytechnique was, by origin, an engineering school. The Revolution needed military engineers to help fight its almost uninterrupted wars. Its scientific orientation was linked to a tradition of practical engineering that has come to be associated with the enthusiastic revolutionary and inventor of projective geometry, Gaspard Monge. Under the Napoleonic empire, young men were enlisted in the army upon beginning their studies at Polytechnique. This was mainly to help enforce discipline, to stamp out the revolutionary tradition that had already taken hold there. Napoleon also moved the curriculum further toward engineering, and indeed military engineering. He shortened courses in advanced mathematics and chemistry to allow more time to study fortifications and related subjects.[50]

A radical reversal of this educational philosophy is often associated with the fall of Napoleon. Terry Shinn argues that under the Restoration government, so exclusive a focus on practical engineering seemed subversive of natural social hierarchies, and that the curriculum was for this reason shifted in 1819 toward theoretical science, and even literature. Laplace rather than Monge became the dominating presence. He wanted to make Polytechnique part of his science empire.[51]

So extreme a shift is implausible. On the one hand, many of the most distinguished scientists to emerge from Polytechnique, among them Biot, Fresnel, Ampère, Carnot, and Poisson, studied there before 1819. On the other, as Jean Dhombres points out, the introduction of new courses in 1819 on social arithmetic and on the theory of the machine suggests that the practical imperative remained powerful.[52] The instruction in social arithmetic, though, was very far from accounting. Evidently there were contradictory influences. Polytechnique students generally prided themselves on an ethos of unrelenting practicality. But this cannot be attributed to the curriculum, which seems not to have been decisive in forming their identities as engineers. If Polytechnique was a school of engineering it was no institution of nuts and bolts, or gravel and paving stones. Engineering at the Ecole Polytechnique was as abstract and mathematical as the study of roads and bridges or artillery could possibly be, and possibly even more so.

Such, also, is the style of mathematics to be found in Cournot's 1838 treatise on mathematical economics. Yet he was not actually a polytechnician, but a graduate of the Ecole Normale Supérieure. This was a more academic, research-oriented institution than Polytechnique, though the contrast became more striking after midcentury.[53] Cournot's model, as Claude Ménard points out, was not engineering, but rational mechanics, and much of his mathematics was translated directly from physics. He did not concern himself with the practice of banking, or the economy of steam engines, and he did not collect empirical formulas relating prices to the quantity of gold, or trading patterns to levels of prosperity. Although he began a book on probability and statistics soon after finishing his great work on political economy, he put no emphasis on empirical statistics in either. He treated practical recommendations as at best the fortunate by-products of a mathematically rigorous formulation of political economy.[54]

It is possible to read Cournot's economic mathematics as reflecting a general commitment to the rationalization of society. But in contrast to the economics of Belpaire, Navier, or Dupuit, his was better suited to provide metaphysical comfort than to furnish a concrete plan of administrative action. His strategy of economic mathematization excluded history, with its irrationality and perpetual disequilibrium. In his philosophical discussion, he insisted that there was an economic art standing outside of mathematical theory, and conversely that there must be space for a pure science separate from practice.[55] Ménard rightly sees this insight, and its preliminary working out, as Cournot's outstanding achievement. He was willing to pay the price of mathematical rationality by excluding the whole domain of *économie sociale*, all the complications that would muddy the pellucid waters of pure economic reasoning. Concrete economic decisions, he argued, involve so many complex factors that practical sagacity must outweigh scientific apprehension.[56]

Still, Cournot was deeply concerned that his mathematics describe something real. Currencies, even gold, fluctuate too much to serve as economic units; he aimed to show mathematically how a "mean price," analogous to the "mean sun" in astronomy, could define a stable reference frame in observational economics.[57] His economics was thus consistent with the commitment to measurement that was so characteristic of nineteenth-century physics.[58] Significantly, it was on just this point that Walras and Cournot parted company. In his letters to the revered older economist, Walras claimed to have gone beyond him mainly in the purity and rigor of his methods. "You," he wrote, "follow a route that takes immediate advantage of the law of large numbers and leads to numerical applications, while my work remains free from that law on the terrain of rigorous axioms and of pure theory."[59]

Walras did not always discuss his work this way. In his letters to Jules Ferry, an old acquaintance who became France's minister of education, he was much more eager to claim practical relevance for his theoretical insights. He urged that the pressing problem of railroad rates could not be solved until economic theory was better developed.[60] And Walras, unlike Cournot, did write on practical issues. He even became active twice in campaigns for economic reform: first, at the beginning of his career, in favor of free trade, and then, near its end, as an advocate of land socialization. But the self-characterization in his letter to Cournot is correct. Cournot framed his theory mainly in terms of macroscopic variables, such as the quantity of money. Walras's originality as a theorist owes principally to his deductions from an abstract model of free exchange, leading to an even more abstract theory of general equilibrium. His microeconomic approach could be used as a language to describe the behavior of a profit-maximizing firm, but Walras did not do so. Although he was genuinely interested in public policy, he did not work out the connections with his theory.

Walras's ties to the Ecole Polytechnique were, like Cournot's, ambiguous. His mathematics was not good enough to succeed in the competition for entry. He did, however, study as an external student at the Ecole des Mines, which, like the Ecole des Ponts et Chaussées, accepted as ordinary students only the most elite graduates of Polytechnique. Significantly, Mines was more aristocratic than Ponts, and perhaps on that account it was more indulgent of impractical knowledge. In any case, he did not work very hard to apply his mathematics to problems like railroad administration. Railroad rates were actively debated during the 1870s, when Walras published his theory. Quantitative solutions were pursued by numerous engineers, and not only in France. Economic liberalism could furnish no answer to the problem of rates, but only suggest that the market would reach the best solution if the monopolies were broken up. This was not what state administrators wanted to hear. They were looking instead for strategies of management and technologies of decision-making. The language of Walrasian theory might have been used to convert the political problem of setting rates into the economic one of finding a maximum of utility or revenue.

Unlike most Ponts engineers, Walras did not mind seeing these decisions reduced to mechanical calculation. But, as he himself insisted, a wide gulf separated his economic mathematics from practical questions of management. This kind of economics did not impress the polytechnicians. He had every reason to woo them, since he was desperate to gain a following in France. For a time, he considered that his best hope lay with the French Circle of Actuaries, dominated by polytechnicians. Its avowed purpose was to apply quantitative reasoning to economic decisions of all sorts.

The history of Walras's relations with them is instructive. In 1873, he presented a paper at a meeting of the Académie des Sciences Morales et Politiques in Paris in hopes of making his work known to the leading French economists. Disappointed, if not surprised, by their incomprehension, he was correspondingly pleased to hear afterwards from Hippolyte Charlon, who had learned of the paper from Hermann Laurent. Charlon informed Walras of the mathematical ambitions of the Circle of Actuaries, and offered its journal as an outlet for his work. Walras, for his part, declared himself pleasantly surprised to discover that he was not so isolated in France as he had thought.[61]

He soon sent Charlon a memoir, the crucial chapter of the *Eléments d'économie pure*, for separate publication, in the hope of drawing attention to his forthcoming book. After a long delay, Charlon reported that the *Journal des actuaires français* had decided not to publish his memoir. Although Charlon had found it "very remarkable and abounding in sound ideas," it was also "off the practical and positive course along which we have directed our Journal. There is a crowd of sciences that, more than political economy, employ or could employ mathematical methods. This is no reason for them to be the object of our publication." There seems, he speculated, to be an unfortunate "incompatibility of humor between economists and actuaries."[62]

Walras had no better luck with the mathematician Laurent. Laurent took the model of the physical sciences very seriously, and he wondered whether economic comparisons over time might be facilitated by using a measure of energy, rather than currency or utility, as the standard economic unit.[63] He appears, indeed, as a bit otherworldly, though in his conscious intentions he exemplifies the urge, typical for polytechnician-economists, to make economics practical. This, he thought, required that it be made mathematical.

In 1902, Laurent published a short book on political economy "according to the principles of the Lausanne school" of Walras and Vilifredo Pareto.[64] Clearly he did not reject their work. He saw it as promising, in contrast to those merely verbal theories that Laurent blamed for the failure of economists ever to agree on anything.[65] Economics divides naturally into four parts, he explained: statistics, "economic facts," the theory of financial operations, and theory of insurance. He honored Walrasian theory by including it under the heading of economic facts. But mathematics could only elevate economics to a proper science if it were closely linked with the study of empirical reality. This for him implied careful attention to statistics: economics without statistics is like physics without experiment. Laurent even wrote a volume on statistics, this "experimental part of political economy."[66]

This book was more about probability than about the empirical findings of census-takers and social researchers. We must allow that Laurent's empiricism was mainly a matter of good intentions. Still, it was real enough to lead the correspondence with Walras onto the paths of incomprehension. Laurent wanted to get outside the narrow constraints of general equilibrium analysis. Not content with economic statics, he sought a basis in economic theory for studying quantitatively the development of economies over time. It was for this purpose that he proposed the use of a unit of energy rather than Walras's ineffable "utility," as the basis of economic analysis. Walras responded that this would be valid only if energy was equivalent to utility at the margin—which he doubted—and that dynamical formulas had no place in his theory. "[I]n my desire to establish patiently the basis of a new science, I have so far more or less confined myself to the study of the phenomena of economic statics." Laurent was unconvinced, and Walras became bitter. There is no "profound knowledge," he concluded, at the Institute of Actuaries.[67]

Cheysson also belonged to the Institute of Actuaries, and his criticism of mathematical economics reflected a similar outlook.[68] The failure of Walras to win influence over these actuaries and economists from the Ecole Polytechnique, or to develop practical economic tools of his own, clarifies the standing of practical quantification in late nineteenth-century France. This was largely an autonomous tradition, cultivated more for administrative than for scientific purposes. The highly abstract models from which Walras built a theory of general equilibrium could scarcely influence the decision processes of engineering administrators. The philosopher Renouvier, another polytechnician, objected to Walras that the gap "between the science and the art of the engineer-economist (if you will permit me this expression)" is much greater than "that between the science and art of the engineer-mathematician."[69] Applied to Ponts-et-chaussées engineers, this claim would be very doubtful. But in relation to Walras, it was fully valid. Even before his disagreements with Charlon and Laurent, he insisted on distinguishing his aims from mere quantification. He refused to recognize Dupuit as his predecessor. Dupuit had written about statistical demand curves; he about utility optima.[70]

ECONOMICS, PHYSICS, AND MATHEMATICS

The pioneers of neoclassical economics depended heavily on mathematical physics for the theoretical structure they imposed on their discipline. Drawing inspiration from statics and energy physics, economists built up a set of mathematical models as impressive and as demanding as are to be found in any natural science. Yet the physicists were generally unenthusiastic, sometimes sharply critical, and not just in France. Simon Newcomb, the American astronomer and influential spokesman for "scientific method," provides one last example. Newcomb was an admirer of political economy, and highly favorable to the project of making it more scientific. He wrote an introductory treatise on political economy, which is full of mechanical analogies to economic processes. Yet, although the works of Walras and Jevons had been available for a decade, he did not employ the calculus, the indispensable mathematical basis for marginal economics. He insisted that a fruitful economics must be closely linked with statistics. And he criticized the British mathematical economist Jevons, arguing that it was useless to make subjective feelings the foundation for economics. One must instead focus on visible phenomena, human actions, which alone can be properly quantified.[71]

Why were physicists so unreceptive to mathematical economics? Certainly they could have understood the mathematics. But they were unable to see the point of a purely theoretical economics. With few exceptions, nineteenth-century physicists took measurement to be more central to their discipline than mathematical deductions. William Thomson, Lord Kelvin, once remarked that "when you can measure what you are speaking about and express it in numbers you know something about it; but when you cannot measure it in numbers, your knowledge is of a meagre and unsatisfactory kind."[72] It is unlikely that those who converted this into a motto and printed it in stone above the social science building at the University of Chicago realized that Kelvin was here complaining about the "nihilism" of Maxwell's physical theory, and would have viewed neoclassical economics with even less favor.

We should not blithely attribute this coolness to methodological commitments alone. Nearly all the critics discussed here were at least close to engineering, and many were professional engineers. The French, in particular, pursued economics as an aid to administrative decisions. Economics was not for them a pure research interest, in the way that physics, at least for some, was. So their objections were in part practical rather than scientific. Significantly, mathematical economics was more appealing to those who were indifferent to, or even opposed, applications of political economy than to those who were looking to rationalize economic decisions. Whewell appears exemplary from this standpoint. Toward the end of the century, Herbert S. Foxwell identified as one of the great merits of the new marginalist theory of Jevons and Alfred Marshall to have "made it henceforth practically impossible for the educated economist to mistake the limits of theory and practice or to repeat the confusions which brought the study into discredit and almost arrested its growth." He even considered that mathematical and historical economics were allies in opposing the misapplication of theory.[73] Mathematical economics had the modest virtue of demonstrable irrelevance, which was morally superior to spurious relevance.

Donald McCloskey has recently written, with no discernible enthusiasm, that the values of theoretical economics resemble those of mathematics much more than those of physics.[74] Modernist mathematics, as Herbert Mehrtens argues, has meant precisely a retreat from the world of space and time, flesh and blood; to a world in which *Geist* is no longer confined to a ponderous, suffering body.[75] Pure theorists have rested their claims for the soul of the discipline rather strongly on their scientific credentials. This is at best very doubtful. The economic writings of physicists and engineers, at least up to the 1930s, suggests that the ambitions of scientists have been more closely allied with ideals of quantification and control than with abstract mathematical formulation. Measurement was not simply a link to theory, but a technology for managing events and an ethic that structured and gave meaning to scientific practice.

CHAPTER FOUR

The Political Philosophy of Quantification

Civil society . . . makes incommensurables comparable,
by reducing them to abstract quantities. Enlightenment
changes whatever does not reduce to numbers, and finally
to identity, into mere appearance.
(Max Horkheimer and Theodor Adorno, 1944)

QUANTIFICATION has not yet become a topic in political philosophy. Not that its political dimension has been ignored. An abundance of seemingly contradictory views have been advanced by moralists, critics, and quantitative researchers themselves. This corpus of writings includes some ill-considered polemics, but also some nuanced and thoughtful discussion. The best arguments are by no means all on one side. Unfortunately, there has been little dialogue. Critics, especially on the left, present the quantitative mentality as morally indefensible, an obstacle to utopia. Advocates have sometimes answered their opponents, but usually by defending the legitimacy of quantification as a way of knowing, not of organizing a polity and a culture.

The intellectualist defense of quantification, to be sure, bears on the ethical issues. A system of demonstrably false or untestable dogmas, the product of state power and not of free persuasion, has obvious moral implications to anyone concerned about individual freedom. This point, indeed, has been at the heart of some of the most influential philosophical defenses of science in this century. John Dewey considered science an ally of democracy, and argued that scientific method means nothing more than the subjection of beliefs to skeptical inquiry. Karl Popper held it up as antidote to the century's totalitarianisms. Science, he argued, "sets free the critical powers of man." It means openness and universalism; scientists "speak one and the same language, even if they use different mother tongues." This is the language of experience, but not just of any experience. Science values experiences of a "public character," observations and experiments that can be repeated, and hence that need not be taken on faith.[1]

While Popper did not stress quantification in his political philosophy of science, his terms could easily be applied to it. A more rigorous language contributes to the project of universalizing experience. But for its technicality, it might be Daniel Defoe's "most perfect style, . . . in which a man speaking to five hundred people, of all common and various capacities, idiots or lunatics excepted, should be understood by them all in the same manner." Yet rigorous definitions and specialized meanings are critical to this avoidance of ambiguity. In John Ziman's more ambivalent formulation, the language of number may be contrasted to "normal, natural language," with its "loopholes such as ill-defined terms or ambiguities of expression," which permit one "to slip out of the noose of a line of reasoning." Scientific claims, like legal documents, "have to be written in a complex, formalized (and ultimately repellent) language."[2] There is a hint of paradox in this alliance of clarity and arcaneness, and appropriately so. Thinking about quantification from the broad perspective of social morality tends to turn contraries into obverses and to emphasize moral ambiguities.

OBJECTIVITY/OBJECTIFICATION

Although it is of course possible to use numbers casually and informally, quantification for public as well as scientific purposes has generally been allied to a spirit of rigor. The ideal calculator is a computer, widely revered in part because it is incapable of subjectivity. Mathematics has long been able to claim a like credibility since it is supposed, with pardonable exaggeration, to involve rules of discourse so constraining that the desires and biases of individuals are screened out. Nature, too, is often cast as the embodiment of what is alien and hence objective, but nature has various guises, and an opposite one has been exalted by Stoic moralists and romantic poets. Nature recorded impersonally by the camera or the illustrator can make a better claim to the image of objectivity, although (as birders know so well) this ideal is not without its contradictions.[3] Strict quantification, through measurement, counting, and calculation, is among the most credible strategies for rendering nature or society objective. It has enjoyed widespread and growing authority in Europe and America for about two centuries. In natural science its reign began still earlier. It has also been strenuously opposed.

This ideal of objectivity is a political as well as a scientific one. Objectivity means the rule of law, not of men. It implies the subordination of personal interests and prejudices to public standards. This has nowhere been more clearly recognized than in the work of the eminent quantifier Karl Pearson. Pearson's argument, indeed, is so clear and so uncompromising that most modern readers draw back from his conclusions.

Objectivity as impersonality is often conflated with objectivity as truth. Pearson, a firm positivist, made no such mistake. He emphasized its moral values even more than its epistemological ones. Always an admirer of religious institutions, if not of religious dogmas, Pearson was scarcely less explicit than Auguste Comte in casting science as the successor to Christianity. He argued in "The Ethic of Freethought" that science admits "no interested motive, no working to support a party, an individual, or a theory; such action but leads to the distortion of knowledge, and those who do not seek truth from an unbiassed standpoint are, in the theology of freethought, ministers in the devil's synagogue."[4] Method was a religious ritual that would permit freethinkers to expel the demon of interestedness.

This, naturally, would be good for science. But an education in science and its methods was just as important for nonscientists. Pearson wanted to reorganize the school curriculum around science, not in order to make technicians, but to provide the best possible moral instruction. The scientific classroom could be a factory for citizens. "The scientific man has above all things to strive at self-elimination in his judgments, to provide an argument which is as true for each individual mind as for his own." Science leads to "sequences of laws admitting of no play-room for individual fancy." "Modern science, as training the mind to an exact and impartial analysis of facts, is an education specially fitted to promote sound citizenship."[5] Science, in short, meant socialism: the elevation of general rules and social values over subjectivity and the selfish desires of the individual.

This exaltation of the objectivity of science is often confused with elitism. As defined here, though, it is anything but elitist. A Pearsonian education should make everyone an expert, and every expert interchangeable. In the event, Pearson found a way to make some citizens more objective than others. But we should not fail to recognize the ethic of puritanical self-denial that pervades his writing. His objectivism would turn even the human subject into an object, to be formed in accordance with social needs and judged according to strict, uniform standards. Charles Gillispie and Donald Worster, from opposite perspectives, argue that the spirit of objectivity in Western science entails no small degree of alienation from nature. Evelyn Fox Keller adds that the control of nature is also the control of self.[6] Pearson's *Grammar of Science* displays this with unexampled clarity.

This challenge to subjectivity has important consequences that are not often recognized. The strong self generally belongs to a social elite. This has been at least implicitly recognized in the educational systems of hierarchical societies, which have almost always conceived their mission in terms of the formation of character and not merely the acquisition of knowledge, still less of technical skills. Nineteenth-century Germans who had received a classical Gymnasium education distinguished themselves from the hoi polloi by their *Bildung*. This was a rich concept, implying culture or cultivation as well as education. Its literal meaning is form, the formation of character. Jan Goldstein shows that French elite education at the same time was fixated on the Cartesian *moi*, the unitary self, which had to be defended against a variety of forces tending to fracture it. Significantly, Karl Pearson followed Ernst Mach in denying any continuity or integrity to the self, whose function could now be replaced by rules and methods.[7]

The educational formation of personal identity was always, implicitly or explicitly, the formation of a culture, usually an elite culture. An insistence on quantification tends to break that culture down, or to compensate for its absence. The American political scientist Harold Lasswell remarked in 1923 that formal expertise was anything but "monarchical." The American political system, he argued, made greater use of quantified, objective knowledge precisely because of its democratic character. By contrast, the British could rely on less formal modes of reasoning and communication because their political and administrative leaders made up a cohesive elite.[8]

The relation of quantification to cultural openness needs to be explored more than is possible here. The current politics of multiculturalism has made scholars more aware than before that scientific methods have a gendered dimension as well as an ideological one. It is often argued that mathematics expresses the special culture of men, or even white men. Yet the situation is surely far more ambiguous, and the net effect of the modern emphasis on quantification has probably been to open up professional cultures to women and ethnic outsiders. Exemplary in this regard is the insistent quantifrenia that prevails in the bureaucratic management of diversity. Affirmative Action offices and courts cannot very easily second-guess every employment and salary decision by a corporate office, university department, or law firm, but they can assemble numbers to establish a prima facie case for discriminatory practices by this or that unit.

It would be worth inquiring into the effect on diversity in American corporate offices of the rise of business schools, teaching highly quantitative management strategies. In Europe and America, mathematics has long been gendered masculine, and this has often worked to exclude women from the sciences and engineering. But the impersonal style of interactions and decisions promoted by heavy reliance on quantification has also provided a partial alternative to a business culture of clubs and informal contacts—an old-boy network—that was and remains a still greater obstacle to women and minorities. Little wonder that the "culture of no culture," to borrow a phrase from Sharon Traweek's study of physicists,[9] is now being vigorously promoted in a variety of contexts by the European Community. The language of quantification may be even more important than English in the European campaign to create a unified business and administrative environment. It aims to supplant local cultures with systematic and rational methods. A revealing French cartoon image depicts a diverse humanity entering the business school at Fontainebleau, and identical white, male, business-suited eurocrats coming out. Its resonances are simultaneously egalitarian and oppressive.

In the quantitative social sciences, the objectification of people has one other crucial dimension. Social quantification means studying people in classes, abstracting away their individuality. This is not unambiguously evil, though of late it has been much criticized. Much, probably most, statistical study of human populations has aimed to improve the condition of working people, children, beggars, criminals, women, or racial and ethnic minorities. The writings, especially private ones, of early social statisticians and pioneers of the social survey exude benevolence and goodwill. In print, though, they generally adopted the hardheaded rhetoric of factuality, which permitted women as well as men to assume the role of a scientific social investigator, and not merely of an agent of charity.[10]

This suppression of moral feeling in favor of rigor and impartiality was refused by many, and came at a high psychological cost for others. Often, though, the moral distance encouraged by a quantitative method of investigation made the work much easier. It is not by accident that numbers have been the preferred vehicle for investigating factory workers, prostitutes, cholera victims, the insane, and the unemployed. This was clear in early industrial Britain and France, and remained true with minor changes in early twentieth-century America. Middle-class philanthropists and social workers used statistics to learn about kinds of people whom they did not know, and often did not care to know, as persons. Counting was not impeded, but encouraged, by their alienness, for averages must always appear less meaningful when drawn from a population of strong and interesting personalities. A method of study that ignored individuality seemed somehow right for the lower classes.[11]

Finally, numbers have often been an agency for acting on people, exercising power over them. Michel Foucault and a host of admirers have on this account dealt harshly with modern social science in most of its manifestations. Numbers turn people into objects to be manipulated. Where power is not exercised blatantly, it acts instead secretly, insidiously. Ian Hacking and Nikolas Rose have been especially acute in recognizing the authority of statistical and behavioral norms, through which an oppressive language of normality and abnormality is created.[12] Those who fail to conform are stigmatized, and most others have internalized the values of an ever more pervasive bureaucracy of experts and calculators. Significantly, their power is inseparable from their objectivity. Norms based on averages advertise a beguiling independence of human choice that enhances their credibility.

TRANSPARENCY/SUPERFICIALITY

The first great statistical enthusiasm of the 1820s and 1830s grew out of a commitment to the transparency of numbers. The London statisticians, most notoriously, resolved that the facts should be allowed to speak for themselves, and that there was no room for opinions in the proceedings of a statistical society. This responded to a fear in the British Association for the Advancement of Science, with which the statisticians had contrived to affiliate themselves, that the statistical section would become too political. It also resonated with the strong empiricism of natural science in early nineteenth-century Britain, and indeed the Statistical Society's motto *Aliis exterendum* (to be threshed out by others) echoed the seventeenth-century Royal Society's *Nullius in verba*.[13]

Naturally this official exclusion of opinion is not to be taken at face value. Of course the British statisticians had opinions. This was a form of self-representation appropriate for particular rhetorical occasions. To appear independent of politics was advantageous not only in the company of the natural scientists, but also of judges. In nineteenth-century England, judicial discretion and personal knowledge were increasingly being hemmed in by rules appropriate for an emerging "society of strangers."[14] Such disinterestedness was especially valued whenever the statisticians wished to present themselves to higher powers in the capacity of unbiased knowers. That is, the statisticians were most inclined to emphasize their objectivity when they were weak, and had to appeal to the strong. But since the statisticians came overwhelmingly from the governing classes, this was by no means always necessary. At least there was no obstacle to using morally charged terms like "shiftless," "degraded," or "honourable," to describe the poor.[15] Still, there were times when opinionated humans were expected to stand aside and make room for the numbers to speak for themselves. And not only in Britain. This push for the openness of demonstration was in the best mathematical tradition; since the ancient Greeks, the idea of a geometrical proof has reflected an "ideal of open knowledge," with legal and political as well as epistemological implications.[16] Americans have shown a particular fondness for the antirhetorical rhetoric cultivated by British statisticians. Perhaps the most interesting discussion of the political morality of statistics, though, took place in France.

MAKING FRANCE A STATISTICAL SOCIETY

The old-regime statistical tradition in France was statist and secretive. Population numbers had obvious implications in the domain of power, and so the monarchy was interested in knowing them, but for the same reason it seemed unwise to permit them to be diffused freely. Condorcet stood for a different, more liberal view of numbers, which he hoped might be put into effect by the revolution. He was himself devoured by it, but circumstances soon became more propitious for his program. The Bureau de Statistique, which flourished around 1800, aimed to gather and publish information to promote an informed citizenry. This ideal unfortunately could not survive long within the Napoleonic imperium. The Restoration government was still less supportive of quantitative research. Even under the July Monarchy and the Second Empire, the French state was not very energetic in its statistical activities. The statisticians were acutely aware of this. "Why not face it?" wrote A. Legoyt in 1863. "Statistics is unpopular. Governments only provide them as a public service under pressure of opinion, and that, alas, of only a very few savants."[17] The desire for reliable public statistics never died out, but it was sustained mainly by energetic volunteers, working privately or taking their own initiative within some corner of the administration. Eric Brian shows how a few liberals and scientists struggled to preserve the statistical tradition in this uncongenial setting.[18]

The ethos of French statistics was on this account rather similar to that in England, where the tone was set even for official statistics by voluntary statistical organizations in London and Manchester. Perhaps the French were even more extreme. Statistics meant the presentation of numbers procured through direct observation. As late as 1876, the committee of the Académie des Sciences in charge of the Prix Montyon (for statistics) expressed doubts that there was much value in the mathematical manipulation of numbers gathered by others. This latter amounted to "economic conjecture" rather than factual knowledge. Statistics was also a decidedly liberal science. The statisticians had very little tolerance for state economic intervention. They believed deeply in the educational value of numerical facts, honestly reported and widely disseminated.[19] For many, public exposure was the only possible route by which their work could become influential.

Thus, for half a century after the Restoration, the dominant rhetoric of statistics in France emphasized transparent factuality. Echoing the London Statistical Society's policy that was declared two decades earlier, the newly created Statistical Society of Paris resolved in 1860 that "statistics is nothing else than the knowledge of the science of facts." It was, their statutes continued, an indispensable science for a liberal state: "It ought to provide the basis upon which society is governed."[20] The Saint-Simonian Michel Chevalier expressed this uncompromisingly: "A well-made statistic is an impassible testimony, above intimidation and seduction alike." For example, the statistics of education and of legitimate and illegitimate births provided "unimpeachable indices of the morality of populations."[21] Some decades earlier, one of Balzac's characters, Des Lupeaulx, pointed to the fetish of numbers as characteristic of the current economic order. "The figure is always decisive for societies based on personal interest and on money, and such is the society which the charter has made for us. . . . Hence nothing is better for convincing the educated public than a few figures. Our statesmen of the left claim that everything is definitively resolved by figures. So let us figure."[22]

As Balzac implied, this faith in numbers was wedded to a belief in progress through public information. A science of statistics based on subtle arguments and requiring long experience was poorly calculated to influence public debate, or to provide a justification for public decisions. Ineffable judgment is a highly undemocratic form of expertise. Statistics was supposed to provide thoroughly public knowledge, suitable, as Chevalier argued, for a democracy. Ideally, democratic statistics would be self-explanatory. Alfred de Foville argued that statistics could teach where to find safety, and where ruin lurked, but that governments were unlikely to listen. The best hope was to give citizens the means to judge the accomplishments of their leaders. Rest assured, he announced, "that wherever the struggle resurfaces between the champions of the general interest and that of private interest, you will find us [statisticians] at our post, armed and ready to march."[23] Chevalier maintained, rather optimistically, that the most reliable and abundant statistics were published by nations with representative institutions, in particular by Great Britain. And why not? For these numbers show their own vast superiority to other nations.[24]

This was an excellent sentiment for public addresses. It rarely worked out so well in practice. As early as 1828, French and British statisticians had been embarrassed by the ostensible conflict between their pet idea that education was a cure for crime and a much-discussed French table of educational attainments and crime rates by département.[25] Whenever something like this happened, the statisticians found reason to distrust appearances, and to probe more deeply. Not until the late nineteenth century, when statisticians had become more confident of their collective expertise, did they begin considering that complexity and confoundedness might be typical, and not exceptional. This happened at least as early in France as anywhere else. In 1874 Toussaint Loua editorialized in the *Journal de la société de statistique de Paris* that, though the government is to be congratulated for replacing "the novels and news of the old *Moniteur* by the statistical information in the [*Journal*] *Officiel*," undigested facts alone do not make a science. It requires instead careful comparison of these facts, to determine their significance and bearing. This can be no mechanical operation. "To ascend to causes, to be able to distinguish them amidst the multitude of diverse elements that act on society, to avoid oversights, requires great sagacity, sustained attention, a profound analytical spirit, and great rigor in deductions—all things that cannot be acquired, even by the most brilliant, except with long experience."[26] André Liesse made the point still more strongly in 1904: "To make a comparison so complex as this demands sustained attention, and a mind accustomed to the relativity of things. For purposes of influencing the general public, an argument loses force in proportion as it takes in more terms and comprehends a wider field. Statistical problems are not questions of elementary arithmetic for the common crowd."[27] In 1893, Fernand Faure called for a specialized school of statistics, to form the basis of a corps like Mines or Ponts et Chaussées. The contemporaneous effort of Emile Cheysson and Hermann Laurent to create a mathematical statistics expressed similar ambitions.[28]

Expert judgment might be acceptable in a close advisory relationship with powerful officials who were themselves authorized to act with considerable discretion. But public opinion was not easily bypassed in the nineteenth century, and the wider public has remained an important audience for public statistics up to the present. For their sake, transparency could not simply be abandoned. Standard index numbers provided the best hope of salvaging it. It was in fact the close relation of social numbers to public action, more than the demands of statistical science itself, that led to the creation of standardized measures and indices in statistics.[29] Although they may sometimes be useful for private consideration, they reflect strongly the public aspect of statistics. They are essential precisely where there is more accountability than authority. They epitomize the social role of objectivity.

Certainly there were earlier cases, but the interest in measures of the value of money that erupted in much of Europe around 1870 was a landmark event from this standpoint. Index numbers could never simply be observed; they normally involved extensive data collection and often difficult or at least tedious calculations. Their credibility required that they be calculated, even if from bad data, and it has never been acceptable to adjust a number on the basis of judgment alone, however expert. To be sure, mathematics counted for little in the absence of institutional power. The history of early efforts to use probability calculations as the basis for reforming the French judicial system is a paradigm of futility, notwithstanding the impressive scientific reputations of its protagonists. Even Condorcet, a political actor of note as well as a distinguished savant, could not get this project off the ground in the absence of solid institutional support.[30] Quantitative arguments had some weight. But they seem more often to reflect the efforts of those with little power to enlist instead the authority of objectivity. To be sure, that authority depended also on institutional power. At a minimum, support from an organized body like the Paris Statistical Society was necessary to create an index of prices or of salubrity. More typically, it would depend on the sanction of the state. For these things nearly always generate controversy, as the following example illustrates.

French statisticians were not slow to recognize the benefits of focusing attention on a few canonical numbers. They were especially alert to the possibilities of directing reform using medical statistics. Assessing the health of districts and institutions was inherently a comparative operation, and for this a measure of mortality, or alternatively of life expectancy (*vie moyenne*), was indispensable. Public-health statisticians were not entirely left to their own resources in measuring salubrity. Measures of life expectancy were pioneered by early writers on mathematical probability, mainly for insurance purposes. But actuarial formulas were not quite adequate for quantifying the health of various départements, much less of orphanages, prisons, and, worst of all, hospitals, where the number of deaths in a year might well exceed the number of patients at any given time. Clearly some more refined index than deaths per thousand per year was needed if statistics were ever to provide the basis for a compelling indictment of unhealthy institutions.

Such at least were Louis-Adolphe Bertillon's objectives in offering the Paris Statistical Society a more adequate set of formulas for mortality and for life expectancy. It is, he proposed, "natural and legitimate that the length of life be taken as a measure of the sanitary conditions in various human collectives." But there were at least eleven competing formulas, which were so much in disagreement that they led to discrepancies in the ordering of départements by health. Thus there was "nothing more arbitrary" than these measures. Arbitrariness is precisely what such measures were designed to exclude. To raise the study of mortality above controversy required a dose of objectivity. Bertillon proposed to replace them with "a truly scientific method," the "only one appropriate for determining the exact longevity of various places."[31]

To rank départements or arrondissements correctly required that gross mortality be supplanted by a measure that took account of the age distribution. On this, statisticians were generally in agreement. To measure the mortality of prisons, schools, or hospitals involved further complications. It was also of vital importance. "The mortality of various human groups is the most certain meter . . . for measuring those conditions, so manifold and complex, that determine the salubrity of an environment. It is thus important to have a method, not only precise but also uniform, and suitable for determining this mortality."[32] He did not think it sufficient for statisticians simply to agree on some conventional measure: "Science knows only one thing, and that is truth." The truth Bertillon sought was one that would take account of high mortality without producing absurdities such as an annual mortality greater than 100 percent. The population of hospitals turns over so often, he decided, that one can only calculate the mortality for the mean duration of stay.

This insistence that only one measure could be consistent with truth was all the more important because others disagreed with Bertillon's analysis. Toussaint Loua was no less convinced of the need for a uniform measure of mortality by which the healthfulness of diverse institutions could be compared. Bertillon's measure, however, he judged to be flawed. He did not like having an index for hospital patients that was calculated in a different way from that for other populations. This would unnecessarily narrow the basis for comparison, when the broadest possible basis was the great desideratum. It would be better, he argued, to calculate mortality per day.[33] Bertillon was not convinced. He responded that the probability of death in a hospital is by no means proportional to the number of days spent there; Loua's methods would permit a hospital to halve its mortality rate by doubling the period of confinement. The real unit of comparison must be the particular malady, and not the day.

This minor debate shows that statistical standardization did not come automatically. Disagreement about new research, after all, is found throughout the sciences. It is more crucial that they saw the importance of reaching consensus. They agreed that the effective administration of hospitals and other institutions required an objective basis of comparison, which could only be quantitative, and that science was the proper basis for establishing such a measure. Science, that is, supported by the state.

Faith in numbers could, of course, be ridiculed. Foville remarked in 1885 that in the theater, "as soon as a statistician comes on the stage, everyone prepares to laugh." An aspiring prefect in Edmond Gondinet's *Le Panache* proposes to put the sexes in balance by marrying (immediately) "one and one half men with three women minus a quarter per kilometer squared." A Labiche comedy has the heroine narrowly escape marrying a certain Célestin Magis, "secretary of the Statistical Society of Vierzon," who can't understand why his rival, Captain Tic, didn't count the projectiles fired by both sides in the Battle of Sebastopol. "Statistics, madam, is a modern, positive science. It casts light on the most obscure facts. Thus, thanks to laborious researches, we have most recently come to know the exact numbers of widowers who crossed the Pont Neuf during 1860." (The answer was 3,498, "plus one doubtful").[34]

This was of course just humor. But it was barbed. The argument that statistical knowledge is inherently superficial, if not ridiculous, was already a common one in the nineteenth century. It is implied, for example by Frédéric Le Play's faint praise of statistics for the benefit of the Paris statisticians in 1885. Statistics, he explained, are not really crucial in states with a hereditary aristocracy, whose members have been raised to govern, and can do so almost by instinct. But since we have experienced a rupture in forms of government, now people who have no practical experience in public affairs can rise to high office. Statistics can help to compensate for this lack of practical experience, and on this account statistical knowledge should be required of those who govern.[35] This need for formal knowledge was widely recognized. Jules Simon argued in 1894, "When there was an aristocracy, a ruling class, one could take for granted that future administrators and future legislators would have received from their family the traditions of their craft. In a republic, where anyone can be anything, the most ignorant may be assigned the most difficult functions."[36]

TWO-DIMENSIONAL CULTURE

This charge of superficiality has two basic forms, one emanating from the left, the other from the right. Le Play's sympathies were manifestly on the right; he preferred the deep understanding of those born to power to superficial expertise. A more recent version, which displays nicely some implications of the statistical constructivism proposed in this book, comes from an essay on rationalism by Michael Oakeshott. The rationalist, and this would no doubt apply a fortiori to the statistician, is for Oakeshott "a foreigner or man out of his social class, . . . bewildered by a tradition and a habit of behavior of which he knows only the surface; a butler or an observant house-maid has the advantage of him."[37] On this account, one might expect the rationalist to be ineffectual. But the tone of Oakeshott's essay was despairing, not superior. Rationalism is a cancerous growth on society, destroying its rich inwardness and leaving only surfaces. By transforming, really negating, a culture, it can become powerful after all. It is an effective tool for understanding a world it has itself helped to construct. It is no less shallow for that, since it has never understood the world we are losing.

The left critique, too, has an element of nostalgia. It comes to us from very nearly the same time, the early postwar period, but now from Frankfurt (and Los Angeles) rather than England. Though advertised as a species of Marxism, a sweeping critique of statistics is almost inconceivable from Marx himself, who spent many years buried in the British Museum assembling numbers out of Parliamentary reports for *Das Kapital.* Max Horkheimer and Theodor Adorno argue in *The Dialectic of Enlightenment* that positivist science replaces "the concept with the formula, and causation by rule and probability." In this form, they thought, knowledge gives up its critical edge. It sees only the linear, not the dialectical. Much better, argued Herbert Marcuse, to attend to Hegel than to the positivists.[38] But the Frankfurt critics were moved to oppose the calculative mentality by more than a longing for the coming revolution. Horkheimer and Adorno deplored the instrumentalist view of nature, with its emphasis on acquisition. Adorno, as we have seen, invoked the quantitative study, and destruction, of culture to exemplify the empty values of capitalism. Mass culture was the enemy. It had not grown up spontaneously, but out of the hollowness of the calculative culture industry. True culture could never be measured, but an increasingly superficial society conceals ever less from those who cannot know except by counting.

It has been urged that objectivity, in its various meanings, is characterized rather by what it omits than by any positive characteristics of its own. Lorraine Daston and Peter Galison write: "Objectivity is related to subjectivity as wax to seal, as hollow imprint to the bolder and more solid features of subjectivity."[39] For them, and for much of this book, the absence in question is the unique, interested, located individual. It involves an ethic of personal renunciation on the part of those who construct knowledge and make decisions. To adopt this ethic neither implies nor presupposes that a person lacks the rich local knowledge exalted by these critics of quantification. But if not, then the self-sacrifice demanded is all the more extreme. Unless you become like outsiders, you shall never enter the domain of quantitative science. The ultimate outsider is the machine, and it is rapidly becoming the greatest in the kingdom of quantification. Mathematics is so highly structured that most computations, and some symbolic manipulations, can be left to computers—that is, can be made independent of anything we would care to call understanding. Inevitably, meanings are lost. Quantification is a powerful agency of standardization because it imposes order on hazy thinking, but this depends on the license it provides to ignore or reconfigure much of what is difficult or obscure. Whenever a reasoning process can be made computable, we can be confident that we are dealing with something that has been universalized, with knowledge effectively detached from the individuality of its makers. As nineteenth-century statisticians liked to boast, their science averaged away everything contingent, accidental, inexplicable, or personal, and left only large-scale regularities.

It is important to add that quantification has the virtues of its vices. The remarkable ability of numbers and calculations to defy disciplinary and even national boundaries and link academic to political discourse owes much to this ability to bypass deep issues. In intellectual exchange, as in properly economic transactions, numbers are the medium through which dissimilar desires, needs, and expectations are somehow made commensurable. The literary technologies of the modern scientific paper are inadequate to convey the tacit richness of experimental technique, or, for that matter, the arcane craft of formulating theories. For most purposes, especially when knowledge crosses the boundaries of community, such intimate knowledge is not particularly desired. The value of superficiality has been argued by Peter Galison, who observes that the interactions among instrumentalists, experimentalists, and theorists in physics are a bit like a trading zone, involving, say, European merchants and South American Indian craftsmen or farmers. Religious, cosmological, and ideological meanings are lost; the traders only need to agree on a price, a number or ratio. Similarly, it is often mainly predictions and measurements that pass between experimental and theoretical physicists.[40] It may even facilitate easy communication if the rich craft techniques of both communities are simply ignored.

Much of what follows will be grist for the mill of those who attack the quantitative mentality as superficial. So it is important to add that there is no fixed limit to what can be quantified, and that a richly nuanced or profound analysis of a large question is never logically excluded by the attempt to quantify parts of it. Often, though, it is politically excluded. For quantification is not an unmovable mover, or the product of a conspiracy, by which a culture has been overturned. It reflected values before it created them, and its massive expansion in recent times has grown out of a changing political culture. Yaron Ezrahi has argued powerfully for a symbiosis between democracy, American style, and a faith in surfaces.[41] This superficiality is called, with some justification, openness, and it is designed to drive out corruption, prejudice, and the arbitrary power of elites. To no small degree it succeeds, though agencies exposed to democratic scrutiny are often also adept at the play of masks. When it does succeed, this nearly always comes at some cost in subtlety and depth. And often, as Oakeshott suggests, their disappearance from discourse may imply their disappearance from the world as well. In no other way has the power of numbers been so impressively displayed.

PART II

TECHNOLOGIES OF TRUST

Whatever some may think of the great advantage to trade,
by this favourite scheme, I do very much apprehend, that in
six months time, after the act is passed for the extirpation of
the gospel, the Bank and East-India may fall, at least,
one per cent. And, since that is fifty times more than ever
the wisdom of our age thought fit to venture for the
preservation of Christianity, there is no reason
we should be at so great a loss, merely for
the Sake of *destroying* it.
*(Jonathan Swift, "The Abolishing of Christianity
in England," 1708)*

CHAPTER FIVE

Experts against Objectivity: Accountants and Actuaries

[T]here is no manufactory for actuaries where you can get one made to order.
(Edward Ryley, 1853)

I HAVE already discussed accounting as an emblem of quantitative practicality, in contrast to the detached and otherworldly outlook nourished within pure mathematics, and sometimes in the scientific disciplines as well. Practicality in the context of accounting is meant to imply a close contact with the world of production or management. In regard to natural science, I use the term broadly to refer to techniques for predicting and controlling phenomena. Obviously it does not follow that theorizing is impractical, even in this sense. What aids in comprehending events will often contribute also to their reliable manipulation. Still, it is time to abandon the identification of scientific knowledge with rigorous, formalized theory. The power of science depends above all on an ability to organize a skilled labor force in seizing hold of the world.

The first two chapters of this book are mainly about how quantification works to project power over large territories and a diversity of objects. In chapter 3 I began to turn my attention to the other side of this problem, how an ethic of exactitude has helped to form the identities of the researchers themselves, and in chapter 4 I proposed that it is linked to an ideal of self-sacrifice. A few scientists, such as Karl Pearson, seem to have adopted it for personal and, in a broad sense, religious reasons,[1] and there is doubtless a pervasive religious component to this spirit of renunciation. In this book, though, I am emphasizing rather its public dimension: objectivity as an adaptation to the suspicions of powerful outsiders. In the three chapters to follow, this climate of suspicion is an explicitly political one. They contain stories of professions that, in varying degrees, abandoned their open reliance on expert judgment in the name of public standards and objective rules. It was never a voluntary sacrifice, but emerged always from a scene of intense pressure or even bitter rivalry. This aspect of the pursuit of quantitative rigor has not often been appreciated. The language of pure and applied science suggests that quantitative professionals pursue rigor and objectivity except so far as political pressures force them to compromise their ideals. But this is exactly wrong. Objectivity derives its impetus, and also its shape and meaning, from cultural, including political, contexts.

And this does not apply only to accounting. Here, again, I am using accounting, along with insurance mathematics and cost-benefit analysis, as an exemplar, in order to clarify processes implicit, though often less manifest, in all knowledge-making. This chapter and the following two are concerned with disciplined practices involving economics and finance, practices whose importance for large organizations—above all the state—is unmistakable. I argue that the pursuit of objectivity in these studies is not undermined by their accountability, but defined by it. Rigorous quantification is demanded in these contexts because subjective discretion has become suspect. Mechanical objectivity serves as an alternative to personal trust.

It deserves emphasis that these contexts are important ones, and that these fields are as significant in their own ways as more academically respected ones like physics, chemistry, and medicine. Historians, sociologists, and philosophers of science cannot afford any more to look down our noses at bureaucratic knowledge-making. The reason for attending to it here is not merely or mainly because quantitative practices linked to accounting are illustrative. Still, I consider that they are illustrative, and in part 3 I will extend the analysis to show how quantification works as a technology of trust in the scientific disciplines as well.

ACCOUNTING AND THE CULT OF IMPERSONALITY

To argue that objectivity is defined by its context, we need to be able to say something about what a form of knowledge would look like outside of that context. The context of modern accounting involves large business organizations or the state, and often both. For a contrast, one might look to premodern bookkeeping, a decidedly less formal and intricate way of keeping track of assets and obligations. Here, instead, I'll begin with a discussion of how accounting might work if the profession were stronger—if the boundaries of the specialist community were less permeable. This is not simply a counterfactual, and it depends on no great leap of imagination. Rigorous objectivity and professional autonomy are opposite extremes on a continuum of possibilities that leaders of accounting have actively debated for at least sixty years.

"A face to face group has no great need of writing," observes Jack Goody, who proposes that bureaucratization was one of the crucial ingredients in creating a demand for literacy. Harvey Graff argues that business literacy first became important in Europe in the eleventh century, as extensive trade networks began to develop.[2] The importance of commerce for the growth of quantification is even clearer, and in fact numbers are often as plentiful as words on Babylonian clay tablets, Egyptian papyri, and early medieval letters. Still, this involved mainly the simplest kind of bookkeeping. "In a world in which businesses were small and managed by their owners on a day-to-day basis and in which no income tax existed, there was little demand for outside accounting services," observes R. H. Parker. The accounting profession got its start in mid-nineteenth-century Scotland and England, where its members exercised the very public function of presiding over bankruptcies and assuring creditors that they would be treated fairly. American and British accountants began just a little later to audit the accounts of public companies such as railroads and gas and electric companies, normally to meet the requirements of new regulations. Their role was to offer independent and expert assurance to shareholders and other interested parties that the books were fair and honest. The crucial ingredients here are independence and expertise. These guarantees of objectivity were indispensable because, as William Quilter told a parliamentary select committee in 1849, auditing is a matter of judgment, and no "dry arithmetical duty."[3]

This view that accounting attains a kind of objectivity through disinterested expert judgment has not lost its appeal. These days, the exclusive identification of accounts with numbers is sometimes criticized by accounting writers in the name of hermeneutics. To anyone who has read and been impressed by contemporary literary theory, it seems plain that an ideology of straightforward facts must give way to a language of interpretations and cultural meanings. The message of accounting hermeneutics is that financial affairs are never sufficiently straightforward to be adequately summarized in a mere table of numbers. A language of inference and interpretation, resting on the discernment that comes with true expertise, could provide much more helpful guidance to stockholders and creditors than a strictly tabular report.[4]

Significantly, this disavowal of mechanical objectivity leads to the vindication of professional expertise. And in fact, rather similar arguments have been made throughout the history of accounting. Accounting professionals have looked to fields like medicine as models of successful practice. Medicine, especially in the early and mid-twentieth century, meant powerful professionals whose expert judgment was rarely questioned. Even in quite recent times, many accountants have tried to claim the same prerogatives. An argument appearing in the *Accounting Re-*

view in 1965, for example, inferred the crucial importance of interpretation from the self-interest of the profession:

> The accounting profession's prime asset is an attribute known as *professional judgment.* Judgment, professional or otherwise, is a product of the mind. If judgment must be made synonymous with subjectivity, we cannot have objectivity and a profession at the same time. Clearly we cannot accept such a view of objectivity. Rather, we must show that the exercise of professional judgment and the desire for objectivity are complementary propositions.

To be sure, this professional judgment had to be free of "perceptual defects." Clear vision, the author continued, should follow from selection and discipline. Accountants properly imbued by a strict training with general principles and goals would command a form of judgment "more effective, more controllable, in attaining a desirable state of objectivity."[5]

A recent and influential historical study, largely untainted by the new constructivism in accounting, laments the increasingly mechanical use of management accounting by uninspired executives trained to manage "by the numbers." The authors, Thomas Johnson and Robert Kaplan, identify two classes of villains who share responsibility for this stultifying drive for rigor. One is the accounting professors, who reside in an ivory tower and yet recognize no disadvantage in their isolation from the real world of production, customers, and contracts. They teach their students to trust in knowledge and action at a distance, which after all is what they are themselves obliged to do, as if the numbers spoke for themselves. The other, still more culpable, is government regulators, who have succeeded in turning management accounting into public accounting. That is, the categories appropriate for managing a company have gradually been supplanted by categories officially specified for calculating taxes and preparing external financial reports. Accountants and executives, trained to revere numbers, ignore this and rely on them anyway. The curriculum for accountants has come to be shaped by research ideals and regulatory demands rather than by business needs.

Like most jeremiads, this one supposes that things were once better. Through the 1920s, Johnson and Kaplan argue, management accounting was mainly the business of engineers, who "invariably relied on information about the underlying processes, transactions, and events that produce financial numbers." Yet they recognize that developments internal to firms have contributed to the new aloofness from reality. In particular, the use of profitability measures like return on investment to evaluate the performance of subordinates encouraged reliance on accounting numbers.[6] Alfred Chandler and his students have shown to what extent the development of accounting was associated with the growth of complex, integrated firms, even before government regulation became so intrusive.[7] From this standpoint, at least, the large capitalistic business corporation is more like a government than it is like a small company. The ambit of accounting is first of all administrative and political. That icon of economic rationality, Robinson Crusoe, could manage his island quite well enough with some basic bookkeeping, as did most small and even medium-sized firms until the demands of income taxation and other forms of public regulation made this impossible.

The drive for rigor and standardization, I argue, arose in response to a world in which local knowledge had become inadequate. Economic concentration meant that people could no longer look their trading partners in the eye. Complex and risky contracts like life insurance, offered by faraway companies, led to demand for government oversight. Banks, which once had been managed locally by and for the benefit of insiders, began exchanging notes and obligations on the open market.[8] Mechanical objectivity was not the only possible response to these changing conditions. Beginning in the late nineteenth century, the elite of accountants in Britain was to be found in independent accounting firms. Independence provided some assurance of impartiality. It was important also for accountants to have a wide reputation for probity and skill, and this came to be guaranteed by the names of a few large firms. These firms spread also to the United States, a more challenging setting in which to maintain a gentlemanly profession.

The shift away from elite disinterestedness toward standardization as the basis for accounting objectivity began in earnest in the 1930s. The occasion was the Depression, but more particularly the efforts of a new regulatory bureaucracy, the Securities and Exchange Commission (SEC), to restore investor confidence. The governmental way of achieving this was through the promulgation of strict reporting rules, so that anyone could read a company financial statement, and so that fraudulent misrepresentation could be easily recognized and punished. Thus it was not for their own reasons that accountants narrowed their vision and identified their craft with rigorous calculation. The shift toward objectivity meant a loss of autonomy, and was a failure of the profession. To fend off an imminent bureaucratic intervention, the American Institute of Accountants established its own mechanism of standardization. In 1934 it voted to establish six "rules or principles" of accounting. In 1938 it established the Committee on Accounting Procedure, which was replaced by the Accounting Principles Board in 1949 and then by the Financial Accounting Standards Board in 1972. These have acted almost like governmental agencies.[9]

There was much disagreement among prominent accountants about the desirability of standardization. The balance of opinion, especially at the most elite levels, was against it. George O. May told the American Institute of Accountants in 1938: "There is no doubt a widespread demand for uniformity. . . . We should regard uniformity only as one of a number of ways in which accounts can be made more valuable, particularly to the unskilled reader. . . . We shall never be able to make superfluous either honesty and skill in preparing accounts, or intelligence in interpreting them." Walter Wilcox explained to his colleagues in 1941 that accountants "have a large audience that expects us to know what cost is. . . . Cost is not a simple fact, but is a very elusive concept. . . . Like other aspects of accounting, costs give a false impression of accuracy." Nobody, not even the chief accountant of the SEC, denied these complexities, though he justified the push for standardization with the argument that they were progressively being overcome through careful research.[10]

The SEC, however, was willing also to overcome them by fiat. It was interested less in accounting truth than in enforceable regulations. A notorious example was a Depression-era ruling that corporate book value should be based on the original cost of assets, not replacement cost. The SEC's reasoning was not obscure: investors were already nervous enough, and direct-cost financial accounting seemed to leave a minimum of room for self-interested manipulation.[11] But few accountants were satisfied by a rule preventing them from revaluing assets to take account of inflation or technological improvement. It led to the awkward situation that assets would have to be revalued, sometimes radically, whenever a company was sold. The rule, in short, seemed to favor expediency and precision over accuracy. It became the focus for an international discussion on the nature of accounting objectivity, pitting, we may say, philosophical realists against political ones. The participants were not idly debating abstract questions of philosophy. They were accountants, working through issues that had real implications for the practices as well as the self-definition of their profession.

OBJECTIVITY IN ACCOUNTING

In accounting as in other sciences, wrote the Australian R. J. Chambers in 1964, we can only claim objectivity when we know what we measure. If our objects are not defined, "it is quite impossible to speak of eliminating known biases and discovering true or estimated measures."[12] True values must be contemporary values; historical cost is meaningless until revised to reflect current conditions. Conventional rules cannot suffice to manufacture objectivity. But what if mortal accountants cannot apply the true standard to achieve consistent results? Chambers, conscious of the political need for accounting uniformity, defined this possibility away. He asserted without argument that agreement can never be obtained except by way of truth. An objective statement is one that any other informed person would make about the same subject matter. In this way he confounded at least two different senses of objectivity: following rules, and reaching truth. It was a handy conflation, and was widely accepted in the profession. The *Accountants' Handbook* characterized "objective" as implying "the expression of facts without distortion from personal bias."[13] An ally of Chambers invoked a Kantian understanding of explanation as the subsumption of particulars under law to ground accounting rationality and at the same time to explain how it could be made invulnerable to "emotive considerations."[14]

Others recognized that there might be a problem here. The bravest of the realists was Harold Bierman, whose article of 1963 had provoked Chambers. He conceded that to abandon accounting according to original costs would oblige accountants to face "a variety of choices"—whether, for example, to adjust financial quantities only to take account of changing price levels, or to base value on expected future cash flows, or even to use liquidation prices. "The accountant's task would be more complex," he warned. "The above suggestions would lead to manipulation of the reports," and would add to the regulatory burden on the SEC.[15] Still, he thought accountants should accept these challenges, in the interest of a better representation of the true state of affairs. He preferred not to dilute rationality with expediency, and he envisioned accounting as a measurement discipline analogous to astronomy and psychology.

Bierman's formulation shows how accounting realism might be allied to a commitment to a strong profession and faith in the discretion of experts. But this had become a minority stance, especially among accounting researchers. More convenient, and more popular, was a strong positivism allied, inevitably, to a quantitative form of behavioral research. The pioneers here were Yuji Ijiri and Robert Jaedecke, who declared against the realists that existence independent of observers has no operational meaning in the context of accounting. The problem faced by accountants is simple: "Accounting is a measurement system which is plagued by the existence of alternative measurement methods." The remedy was equally simple: "If the measurement rules in the system are specified in detail, we would expect the results to show little deviation from measurer to measurer. On the other hand, if the measurement rules are vague or poorly stated, then the implementation of the measurement system will require judgment on the part of the observer."

Objectivity, for these accountants, was a mechanism to exclude judgment. It could be "defined to mean simply the *consensus* among a given group of observers or measurers," and hence measured (inversely) as a statistical variance. That is, if several accountants give nearly uniform figures for book value according to one measurement scheme, and rather diverse ones according to another, the first is by definition more objective, whether or not it seems plausible. The importance of this kind of objectivity was not so overwhelming as to exclude consideration of "reliability," meaning accuracy. But it could not be neglected, for without consensus there could be no reliability either.[16]

Practicing accountants and researchers alike found this reasoning plausible, even compelling. Practitioners were acutely conscious that reaching agreement by following rules provided their most powerful defense against government bureaucrats and other meddlesome outsiders. The paramount need to minimize the appearance of subjective discretion—"managerial whim"—in financial reports was stressed in almost every accounting discussion of objectivity. Researchers also applauded the quantitative form of objectivity for its amenability to empirical—meaning statistical—research, and on this account it became the consensus concept of objectivity in accounting.[17] Nor were research accountants immune to the allure of "perfect operational objectivity," which could be realized "only where the entire accounting process was reduced to programmable sets of procedures."[18]

Still, a gap between methods, however constraining, and theoretical reasoning was an embarrassment. In general, uniformity and standardization were probably enhanced by the ostensible rationality that comes from explicit theoretical reasoning. Certainly it is a threat to normal procedures, however well standardized, if they have no credibility as true measures. Participants in this discussion over accounting by direct costs or present value tried to bring the ideals of truth and uniformity together. Robert Ashton, who subjected a population of accountants to a sample survey in order to measure the objectivity of rival accounting methods, was pleased to find that the theoretically preferred measure, present value, was indeed more objective (i.e., showed a lower standard deviation over different measurers) than its rival, accounting by original cost.[19] Most accountants, though, have assumed that uniformity would be possible only with clear and relatively inflexible rules. Their willingness to insist on standardizability, even where it violates the best judgment of expert practitioners, will rarely be found except in fields that are highly vulnerable to criticism from outsiders. The reluctance of accountants in many situations to admit the exercise of personal discretion is evidence of their public exposure, and in this sense their weakness.

That weakness derives from an absence of trust, which seems inevitable in this domain. Fortunes have been made and lost through the reinterpretation of financial categories; heroic entrepreneurship and criminal embezzlement may be distinguished by no more than a subtle point enunciated a few years back by the regulatory agencies. Income taxes mean nothing if definitions of investment income, depreciation, necessary business expenses, and capital gains are not defensible in courts of law. The strongest profession would be hard pressed to maintain the public credibility of expert judgment in the face of such challenges and temptations. The preferred bureaucratic and legal way of dealing with these issues is the promulgation of rules. As is the case with scientific laws, art and judgment are required to connect those rules or laws to the actual phenomena of experiment, observation, or economic life. But whereas scientists generally benefit from the order that this shared culture makes possible, economic actors strive perpetually to undermine it. Hence the presuppositions of accounting rules must themselves be codified and published, and so on until the whole Malthusian cascade presses up against the supply of paper and patience.[20]

It is important to note that the form of knowledge resulting from this relatively rigid quantitative protocol is decidedly public in character. It embodies, and responds to, a political culture requiring that as much as possible should be brought into the open. Judgment and discretion, normally the prerogatives of elites, are discredited. Anne Loft's historical study of cost accounting captures nicely the political resonances of quantitative objectivity. Though developed by American corporations in the late nineteenth century, cost accounting became important in Britain during the First World War. Economic mobilization upset private markets, especially for items needed by the military. How were prices to be determined? The government and industry might have simply negotiated a price. But a private agreement resting on no more authority than administrative judgment lacked credibility. An especially untrusting party was the trade unions, whose members were being asked to hold down wage demands in the national interest. They regarded price negotiations between the companies and Whitehall as an opportunity for collusion. They insisted on objective evidence that they were not sacrificing wages for the benefit of profiteers. So cost accounting, rather a new and undeveloped technology, was mobilized during the war to establish quantitatively that manufacturers were taking only a small profit above their actual costs of production.[21] Any economist can give good arguments why pricing according to cost plus profit is an inefficient way to run an economy. But in a situation of distrust it may be the most credible way to run a polity.[22]

Modern scholars almost instinctively regard this kind of quantification as a ruse, and the poor workers as dupes. But our habitual suspicions may go too far. If the bureaucrats and industrialists had the power to do whatever they wanted, they would not have had to seek refuge in quantitative rules. We are, after all, talking about public knowledge. Whenever such calculations are exposed to unfriendly eyes, deviations from standard practices can be noted. Unless the relevant expertise is completely monopolized by the parties actually performing the calculations, the rules become genuinely constraining, even if some room always remains for creative manipulation. The civil servants who are the heroes of this story had to become faceless, to let themselves be standardized by a quantitative protocol in order to minimize conflict and avoid stalemate. The authority of public bureaucrats and private contractors was suspect; the reaction of the trade unions clearly mattered. Since it was difficult, perhaps impossible, to coerce them, they had instead to be persuaded to acquiesce. What went on was not quite reasoned democratic discussion, but neither was it simple coercion or trickery. This is power not in Stalin's sense, but Foucault's. Potentially, at least, it can constrain the administrators almost as much as it constrains the workers. Quantification provided authority, but this is authority as Barry Barnes defines it: not power plus legitimacy, but power minus discretion.[23]

The drive for objectivity in accounting did not follow naturally from a logic of finance. Neither was it the result of unchecked power on the part of professional experts. It was a consequence of self-aggrandizing self-effacement, the methodological equivalent of gray suits, adopted by men who would otherwise have had even less chance of acting autonomously. Accounting is much less bound by rules than outsiders generally suppose. Still, its current status as the paradigm of impersonal rule-following is not groundless. Accounting embodies an ethic of self-denial that provides a model for the sciences. A guidebook for psychologists, for example, urges researchers to put aside hypotheses that approach but do not reach statistical significance: "Treat the result like an income tax return. Take what's coming to you, but no more."[24]

HIERARCHY AND DEFERENCE: THE BRITISH CIVIL SERVICE

Not everybody has found it necessary to give in to the view that what's coming to them is given by quantitative rules. There have been two main bases for resisting the advance of mechanical objectivity. One is the right to privacy, usually meaning private property. The other is a reasonable claim to the prerogatives of an elite. This is not mainly a matter of naked power, but rather of mobilizing a discourse that sanctions expert discretion. British actuaries in the mid-nineteenth century had only a new and badly divided professional organization. Neither could they claim gentle birth, or consistently high educational attainments. But the potential regulators they faced were not very insistent, in part because they feared they were intruding into the proper domain of free enterprise. Also, the actuaries' argument that they deserved to be trusted as experts and gentlemen had then a force that is now almost lost to memory. A few remarks about professionalism and bureaucracy in Britain will help to appreciate what follows.

This story centers on a set of parliamentary select committee hearings in which the English actuaries ably defended the intricacies of their craft. It took place in 1853. In 1854, a report known by the names of its principal authors, Northcote-Trevelyan, called for recruitment into the civil service by examination. Those recommendations were not implemented until 1870. Until about that time, the British had still a rather rudimentary bureaucracy, recruited unsystematically but mainly on the basis of patronage. Its weakness was especially notable in regard to the regulation of companies. The Joint Stock Companies Act of 1844, inspired largely by evidence of insurance fraud, led to the creation of a staff of two, a registrar and his assistant, to try to keep track of many hundreds of companies. This was not a formidable regulatory presence. The actuaries did not answer to a permanent bureau, but directly to Parliament, which was scarcely in a position to intervene forcefully on its own. Unless it was prepared to create a large bureaucracy, it had to rely on the friendly cooperation of the actuaries themselves.

The Northcote-Trevelyan reform is relevant to this story mainly for the attitudes it revealed. It made few concessions to the practical applicability of formal knowledge. The upper grade of the civil service was to be recruited from those with a classical education, normally from Oxford or Cambridge, based on their performance on an examination emphasizing dead languages and geometry. This elite was to be made up of generalists, whose intellectual training and cultural background would, it was supposed, enable them to pick up in short order whatever technical knowledge they might need for any particular position. They were highly mobile, moving from department to department.[25] Specialized knowledge fit a man only for the lower levels of administration. And there were serious doubts whether schools were the proper place to acquire appropriate expertise. When esoteric knowledge was called for, British administration was rather inclined to respect the information and skills picked up on the job by businessmen and technicians. Even in fields like electrical engineering, the value of scientific knowledge was sometimes powerfully doubted up to the end of the century. Jose Harris notes that as late as the interwar period, it was normal for inquiries on social questions, and even on the income tax, to ignore the academic economists, or to include one only because of his practical experience or political position.[26]

This administrative style proved remarkably durable. In 1974, Hugh Heclo and Aaron Wildavsky explained how a few top officials in Treasury were able to negotiate the government budget without relying on fancy accounting and without much help from expert staffs. Two factors were crucial in permitting the highest civil servants to dispense with formal knowledge and explicit procedures. First, the government was able to keep secrets, to maintain privacy in regard to the public business. Decisions were made known only after they were made. This owes partly to British law, which provides protection for official secrets.

It owes even more, though, to the second factor. The British government, Heclo and Wildavsky found, was comprised "of people whose common kinship and culture separates them from outsiders." That culture was made possible by a shared high socioeconomic status and elite educational background, but even more by a career pattern that, near the top, involved frequent movement between departments. This created unity in the upper civil service, and promoted trust. "The one inescapable theme in virtually every interview we conducted is the vital importance participants place on personal trust for each other," write Heclo and Wildavsky. "Treasury officials are able to do their job because there are relationships of trust." It was not blind, but highly nuanced: "By their own account, the most important skill Treasury people learn is 'personal trust and where it should be put.'" Of course, personal trust is an important factor in every form of human organization, including the much more open and vulnerable American form of bureaucracy. So this is a matter of degree. The British administrative elite was sufficiently closed and cohesive to rely more on people than on impersonal knowledge, and to depend minimally on formal expertise.[27]

Economics was anything but unimportant to Treasury officials and other administrators. It was in fact so important that they were unwilling to leave it to the academic experts. Economics was no specialty, like law and medicine, but the shared property of educated generalists, like moral philosophy and politics.[28] In recent decades, the British government has found some use for quantification in complex and highly factual inquiries. But the conclusions might in the end be simply set aside, as happened in the case of the monumental Roskill cost-benefit analysis in choosing the site for a new London airport.[29] Since about 1960, and partly in imitation of Americans, the British government has sometimes relied on formal quantitative analysis to plan highways and underground lines. Margaret Thatcher assigned accountants and cost-benefit economists a central role in gaining information that might help penetrate the relatively autonomous National Health Service.[30] It is still not clear that there has been a fundamental realignment at Whitehall. In some respects, quantifiers have benefited from the British system of administrative privilege, since opponents of airports and power plants may be denied the opportunity to challenge government studies unless they can match the official experts and produce comparably full and detailed analyses of their own. Still, the economist Alan Williams was not wholly off the mark in calling cost-benefit analysis a challenge to the "authoritarian" and "paternalistic" assumption that leaders already know what is best for society, even if it does not much promote the forms of public participation that many critics would prefer.[31]

The British administrative elite survived so long that in retrospect its success seems to follow inevitably from long-standing cultural patterns. In fact there were real alternatives, especially in the mid-nineteenth century. Northcote-Trevelyan was a brilliant triumph for Benjamin Jowett's Oxford and the Coleridgean clerisy over the Benthamite ideal of practical education. Benthamism had won some successes earlier in the century, and was installed at such sites as Haileybury, where future officials in the East India Company learned Indian languages and political economy. But even in the 1840s and 1850s, it was not at all easy for mere experts, without family connections or close ties to the interests affected, to contribute much to the making of public policy.[32] The British political order was sufficiently hierarchical to rely more on trust and deference than on objectivity or precision. Incipient professionals recognized this, and understood too that as mere technical specialists they could never become proper elites. The reigning sense of moral order easily accommodated gentlemanly professionalism. If, for example, actuaries could present themselves as trustworthy gentlemen, a parliamentary select committee was unlikely to demand that they behave like calculating engines.

GENTLEMANLY ACTUARIES

None other of the human sciences acquired the discipline of mathematics so early as did the business of the actuary. By the early nineteenth century the best life insurance offices depended on extensive calculations to set their rates. But the processing of numbers by actuaries is not the place to look for an ideology, or even a practice, of faith in mechanical objectivity. The ability to apply mathematical formulas was a minimal requirement for the novice actuary. These calculational skills were to be used for the preparation of life tables and the determination of premiums. Actuaries were unanimous in recognizing the importance of reliable statistical records. But they did not believe in the possibility of precise measurement, of reducing their work to calculational routines. Very few actuaries aspired to see their craft made perfectly mathematical. The push for objectivity came instead from Parliament and from regulatory authorities in pursuit of political and administrative ends. British actuaries conceived themselves as gentlemen whose integrity and judgment had earned the public trust. A strict regime of calculation would have implied the denial of that trust, in the name of democratic openness and public scrutiny.

Midcentury actuaries believed firmly that slavish adherence to formulas was inconsistent with sound business practice. They gave a variety of reasons, but the crucial one had to do with the selection of lives, which I discussed in chapter 2. This required skilled personal attention, or else a company would suffer adverse selection, meaning mortality rates higher than those for the general population. Nearly all agreed with Edwin James Farren of the Asylum Life Office that a system of life contingency calculation based on rigid mortality schedules and fixed interest rates is at best suitable for the infancy of life insurance. He wrote that the neophyte actuary, trained in this "exquisite logic," must be surprised when he confronts the real world, marked by the inescapable variability of the "so-called law of mortality." The "assumption of absolutism" in this domain is no longer useful.[33]

Arthur Bailey and Archibald Day argued in 1861 that no general law of mortality was yet known, and that if one is someday discovered "it will represent the law that really prevails among the living, moving, thinking men that inhabit the earth, much in the same way that the statue of the Apollo Belvedere represents their bodily form. Such a law will never supersede, in our pursuits at least, the exercise of . . . careful judgment and sound discrimination." Insurance, after all, was always a local problem. Actuaries must calculate on the basis of "dependent risks," which apply only within a given company at a given time. The tables apply to lives actively selected, whose quality may determine the viability of the company. William Lance of Lloyd's explained: "It is well known to actuaries that the success of a Life Assurance Office does not necessarily follow from accepting assurances at rates determinable upon tables of mortality, but according to the judgment with which lives are selected, and the premiums improved at interest."[34] Tim Alborn suggests that the actuaries' preferred subjective interpretation of probability reflected their preference for expert judgment over mere mechanical calculation.[35]

Since actuaries denied so stubbornly that their profession could be reduced to the performance of calculations, it is important to ask just what role mathematics did play in the British insurance industry. Notwithstanding the abundance of testimonials to expert judgment, life insurance was a thoroughly quantitative business. The amount of space in the *Assurance Magazine* devoted to probability mathematics and to the presentation of statistical tables makes this clear. Preparation of life tables and rate structures was the principal business of the actuary. Mathematical reasoning was indispensable. What it could not provide was objective measurement. Because of the essential heterogeneity of company practices and of the populations insured, no mere process of collecting and tabulating results from the population at large could yield numbers that would be valid for any particular company.

This was all conventional wisdom for practicing actuaries. Purely mathematical solutions to insurance problems in the *Assurance Magazine* often provoked skeptical or satirical responses, which the editors did not hesitate to publish. A mathematical piece on the value of an inheritance in perpetuity conditional on the earlier death of some elder brothers, for example, was answered twice in the pages of the journal. The problem could be solved mathematically by presupposing the validity of the tables, the critics conceded. "That would not be the view taken by an actuary before whom the case came in the ordinary way for an opinion." A proper solution would depend also on factors ignored by the tables, especially the health of the various parties. Hence, "actuaries would, in stating a value, be guided more by their own judgment than by any tabular or mathematical value, which can only, without great trouble, be approximate."[36] In the mid-nineteenth century, nearly all insurance offices remained small, having at most ten thousand policies and perhaps a few dozen clerks. Candidates for insurance could expect individual attention, which was not necessarily to their advantage, since their health or morality might be judged inauspicious. Large-scale data processing in insurance was not worked out until the 1850s and 1860s, when the Prudential began marketing policies with a small face value to working-class people.[37]

The reliance of actuaries on quantitative analysis is most obvious not where tables were available, but where they weren't. While actuaries held devoutly to the view that the variability of phenomena made judgment indispensable, none believed that insurance could be managed by judgment alone. Only through extensive calculations, they pointed out, could it be demonstrated that an insurance company or friendly society with large reserves faced insolvency when its members became older. Laymen and especially working people without much knowledge of mathematics could never be convinced.[38] Actuaries urged their brethren in new companies to rely on tables from comparable firms (but not on national statistics) until they had acquired sufficient experience of their own. New types of policies, such as insurance on the lives of older men with barren wives that would take effect only "in the event of issue" (presumably after a death and remarriage) inspired efforts to construct tables to define the appropriate risks.[39] It remained common in mid-century for the more conservative companies to require insured persons traveling abroad to surrender their policies, because nobody knew how much the risk was increased in India, Africa, or the Caribbean. Others imposed somewhat arbitrary, and generous, surcharges. At the same time, actuaries worked diligently to gather information on the experience of Europeans abroad, so that insurance contracts could be maintained on military officers and colonial administrators without increasing the risk to the company. Their inquiries produced some of the best evidence we now have concerning European mortality in tropical colonies.[40]

The unease of actuaries in the absence of systematically gathered quantitative data on risks is evident also in their writings on fire and marine insurance, which were supported by nothing like the mortality tables of life insurance. Samuel Brown, one of the more vigorous champions of mathematical probability as a tool of insurance, complained of the failure of the companies to gather up and share their experience with risks to buildings and shipping. Others argued that losses in these categories display regularities from year to year comparable to those that govern human mortality.[41] Underwriters involved in marine and fire insurance, though, were less easily convinced. Buildings and technologies changed too rapidly for past results to be generalized into the future, they held. J. M. McCandlish wrote on fire insurance in the ninth edition of the *Encyclopaedia Britannica*: "The slightest observation reveals an endless diversity in the risks undertaken, and even if an absolute law could be reckoned on, the risks would require careful and accurate classification before the law could be deduced. But, in point of fact, the risks are always changing."[42] Clearly the companies relied on experience, but they used it in a more secretive and informal way than did life insurers. In the absence of regulatory interference, there was never sufficient incentive to systematize and rationalize that experience. When losses of a certain type or in a particular city became noticeable, the companies colluded to raise prices and restore profitability.[43]

Life insurance companies could not deal with risk so casually. They offered long-term, whole life contracts, in which the premiums paid by each policyholder greatly exceeded the value of risk on his life for several decades, but then fell considerably below it. Rates could not be readily adjusted if experience showed that they were too low. Sound instinct or seasoned judgment by itself provided little guidance to the pricing of so complex a contract. Calculation, if not definitive, was at least approximate. Cautious, responsible companies normally based their calculations on low interest rates and conservative life tables, then added a percentage for extra confidence. They might also make adjustments for the expected quality of lives or for unusual investment opportunities. By 1850 most companies were operated at least in part on mutual principles, which meant that some of the profits were returned to the insured. The measure of profitability was anything but self-evident, and an actuary testifying before a parliamentary select committee in 1853 admitted that what is called a return of nine tenths of the profit might in reality be no more than one half.[44] The point here is that reliance on calculation was by no means inconsistent with the exercise of discretion, provided that nobody pretended to perfect quantitative precision. In life insurance, judgment entered not as a fundamental alternative to calculation, but as a set of strategies for setting up the computation and then adjusting its results.

Mathematical precision was regarded as suitable above all for novice actuaries. It was argued, somewhat paradoxically, that exacting calculation helped to form a mature judgment. In 1854, Peter Gray explained that the construction of life tables from survivorship data can be accomplished according to at least two distinct methods. The "logarithmic method" has the disadvantage of yielding "no more than seven figures" in its results. Seven figures may or may not be sufficiently precise for life tables. "On this point different computers will entertain different views." It was an advantage of his other method, "construction in numbers," that it permitted the calculation of an arbitrary number of digits. He admitted that such calculations would quickly go beyond the reliability of the data. But he considered that the act of computation had its own value. While "practised computers" might find his discussion excessively minute, "younger members" would confer a great benefit upon themselves as well as the profession by calculating tables according to this more exhaustive method. "They would find that they had acquired such an intimate acquaintance with the structure and properties of the tables, that they could apply them to practical purposes with a facility and confidence which without this preparation long experience alone could have imparted."[45]

Henry Porter explained the value of mathematical study for actuaries mainly in moral terms, as promoting diligence and care. Those qualities, he conceded, were not universally admired, and insurance directors had been known to ridicule "the prominence, phrenologically speaking" of their actuary's "organ of caution." Certainly Porter was not prepared to view insurance as a preeminently technical discipline. "I believe that it is now generally considered, that a very abstruse mathematical knowledge is not absolutely requisite for the general business of an actuary." Switching to (faint) praise, he spoke against the opinion of some, "that it is rather detrimental than otherwise, as we know of many instances which prove that the highest scientific knowledge and perfect business habits are not necessarily incompatible." Mathematics, he determined, is essential for actuaries, even if "the prosperity of some Companies has been sacrificed to the closet meditation of the profound theorist." No such ambivalence was attached to Greek and Latin, which Porter deemed necessary as aids to the mastery of the relevant technical vocabulary, or to physiology, which would assist the actuary in judging doubtful lives and deciding on an appropriate supplement to the premium.[46]

It is necessary to add that Porter began his lecture by calling mathematics "the foundation of all actuarial knowledge." Statistics, meaning numerical data, provided "the very foundation on which the superstructure of life assurance is raised." But these foundations were wobbly. "[I]t is not sufficient at once to adopt the numerical result arrived at by calculation." Mere calculation can lead to absurdities. A senior wrangler, "without experience, would be helpless in a Life Office." Men of experience recognize the crucial importance of "*judgment*" in actuarial practice. Porter's lecture was a paean to "judgment and experience," which "cannot be taught" but only acquired through an apprenticeship, as in all professions. He considered that the license to practice should be conferred only after a "searching examination, by gentlemen of undeniable attainments" in this refined art.[47]

By coincidence, Porter himself had been in the first cohort of successful candidates to undergo such an examination, under the auspices of the new Institute of Actuaries. If, however, self-interest is to be invoked to explain his pronouncements, that interest was at least a collective one, widely shared among English actuaries. Edwin James Farren deemed it "a well-known feature of the advancement of learning, that the more knowledge becomes prevalent, the less positive do opinions become." Actuaries had too much experience to believe any more in "abstract truths" and "fundamental axioms" from which all practical conclusions could be derived.[48] This sensibility that knowledge is local, and that even general rules are useless except to those who understand the conditions under which they should be applied, seemed most compellingly true when there was threat of bureaucratic intervention. It was expressed most persistently in response to an 1853 inquiry into "assurance associations" by a select committee of the House of Commons.

A SELECT COMMITTEE SEEKS EXACT RULES

Charles Dickens chose life insurance as the exemplar of fraudulent enterprise in his novel *Martin Chuzzlewit*. The Anglo-Bengalee Disinterested Loan and Life Assurance Company disguised its rapaciousness with a veneer of solidity and trust. That was in 1843. The novel drew on stories of shady insurance operations taken as evidence in 1841 and 1843 by the Select Committee on Joint Stock Companies. In ensuing years, speculative investment in life insurance reached new heights. Despite attempts to regulate insurance through the Joint Stock Companies' Act of 1844, Parliament had almost no control over this activity. The Select Committee on Assurance Associations, which carried out its inquiries in 1853, was able to learn from Francis Whitmarsh, Registrar of Joint Stock Companies, how many companies had been provisionally or completely registered in the intervening years. It seemed that many, probably more than a hundred, had already failed, though the bureaucratic machinery was not adequate to offer any certainty.

The word on the street was not encouraging. James Wilson had written on the problem for the *Economist* newspaper, and was as well informed as anyone about business practices in life insurance. As chairman of the select committee, he looked to actuarial precision to provide readily understandable information so that people would be able to learn for themselves which companies were solid.[49] Nearly all testimony to the committee came from professional actuaries, most from the older and more respectable companies rather than the new, possibly shady ones. These actuaries were polite but unmovable. Precision is not attainable through actuarial methods. A sound company depends on judgment and discretion. Actuaries are gentlemen of character and discernment. Trust us.

Wilson set forth his view of the matter plainly in the form of leading questions. Life insurance is a long-term proposition, and the insured need some basis for confidence that the company will still be there when, at their death, a claim at last comes due. Premiums depend on age of admission to a company, but are set at a rate that is to remain fixed over the life of the insured. Death rates for those recently insured are relatively low, both because of their youth and because they are select lives. Hence a new company will accumulate a lot of capital in its early years. If it is responsible, it will set most of this aside in anticipation of an increasing rate of claims twenty or thirty years later. But many companies seemed not to be responsible. They paid their officers and directors high salaries. They expended vast sums on advertising to bring in new lives and new revenue. And some did not exude solidity. A certain Augustus Collingridge had in just a few years set up and closed down several companies. An investigator reported to the committee that the Victoria Life Office consisted "of one room, over a milliner's shop, in New Oxford-street, containing only two chairs, a broken table, and a large number of prospectuses, printed on foreign note paper." The Universal Life and Fire Insurance Company was lodged in a room in a very small house, "which the parties never occupied, some person calling there for the letters; no one had called for letters during the last six or seven days, in consequence of the police having been making inquiries" about the proprietor.[50]

John Finlaison, actuary for the government, was among the more obliging witnesses. Insurance mathematics, he explained, "is extremely simple: there is no difficulty whatever in determining the actual condition of an office at any given time." An insurance company is solvent if it has resources to pay the present value of all outstanding insurance contracts. Would it be possible, Wilson inquired, to display all this in standard, published accounts? No, replied Finlaison, because actuarial judgment is involved both in predicting the rate of interest at which assets will grow and in choosing a life table. His own practice was to calculate interest at 3½ percent and to use a life table he had prepared, whose validity he knew from personal experience. Of course if the companies do not exercise due care in admitting lives, the tables will not predict actual mortality. Also, the value of fixed assets will vary as interest rates fluctuate, and only long experience prepares one to judge their value.

Wilson soon picked up the drift of Finlaison's arguments. "If I understand you rightly, you are of the opinion that so much depends upon the discretion and good management of the office, apart from anything that can be shown on paper, that you would not be disposed to place much confidence in any check that would be obtained by accounts?" The reply was affirmative. Finlaison also worried that publication of accounts would lead to invidious comparisons which would work to the disadvantage of new companies. Even where there was reason to suspect insolvency, the appropriate remedy was a discreet, confidential inquiry by an independent actuary working with the company's regular actuary, not a loud public investigation. Insurance fraud was very rare, he explained. A respectable board of directors guarantees the integrity of the office. Moreover, actuaries calculate conservatively, by adding a margin for safety and profit on top of calculated premiums.[51]

The record of testimony before the select committee provided little comfort to those who hoped that accounts could be rendered uniform and precise. The witnesses were nearly unanimous. It is impossible to judge companies against any single set of life tables, because they are run according to different principles and insure lives of varying quality. Income and assets cannot be fixed from a single interest rate, because investments are so diverse. Some actuaries are optimistic, others pessimistic, and there is no legislating away these differences. Edward Ryley, a decidedly unfriendly witness, put the argument most bluntly. Actuaries disagree. "If you appoint a Government actuary, you must necessarily select him from existing actuaries; there is no manufactory for actuaries where you can get one made to order." Uniform rules of calculation, imposed by the state, might yield "uniform error."[52] Charles Ansell, testifying before another select committee a decade earlier, argued similarly, then expressed his fear that the office of government actuary would fall to "some gentlemen of high mathematical talents, recently removed from one of our Universities, but without any experience whatever, though of great mathematical reputation." This "would not qualify him in any way whatever for expressing a sound opinion on a practical point like that of the premiums in a life assurance."[53]

Wilson and his fellows on the 1853 committee did not simply capitulate before this barrage of expert testimony. Can we not apply "some general average rule," he asked Ansell, "so as to work out the result upon some given general principle?" Ansell said no. How about using John Finlaison's life table, applying a uniform interest rate of 3½ percent, and adding 10 percent for contingencies, in order to determine whether a company has adequate resources to meet its expected liabilities? There would be great difficulties and valid objections, replied Ansell. All these things depend on the particular circumstances of each company. And the best calculated scale of premiums will avail a company nothing if it has been careless in selecting its lives.[54]

The actuaries were by no means uncompromising in their hostility to general principles. Often, their responses to Wilson's initial queries were favorable. Without fail, though, they added qualifications and stipulations that undermined the principles. Many believed in general maxims; none would concede the possibility of precise, standardizable rules. Samuel Ingall, for one, had a ready reply when asked for a general test of the solvency of a company. It should have one-half the premiums received on existing policies in hand, as capital. This was more informative than the usual answer, that a company should have funds to buy off its policies. Ingall further declared that during its first twenty years a company should accumulate each year, on average, one percent of its potential liabilities. But soon he admitted that these maxims only applied if business was approximately steady, and moreover that actuaries disagreed about how the profits of a company should be calculated. This last reservation sounded rather crucial to Wilson, who then asked how one could test the accuracy of the returns. Ingall replied: "I think the best security is the character of the parties giving them."[55]

This sentiment was repeated like a refrain throughout the hearings. Although actuarial principles are well established, said Ansell, there remains much uncertainty "in applying known principles to different scales or premiums." It takes as long as two years for an outside actuary to assess the economic health of a company, argued James John Downes. Even to understand the published reports of insurance companies requires minute knowledge, so that the actuary of an office inevitably understands its position better than any other person, however clever. Since we must depend on the skill and integrity of the actuary to prepare the data, he might as well be trusted to make the final calculation. Actuaries are "gentlemen of character," reported William Farr, and the government should leave the preparation of accounts to them. No quantitative measure of solvency can be adequate, insisted Francis Neison. A considerable expenditure of funds for publicity when a company is founded may be the best way to secure its future. There is always "a special knowledge beyond the accounts, not appearing in the books of the institution." The success of a company depends ultimately on skilled management.[56]

These were brave claims from a profession so new. The Institute of Actuaries had only been founded in 1848. While the office of the actuary had been officially recognized as early as 1818, the identity of this being remained notably murky. An act of that year permitted friendly societies to be enrolled by the government and to receive certain benefits if their life tables were evaluated by a committee including at least "two professional actuaries, or persons skilled in arithmetical calculations."[57] William Morgan of the Equitable complained in 1824 that some people "call themselves actuaries, who are nothing but schoolmasters and accountants." A clergyman and magistrate at Southwell remarked that he had been unable to put this law into effect, since "how to define who is an actuary exceeded my power."[58] In 1843, the actuary John Tidd Pratt testified to another select committee that mere schoolmasters were often asked to certify tables because "no one knows who an actuary is."[59] The main object of the new institute of actuaries was to secure the recognition befitting a profession. This entailed more than a demonstration of technical competence, and the oft-expressed disdain for mere calculation must be understood in part as a strategy of legitimation in a society that gave little respect to mere technical experts.

In any event, British actuaries had no faith in quantification according to a rigid protocol. Argued Downes: "We may have a person who is a good theoretical actuary, who may apply formulas with very great facility, but who has had no experience in the working of companies, and, therefore, he would not be able to apply those forms and theorems usefully to the business of a life insurance company." Charles Jellicoe, testifying for the Institute of Actuaries, explained that subtle shades of difference can be decisive on such issues as the margin to be set aside for future risks. "Then you do not differ in point of principle, but you differ in the mode of applying that principle?" he was asked. He did. Indeed, he volunteered, an actuary could make a company's books look good by overvaluing assets and undervaluing risks. But this would be fraud, proposed Wilson. Perhaps, Jellicoe replied, but the company could always argue "that the lives were particularly good ones, that the mortality, therefore, would be small," and the like. For such reasons, you cannot effectively legislate a minimal guarantee fund; it is a matter of detail in each office. Parliament should simply "do the best they can to get persons whose judgment and discretion enable them to do the duty properly," by authorizing the Institute of Actuaries to grant a license or diploma to those who meet its standards.[60]

This was very far from what the select committee wanted to hear. Wilson aimed to clear up the problem of proliferating, possibly insolvent life insurance companies with a minimum of government interference. The committee never disputed the truth of what it was told again and again: that government meddling in the business of insurance would cause far more trouble than it would relieve. Wilson hoped that the mildest intervention would suffice. His ambition depended only on the possibility of making insurance readily interpretable, of standardizing the calculations sufficiently that potential purchasers could judge the companies for themselves from a few crucial numbers. But the actuaries resisted his suggestions in the name of judgment—of subtle shades of meaning and innumerable points of detail. The government sought public knowledge, while the actuaries denied its possibility. The government sought a foundation for faith in numbers, while the actuaries demanded trust in their judgment as gentlemen and professionals.

Such actuarial skepticism was heard whenever someone on the select committee suggested that the government might tell the companies how to keep their books. None of the actuaries could countenance such interference. The most sympathetic view of the committee's suggestions was expressed by a Scot, William Thomas Thomson, who assented to the proposition that a single form of balance sheet would be appropriate for all companies. He added that in any event the companies could easily maintain their existing form of accounts and with modest effort translate them into a standard form for public purposes. He even spoke favorably of American-style regulation, as enacted recently in New York. None of this appealed in the slightest to the other expert witnesses. When Thomas Rowe Edmonds was asked if Mr. Thomson's forms would provide the kind of information the public needed, he responded angrily: "Not in the least."[61]

To reject quantification according to a mandated protocol was not quite the same as denying that useful information could be conveyed by numbers. Standardization was what the actuaries opposed. Precision, in any useful sense, entailed a large measure of centralized control, to which they were firmly opposed. They did not on this account refuse to report quantitative information. Without exception, the actuaries claimed to regard clear, accurate reporting as good business practice.

This was another argument against the need for new legislation. One strong opponent of regulation went so far as to lay the blame for unintelligible reports on government interference. Companies naturally become secretive when surrounded by "lynx-eyed" inspectors, "searching out any kind of seeming irregularity in these accounts, and making the most of it, to the prejudice of the institution."[62]

Wilson and the committee responded favorably to the idea that an experienced public official could help to assure that the accounts were clear and accurate. The only actuary who favored anything like this was Thomson, the Scot, who however wanted independent auditors rather than government ones to certify the accounts. Either possibility seemed far too meddlesome to the others. Public investigation could be justified, they argued, only where there is reasonable suspicion of fraud. Routine public inspection of insurance companies would weaken the self-reliance of citizens and thus aggravate the very problem the government had set out to solve.

Still, several witnesses allowed that the government might reasonably require publication of financial records. This, the actuaries argued, should take the form of a straightforward factual presentation. It should not include summary numbers standing for assets and liabilities, which would imply that solvency could be determined by anyone capable of noticing which was the greater. Permitting interpretive accounts in public records would make the government "a medium for advertising all sorts of opinions and fallacious valuations, . . . publishers of puffs."[63] The only thing suited to the public domain was demonstrable facts. By indicating simply the quantity and nature of each holding, or presenting their own figures along with an explanation of the principles by which they were calculated, the companies would provide potential customers all they needed to evaluate the security of a policy. Not that everyman could interpret this information for himself. That would presuppose the impossible—a highly standardized document, with everything reduced to a few categories and expressed in the same terms. Rather, customers could assess the security of a company by consulting a personal actuary, much as they might consult an attorney about matters of law. The public could not learn much from a balance sheet, said Thomson, but it would permit professional actuaries to form "a very decided opinion." Would it be a correct opinion? asked Wilson. "I should say if it is decided it must be correct, so far as depends on the individuals' judgment."[64]

The underlying issues emerged here very clearly. The validation of expertise meant that standardization was unnecessary. If the key entities at issue resisted precise measurement, then trust, or intrusive regulation, was required to fill the gap. To be sure, the preferred solution of the less militantly antigovernment actuaries required exact quantitative reports, but this was basically a matter of a modest opening of their books, not of precise measurement of anything. In Victorian Britain and twentieth-century America alike, the campaign for objectivity was led by the government, and opposed by mathematical actuaries and accountants. Objective knowledge meant public knowledge, which they considered to be neither possible nor desirable. In place of precision they offered a profession.

CHAPTER SIX

French State Engineers and the Ambiguities of Technocracy

We err if we assert that science has supplanted routine.
It replaces old routines, but it requires new ones, and so far as
these are not born, science remains powerless.
(Auguste Detoeuf, 1946)

THE UNITED STATES gave us the word "technocracy," but France seems to have some claims on the thing itself. The Ecole Polytechnique, product of the French Revolution, is often taken to epitomize technocratic culture in France. Polytechnique, with its emphasis on mathematics and science, was central to what Antoine Picon calls the invention of the modern engineer. Quite unlike its imitators, it educated the highest stratum of elites. Where else has administrative power been so closely allied to technical knowledge?

This alliance helps to explain the French tradition of what would now be called applied economics, discussed in chapter 3. Ponts-et-Chaussées engineers brought to economic issues a level of quantitative sophistication unmatched in other lands before the twentieth century. I was concerned in the earlier chapter with published works, written mainly by engineers for each other. But a research literature, even carried out among persons with administrative responsibility, does not make a technocracy. This chapter is about economic calculation in action. Its social and administrative role depended on a shared mathematical culture, and also on bureaucratic organization. This was never a matter merely of abstract economic knowledge, but always of an interaction between quantitative methods and administrative routines.

Bureaucratic uses of economic quantification, inevitably, were closely allied to accounting. So is much of economics itself, especially those parts of it that have been created or mobilized to aid in management, planning, and regulation. Accounting means, among other things, placing monetary values on goods and services that contribute to production or sales but cannot themselves be readily exchanged in the marketplace. Nineteenth-century French engineers went one step further, attempting an analysis of the (often unpriced) benefits of public goods to balance against their monetary cost. In this context, values had to be placed on objects, services, and relationships for which there was no proper market, or whose prices could give no adequate measure of their value to users. This "cost-benefit analysis," to introduce the anachronistic term, remains an elaborate form of accounting. These engineers refused to depart so far from the market as to assign values where there was no contribution to the making and distribution of goods for sale, and in the end to the total production of France.

It is sometimes implied that the drive to make decisions by the numbers is just a matter of engineers doing what comes naturally, the consequence of a marriage of technical knowledge and political power. I have already suggested in relation to American accountants and British actuaries that it was not. Numbers were of course important to both, but in each case the profession insisted on the legitimate and necessary role of expert judgment. Not the experts themselves, but powerful outsiders, worked to simplify regulation by reducing judgment to rules of calculation. The Corps des Ponts was also subject to such pressures. Decisions about the location and pricing of canals, bridges, and railroads inevitably became mired in intense local political debate, and sometimes were vigorously debated on the national scene. As an agency of the French state, its public responsibility and dependence on higher authorities were incontestable. In the twentieth-century United States, as the next chapter shows, such pressures would inspire a monumental attempt to reduce cost-benefit analysis to firm rules. At the Corps des Ponts, in contrast, this never really happened.

I explain this not in terms of its weakness, but of its strength. It was extraordinarily secure and prestigious. It was capable of taking decisions in relative secrecy. Also, it was a highly coherent and centralized body, with the power to regulate the private as well as the public lives of its members. Nineteenth-century French engineers did work out routines of economic calculation. As in the more recent United States, these methods were a tribute to public accountability. They developed in quite specific contexts, to which they responded in considerable detail. But Ponts engineers never had to pretend that calculation was simply a matter of following unambiguous rules. Given the institutional autonomy and elite standing of their Corps, it was quite inconceivable that these engineers could have been deprived of the ability to exercise discretion. The authority of numbers in public life has depended on the growth of science and engineering, but is no mere by-product of it. The public role of quantification reflects social and political developments that cannot be reduced to scientific and technological ones.

THE CONTEXTS OF ECONOMIC QUANTIFICATION

As François Etner shows, economic calculation in France was long centered at the Corps des Ponts et Chaussées. Etner is an economist, on which account historians might not expect him to give much attention to the bureaucratic and political pressures that drove engineers to quantify. In fact he is acutely conscious of them. He interprets economic quantification as an agent of "la lutte [struggle] contre l'arbitraire." Ponts engineers "had to distribute funds, supervise and choose among alternative projects, all in the name of the general interest, according to rules that are written, public and nondiscriminatory."[1] Etner describes here an ideal, one that animated much of the corpus of published economic writings on which his book is based. Documents closer to the scene of bureaucratic action show the limits of quantitative rationality. There was always frank political negotiation. Although various quantitative indicators of the worth of projects were widely accepted, no single standard or hierarchy of standards won general approval even within the Corps. Its top decision-making body, the Conseil général, often had to decide between rival programs, each supported not only by local interests but also by the responsible engineer. The council reached its decisions in closed session. Many of its recommendations were approved almost automatically by higher levels of government.

Given the manifest reliance on expert judgment and administrative authority in such cases, one might doubt the importance of the ritual exercises in quantification. Such parades of rationality are now commonly dismissed as a smokescreen, behind which well-connected interests struggle to get what they can. Of course it would be naive to ignore these political struggles. But it is also a bit naive to dismiss the formal processes of decision as mere illusion. In decisions about public works, interests are always powerful, but often they are nearly in balance, so that any decision will have political costs. When state engineers were planning the trunk lines of the French rail system in the late 1830s and early 1840s, they typically had to decide among several possible routes, each of which was strongly supported by the affected towns and départements. The numberless proposals for local lines in the 1870s and 1880s required similar choices. Even supposing, rather implausibly, that engineers did not care whether the French system of canals and railroads contributed to the nation's prosperity, they at least had an interest in orderly planning. Otherwise their Corps, and indeed the state itself, would be reduced to pawns in a game played by private interests. A bureaucracy that knows no higher purpose than to control its own turf may yet develop and observe a rigid set of rules for that very reason. Quantitative decision criteria may often be overwhelmed by politics, but at times they can be politically indispensable.

The Corps des Ponts et Chaussées did not quantify according to rigid rules. The decisions it made were ordinarily so complex that no such rules could ever have gained general assent. Moreover, the engineers enjoyed the prerogative of an elite: to exercise judgment even in regard to issues of public importance. Theirs was a distinguished agency within the French administration, and they derived great prestige from their connection with the state—prestige that was denied to "civil" engineers. They did not think of themselves as mere calculators.

They were also a meritocratic elite, as attested by their performance on the largely mathematical *concours* to gain entry into Polytechnique and their success as Polytechnique students. The ability to deploy mathematics skillfully was an important component of their professional identity. Just what this meant had to be worked out in the half-century or more after 1795, when Ponts-et-Chaussées engineers began to be recruited exclusively from those who had completed the highly mathematical Polytechnique curriculum. Charles Gillispie argues that the abstract analysis emphasized at Polytechnique was taught for reasons having more to do with science and mathematics than engineering. That style of mathematics was, he suggests, almost useless for roads and canals. Eda Kranakis proposes that it was worse than useless, that it gave form to an ethos of disdain for skills and materialities—the art of construction.[2] The distinguished engineer and physicist Louis Navier, Kranakis's paradigm case, did put analysis to work, especially for designing bridges. He argued vigorously during the 1820s that the Ecole des Ponts et Chaussées should prepare engineers to use the most sophisticated forms of analysis, the tools of mathematical physics.

Since he was inspector general of the Corps, his views mattered. But they were by no means unquestioned. Rivals such as Barnabé Brisson, inspector of the Ecole des Ponts, argued for a greater emphasis on descriptive geometry and political economy. This was in the tradition of Gaspard Monge, revolutionary champion of a practical Ecole Polytechnique accessible to talented men of all social strata.[3] It is tempting but false to suppose that abstract analysis was destined to emerge triumphant. Neither style of mathematics won exclusive dominance at the Ecole des Ponts. Even Polytechnique had begun to move some way toward a more practical mathematical education by the 1850s.[4] Perhaps more crucially, as Antoine Picon observes, students of the Ecole des Ponts spent most of their time in a kind of apprenticeship, and took course work only from November to March. They inferred, correctly, that this was not of the highest importance for their careers.[5] The issues at stake between Navier and Brisson were real, but their positions were not irreconcilable opposites. Teachers and student engineers alike welcomed forms of quantification they could actually put to use. This included economic measurement, whose appropriateness for the planning of routes was never fundamentally questioned from within the Corps. Picon suggests that the prestige of pure mathematics was waning among Ponts engineers by midcentury. A more applied form was meanwhile becoming more central to teaching the art of construction.

Mathematics helped to form the identity of Ponts engineers, though it had to be made consistent with their sense of themselves as men of action. It provided evidence of disinterestedness as well as expertise. Opposition to the Corps, explained one engineer, came from advocates of false systems who would "substitute for industrial conceptions based on science . . . some narrow combination of aggressive ignorance and personal interest."[6] Quantification was never merely a set of tools. Making up numbers in deference to political necessity was unacceptable to these engineers. It compromised their status as a disinterested elite and violated standards of mathematical integrity that they took seriously. Negotiating mutually acceptable numbers was, however, another matter.

THE QUANTITATIVE MENTALITY IN ACTION

The Conseil général des Ponts et Chaussées[7] decided in closed session about projects to be undertaken, routes to be followed, contracts, subsidies, and rates. Much of its business, inevitably, was routine. But when the written record was unclear or contradictory, the top officers who made up this body had the authority to decide whom to believe and how to act. The appearance of near-absolute power here has a grain of truth, yet it is doubly misleading. The council could speak for the Corps, but the Corps could do little on its own. It could only recommend to the minister of public works, who in turn made recommendations to the national legislature. On the other end, the law dictated an elaborate process of inquiry before a recommendation could be made at all.

First a project outline, or *avant-projet*, was prepared. An ordinance of 1834 required that a hearing, or *enquête d'utilité publique*, be held in every affected département. For this purpose a commission would be assembled, made up of nine to thirteen principal merchants, factory chiefs, and proprietors of lands, woods, and mines. The form of their findings was prescribed by a Latin phrase: *de commodo et incommodo*. Or, in French, they were to identify *les avantages et les inconvéniens*, advantages and disadvantages. For this they would consult with engineers and invite testimony from interested parties. Chambers of commerce in affected towns would also be invited to comment, though ambition and envy were more familiar currencies to them than advantage and cost.[8] Their proceedings and conclusions would be transmitted to the prefect of the département and then on to the higher administration.[9] The Corps des Ponts had to negotiate with the prefects and even with towns. A law of 1807 required costs to be allocated in proportion to the "respective degrees of utility." In practice the state would provide one-fourth, one-third, or one-half of any subsidy offered the concessionaire, and local government would be required to come up with the rest.[10] Until agreement was reached on the allocation of costs, no project could be undertaken.

The results of these inquiries were predictable. Each département favored the proposed line that best served its own population. Often the commissions and prefects called for diversions or branch lines to serve other important towns or economic establishments in their jurisdiction. Engineers were normally responsible for the proposed route through their département, and generally favored it.

Public works, then, were highly charged politically. This will come as no surprise. But its implications have not always been properly appreciated. Contests in the political and administrative spheres provided the main incentive to the formalization of economic rationality. A favorite rhetoric surrounding the measurement of benefits and costs naturalizes it as the form of analysis spontaneously used by rational economic actors. Even the council of the Ecole Polytechnique, in setting up a course in social arithmetic, explained its need in terms of the growth of private enterprise (see chapter 3). They were not completely wrong. Private companies seeking to sell railroad or canal bonds on the open market would print a prospectus, which might well include estimates of revenue. At least this was standard practice among the British, whose capital markets were better developed than French ones. Ponts engineers paid close attention to the financial as well as technical aspects of British railroad construction.[11] Still, the more elaborate forms of economic calculation were nearly always associated with public projects, and with the political processes by which they were approved and regulated. They were as much a form of political as of economic management.

While Ponts engineers made strenuous efforts to minimize their political exposure, they otherwise felt little ambivalence about the public dimension of their craft. Their identity was bound up with an ethic of state service rather than a desire for profits. The terms of their economic quantification reflected this. Although budgeting required them to calculate costs and revenue, they preferred to plan in terms of a logic of public utility. No private company ever worked this way, unless as a defense against hostile political forces. Ponts engineers asked under what conditions a railroad or canal will bring more benefit to the public than it costs the state. They sought a structure of charges that would distribute expenses justly among users, or (alternatively) maximize the public benefit. They undertook to show that the state was justified in constructing roads, canals, and railroads even where private investors could never be induced to build them, because the profits would accrue to users rather than the entrepreneur. Since these would usually involve a monopoly, with operating costs much lower than fixed costs, the state also needed some basis for deciding how to charge for their use.

This perspective on public works was by no means unique to engineers. Public utility was a standard term of political discourse, and even of law. The départements, as we have seen, evaluated projects through hearings called *enquêtes d'utilité publique.* Final approval of a project, when the emperor or National Assembly granted a concession to build and operate a rail line to a private promoter, took the form of a *declaration d'utilité publique.* Negotiating these was as central to the job of the state engineer as surveying routes, planning networks, and letting contracts.[12] As I will show in detail below, the "public utility" in these grand phrases had no specific meaning, and was often construed as wholly nonquantitative. Still, it was a convenient phrase for a public agency. The only form of economic calculation specifically required of the Corps was budgeting. In the French canal boom of 1821–51, the state created a precedent of raising money by guaranteeing the safety of capital plus a modest return. Budgeting for this depended on Corps estimates of costs and revenues—which proved disastrously optimistic.[13] Estimates by private English companies were just as bad, according to a table prepared by Charles Joseph Minard.[14] Still, the language of public utility could almost always be construed more favorably to the Corps than could the accountant's bottom line. Measures of public utility had the obvious virtue that they couldn't prove simply wrong in the same way that revenue estimates could. More crucially, transforming the legal and moral term "public utility" into a quantitative one might provide the Corps some protection against the buffetings of day-to-day politics.

For public utility was, especially in its quantitative meaning, a universalizing concept. Henri Chardon, writing early in the twentieth century about public works, remarked that public utility meant "utility for the whole nation." Like many Ponts engineers, he considered that this ideal was often betrayed because politicians rather than experts were put in charge of such decisions.[15] André Mondot de Lagorce acknowledged in 1840 that not all relevant considerations could be reduced to numbers. "Must we, then, renounce economic calculation?" he asked. "No, because this would abandon the legislature to 'inopportune solicitations.' It is much better to adopt a synthetic formula, even if it is not mathematically perfect, so that there will at least be some coherence to budgetary expenditures."[16]

This rather unsubtle allusion to corruption gets at the heart of the appeal of calculation. Numbers meant coherence and generality, a defense against the forces of parochialism and local interest. It helps in this context to recall one central mission of the Corps that remained unchanged, whether it was maintaining roads, building canals, or laying out rail lines. It aimed to unify and administer the French territory, and even to civilize the French peasantry.[17] The basic design of the French rail system, worked out in the late 1830s and early 1840s, is emblematic. The Corps envisioned five or six main lines, spreading out from the capital to all edges of the French hexagon. This very sensible plan was, alas, threatened by the disorderly existence of towns and rivers. Population and industry were concentrated in both. The Corps faced infinite demands to divert lines from the geometrical pattern that many of its officers preferred. "It pains us to see that passions of locality work to embitter a discussion that ought to concern only the general interest."[18]

Some of the more extreme rationalistic ambitions were expressed in 1833 by the suitably named engineer Charlemagne Courtois, using, however, a form of quantification that was completely ad hoc and that never caught on. There is no reason to select routes "more or less arbitrarily" when this problem "admits a rigorous solution." The trick is to maximize the "effect" or "advantage," which is equal to quantity of transport divided by costs, n/D. Some manipulations, more verbal than mathematical, converted this into advantage per unit cost, a quantity that diminished with the *square* of costs. This remarkable result suited Courtois nicely, since any lengthening of the line to pass through intermediate cities would now decrease "advantage" with a more than doubled effect.[19] By 1843 the cost factor in the denominator had grown, still more implausibly, into a cubic term. If a line is lengthened by 10 percent to pass through a certain city, this somehow increases the mean annual expenditure per kilometer by a factor of 1.33. The difference, which for a line of 322 kilometers extended to 355 would amount to 3,503,360 francs, should be charged to the city and département for whose sake the line was displaced. But they would never consent to pay so much; to state the figures is to reveal the absurdity. "If this rule we have just determined were applied rigorously, how many persons defending the interest of their locality would cease to fatigue the administration" with their endless entreaties? The "general interest" is to achieve "the greatest possible effect with the available means. The highest goal of public economy is to determine this maximum."[20] It must not be compromised for mere "financial success."

Courtois's general formulation had particular purposes. By 1843 he was at work as a chief engineer on the proposed line from Paris to Strasbourg and on to Germany. The original plan for this line had been sketched by Navier just before he died in 1834. Navier wanted the track to make only minor deviations from a straight line, avoiding mountains but ignoring towns and rivers. Courtois, who in 1843 still thought canals far superior to railroads, favored a trunk line toward Strasbourg and Lyon, branching at Brienne. Brienne was the solution to a mathematical problem of minimizing a certain combination of construction expenses and distance. He calculated, using his own special formulas, that if instead the trunk line meandered along the Seine to Troyes, this would involve an "equivalent loss" of 27 million francs. But his proposed line to Brienne, like Navier's direct one to Strasbourg, followed no great rivers and passed through no major towns. Courtois calculated that it was advantageous to deviate from the shortest path to go through a town only if a branch line would require at least six times as much track as the added length of the diversion.

Jouffroy, in his history of the construction of the Strasbourg line, remarks that Courtois's reasoning was geometric, not economic.[21] But this is not quite right. A straight track was designed for long-distance commerce, international as well as domestic. In particular it was to provide a better connection between England and Germany than any lines on Belgian soil. Courtois believed also, with Navier, that a rail line would bring prosperity with it, so that building through poor or thinly populated regions might contribute more to the economy than a line through the main towns. Finally, what point was there in building a rail line where much cheaper transportation was already provided by a river or canal? Not all engineers accepted this reasoning. Minard, who favored the line along the Marne that was eventually chosen by the council of the Ponts et Chaussées, printed graphs and statistical tables showing that most traffic is local, and that few passengers travel the entire length of a line. A really useful line, then, must travel through as many densely populated places as possible.[22] The successful designer, Marinet, provided detailed figures for the current road traffic along his chosen route, then multiplied by three to estimate rail use. His result for traffic on the first leg, from Paris to Vitry-le-François, was precisely 4,230,501 travelers going a total of 117,809,796 kilometers and saving two centimes each per kilometer. This amounted to a total saving of 2,356,075 fr.92, an annual return of 4.46 percent on capital.

These lines to Strasbourg and Lyon were controversial in proportion to their importance. Some engineers lamented that politics favored tracks wandering aimlessly through every town with an effective political voice. Calculation was one scheme for neutralizing politics. The other was impartial commissions. A prominent one led by the Comte Daru took a broad look at transportation policy. "The most contradictory assertions are made public; utterly inconsistent figures are produced, for nothing is more elastic than figures." Yet the most important rule it could propose for escaping the quandary was to compare expenses of each project with probable receipts.[23] Daru, in the end, agreed with Minard that the lines should travel through regions of dense population. Another commission, studying the line from Paris to Dijon, from whence it would proceed to Lyon and to Mulhouse, had less to say about receipts but emphasized that traffic volume would be enhanced by near proximity to navigable waterways.[24]

These appeals to the quantitative had a certain rhetorical effect, but so long as they were not incorporated into routines they had to remain relatively weak. Everybody professed opposition to the corruptions of politics, and many were willing to view the choice of routes as some kind of maximization problem. In 1843, though, there was no hint of a consensus about what should be maximized. From a budgetary standpoint, it would be nice if receipts were sufficient to pay the interest on bonds. But was traffic volume, or revenue, an adequate proxy for utility? Most engineers and other commentators insisted that it was not. Edmond Teisserenc, a polytechnician who later became minister of public works, argued against Daru: "If the interest of the community in the establishment of a transportation line could be measured, as the subcommission says, by the revenue one can anticipate from its exploitation, by the amount of traffic where one proposes to put it, then nothing would be easier than choosing the best routes." But Daru's figures did not distinguish whether a unit of freight to be carried by rail could make the same trip for the same price by river or canal. If a good navigable river runs along the same route, a rail line brings no real benefit except the saving of passengers' time. Where there is no water transportation, a train will also bring much cheaper fares for passengers and, especially, freight. Teisserenc illustrated this by a hypothetical example. A rail line far from navigable water carrying a certain quantity of passengers and freight might contribute 3,463,000 francs to the public utility, whereas another of the same length receiving the same revenue from the same loads, but traveling along a good river, would contribute only 263,000.[25]

ASSESSING PUBLIC UTILITY

Teisserenc's illustration shows how a quantitative reading of public utility could be deployed to subdue the politics of particular interests. The requirement that projects receive a declaration of public utility, itself designed to reduce the play of politics in public works legislation, would seem to invite such a reading. When we consider further that a shift to a quantitative language would have been to the advantage of these very numerate engineers, it may seem surprising that measures of utility were not deployed routinely in official reports or recommendations. Although numbers did have their uses, it was quite possible to assess public utility without any attempt at measurement and without any comparison of expected benefits with costs.

If a dock or bridge might fail in storms, or a canal dry up part of the year, this had an obvious bearing on its worth, and issues of safety and reliability were fit matters for discussion in an evaluation of public utility. Military usefulness and vulnerablility to an invading army came up frequently, and sometimes overbalanced purely economic advantages that were uncontested. Even within the economic domain, engineers and other inquirers generally did not attempt to reduce all factors to common, financial terms. A proposal to construct docks in Marseille modeled on those of London and Liverpool was found advantageous by a special commission in 1836 because of its convenience to the city and the reliability of access to the sea. The chief engineer of Finistère reported in 1854 that a commission d'enquête had concluded in favor of a trunk line through central Brittany, even though it would pass through more rugged terrain and would require steeper grades than a more southern route. Its main advantage was economy: such a line could, with branch lines, serve the entire peninsula. This would be cheaper than trunk lines to the north and the south, and an extravagant proposal involving both might never get built at all. It probably didn't hurt that the destination of this central line was to be named for the dynasty: Napoléonville.[26] This last advantage, certainly, could not very well be quantified and compared against costs.

Planning discussions within the Corps des Ponts typically wandered freely through technical, economic, and broadly political considerations. Chief engineer Jean Lacordaire, who disagreed with his colleague Auguste-Napoléon Parandier about the best route from Dijon to Mulhouse, sent the inspectors general of the Ponts et Chaussées a double-column manuscript critiquing his rival point by point. At issue were relative degrees of public utility. Parandier's route along the Doubs, he wrote, was not at all preferable from the standpoint of steepness of grades and number of required crossings to Lacordaire's own along the upper Saône. Lacordaire's tunnels were not so difficult or expensive as certain persons charged. And Parandier's comments on the difficulty of Lacordaire's soils only showed that he hadn't properly studied them. Notwithstanding this argument for the manifest superiority of his original proposal, Lacordaire reported less than two weeks later on a mixed line, "of conciliation." It was a little longer, and had to climb a bit more, but it would be cheaper than following the sinuous valley of the Doubs. He found it quite inexplicable that the city of Besançon should want a line up this difficult, costly, and dangerous gorge, especially since the river and canal provided perfectly good transportation there already. Besançon would back the conciliatory line if it understood its own interests. Another advantage was that this line wouldn't draw its traffic away from the canal between the Rhône and the Rhine. Finally, it would make the town of Gray prosper as an entrepôt, whereas the Doubs line would destroy it.[27]

This is how one generally assessed public utility. It did not mean a quantitative surplus of benefits over costs. Declarations of public utility at the national level were used to distinguish the general interest from local interest, and hence to exclude state subventions to small branch lines. Military need was often decisive, as was territorial unity, and in 1878 when Charles de Freycinet proposed a vastly extended local rail system, he mentioned administrative centralization among its crucial advantages.[28] Public utility had something to do with practicability. The committee of the National Assembly charged to examine a channel tunnel that was proposed in 1875 by a society under the presidency of Michel Chevalier considered it manifestly useful, but was reluctant to issue a formal declaration for a project whose possibility remained in doubt. The council of the Corps des Ponts advised that the project be granted only a *concession eventuelle* until the difficult geological and diplomatic questions were more nearly settled.[29]

Above all, public utility had to do with rationality of planning, the avoidance of unnecessary competition. In the early decades of railroad building, this often meant that a rail line should follow a new route, and not duplicate a service already provided by canals. Canal defenders were still active at the turn of the century, as attested by the embattled tone of Henri Chardon's remark in 1904 that statistics had incontestably shown railroads to be twice as cheap as canals. If certain planets have massive canal systems, he added, alluding to the contemporary astronomical debate about Mars, they must be a lot flatter than this one.[30] Still, by the 1850s a proposal was most likely to fail the test of public utility when other rail lines were deemed to provide adequate service already. The French government was little inclined to subsidize competition to tracks whose bonds it had guaranteed. After 1852 the government gave special standing to six large regional railroad companies, with which it gradually developed a kind of partnership. This did not quite exclude entrepreneurs from laying new tracks, or even receiving a public subsidy to do so, but the formalities of assessing public utility were used to block lines that would draw traffic away from the established companies.[31]

A glance at the disposition of any moderately complex case reveals issues that could never have been settled by a simple comparison of costs and benefits. In December 1869, the council of the Ponts et Chaussées considered a report from the engineer Kolb on a local line in the north of France, 66 kilometers long, running from Alençon along the Huisne River to Condé. Kolb forecast a net return of 6.8 percent on capital, a very good number. Unfortunately, another line was under consideration from Orléans to Lisieux. The municipal council of Nogent le Rotrou feared that this new proposal would lead to delays in the Orléans-Lisieux line, and the commune of Bellême complained of potential competition. Kolb disagreed, arguing that the two lines might even benefit each other, especially if the concessionaire could be persuaded to abandon the valley of the Huisne for a route through Bellême.

Alas, he could not. Meanwhile there were complications in planning the line from Orléans to Lisieux. In fact, several proposals were in play, and the line might go to Bernay or L'Aigle rather than Lisieux. Each of the ordinary engineers who had been assigned a particular direction to study had concluded in favor of it. The various enquêtes yielded predictably chaotic recommendations. In the Loiret they favored a declaration of public utility for a line to Lisieux by way of Ormes, Patay, and Chateaudun, which indeed was the proposal of the chief engineer. In the Eure et Loire they specified additional intermediate cities: Brou and Nogent le Rotrou. The départements of the Orne and of Calvados were content with this, but in the Eure they proposed to build along the valleys of the Charentonne and the Calonne rather than the Vie. The chief engineer conceded that this would save money, and that the Charentonne had much industry, but the Vie too was very rich, and in any case a route along the Charentonne would be too close to another line just to the north, from Orléans to Elbeuf. He was able to negotiate subventions from the départements and communes slated to benefit from his proposal, and the council of the Ponts et Chaussées accepted it.[32]

The reports on projects prepared for the National Assembly during the Third Republic interpreted public utility in similar ways. Often they appeared under the name of Ponts-et-Chaussées engineers who were also members of the assembly. Ernest Cézanne, writing for a commission to consider a gloriously Saint-Simonian line from Calais to Marseille—the English Channel to the Mediterranean—asked whether this was needed. He dismissed competition as a reason for building a new line, since the companies would inevitably collaborate and conspire. From Calais to Amiens there were already two lines, with revenues per kilometer of 62,000 and 43,000 francs, so evidently there was no need for another. From Amiens to Creil there was only one, with revenue of 122,000 francs per kilometer, which was why another had already been authorized. There was enough track already from Creil to St. Denis. Beyond Paris, there was need for a good double track with gentle grades and curves from Nîmes to Lyon, or for a really excellent one from Paris to Marseille costing a million francs per kilometer and permitting speeds over 100 kilometers per hour. The proposed line didn't provide this. It would not contribute to the public utility, but amounted to a wasted investment of 600 million francs. And it would cause the further loss of 20 million francs per year on existing lines.[33]

It is clearly relevant, but difficult to know, what was the relation between decisions taken by the council of the Corps des Ponts and projects recommended to the National Assembly. Much information was not controlled by the Corps; legislative proposals could be based on the dossiers of the projects, the record of testimonies at enquêtes, the opinions of commissions formed by départements, and reports by prefects, by the Corps, and by the Conseil d'Etat (the top of the French administration), which often relied on a special commission. One could be sure that the Minister of Public Works and the affected companies would take an active interest.[34] Still, proposals only reached them after most of the important decisions had been provisionally made. This included planning routes, negotiating subventions, and drawing up contracts governing the concession. On at least one occasion when the enquêtes were unanimous in favor of a plan that the Corps council contested, the higher administration agreed with the council and it was sent back for further study. In this case, though, when the Corps finally negotiated a plan whose public utility it could endorse, the Conseil d'Etat rejected that one too.[35]

Alfred Picard's detailed chronological history cites some instances from the 1840s when routes were debated at length in the Chamber of Deputies. In this forum, the intervention of legislators seems generally to have failed. Occasionally political pressure obliged the minister of public works or the Corps to consent to further studies.[36] They preferred to settle these issues quietly, and did so very effectively. Proposals that did not gain approval from the Corps were normally withdrawn. Entrepreneurs sometimes complained of its despotic power.[37] Of course it could act despotically only against weak outsiders. Still, its power in regard to routine planning decisions was very great. And, crucially, that power was exercised more through the channel of private negotiations than public reports.

PREDICTING REVENUES AND ESTIMATING BENEFITS

A project description by Charles Baum provides a useful guide to the forms of economic quantification that figured in project planning. In 1885, Baum was chief engineer in the département of Morbihan, on the southern coast of Brittany. The project description incorporated various features of his own economic writings, published in the *Annales des Ponts et Chaussées.* It was not the custom to publish project descriptions, and we may infer that Baum regarded his as a model for the genre. He introduced it with an argument for building rail lines, particularly in Morbihan, despite the insufficiency of revenue to cover costs plus interest. He pointed out that Morbihan had 267 kilometers of rail line, only 0.392 meters per hectare (versus a French average of 0.586), and only 511 meters per thousand habitants (versus 815). He had five proposed new lines to consider, all of which he found to possess, though in different degrees, a "well-defined utility." This utility couldn't be measured by revenue per kilometer, but was rather the sum of benefits to all users. Judged from the differences in transport costs between roads and railroads, the benefit over costs amounted to about 24 centimes per ton-kilometer, as compared to actual freight charges, and hence revenues, of 12 centimes.

This was the only point at which Baum argued in the language of public utility. In his detailed study of the proposed lines, leading to an assignment of relative priorities, he wrote instead in strict accounting terms about costs and revenues. First he described the route, kilometer by kilometer, indicating stations, bridges, curves, grades, and other special features. He estimated costs of construction per kilometer, without the rounding that was by then increasingly common in these documents, at 59,845 fr.44. In estimating the cost of operation, he used a special quantitative concept that he had adapted from the Swiss, "virtual length." This was the length of flat, straight track that would consume the same quantity of mechanical work in the passage of a train over it as the actual line in question, with its grades and curves. For this purpose he calculated *coefficients d'allongement*, or multipliers, to apply to each unit of track with a given steepness and radius of curvature.[38] On the difficult terrain between Vannes and La Roche-Bernard the mean coefficient was 3.323, so that the equivalent length of his 45-kilometer track would be about 150 kilometers.

Having finished his discussion of costs, he moved on to revenues. Because the line was rather expensive, he proposed charges slightly higher than the average for trains of local interest, on condition that steep discounts be offered for round-trip passenger travel. There remained the thorny problem of estimating traffic. The annual revenue per kilometer (*recette kilométrique*) on ten "comparable" lines varied from 2,500 to 5,700 francs. The *recette probable*, or expected revenue, should be somewhere in this range. The preliminary planning document (*avant-projet*), however, had estimated revenue from the mean rail usage per inhabitant in the west of France, using the records of the Compagnie de l'Ouest. It proposed a *recette kilométrique* of 7,088 fr.77. This seemed too high.

But what is the correct value? Fortunately Baum had published extensively on just this question, including a paper concerned specifically with trains of local interest.[39] The best comparison, he judged, was with the two lines currently serving the city of Vannes. They showed revenues per kilometer of 2,321 and 5,624 francs, or a mean of 3,962 (actually 3,972) francs. This, happily, was toward the middle of the range defined by the larger group of comparable lines. It was also in good agreement with a figure of 4,400 francs obtained by tallying the population served, taking account of its distribution along the line, making an appropriate reduction from mean per capita figures because it was mainly agricultural, then adding a bit for maritime traffic.

None of this was simply ad hoc. Ponts engineers had published in some detail on the problems of choosing comparable lines, on counting the affected population, and estimating freight and passenger usage per individual. These depended on certain assumptions about uniformity of behavior, but the engineers had ideas about how to adjust the standard figures for agricultural and urban, wine and wheat, and northeast or southwest, as well as to take account of large industries or mines. This had all been laid out, with formulas and approximations and advice for identifying exceptional cases in an 1868 paper by Louis-Jules Michel.[40]

Baum's project description proceeded finally to the crucial comparison of costs and revenues. Taking into account that local lines should be built cheaply, and that he was recommending a narrower than standard gauge, the formula for operating costs was D (*dépenses*) = 1,500 + R/3, where R was revenue (*recettes*). For R estimated at 4,400 francs, this implied expenses of about 3,000 francs, hence a net revenue of 1,400 francs. To pay 5 percent on capital would require a net revenue per kilometer of 3,000 francs, so this line required an annual subvention of about 1,600 francs. Assuming that revenue grows according to a "loi de progression naturelle" of 2 percent per year, this subvention would diminish and then vanish after about sixteen years. Was it worth it? Baum had no doubt that the surplus of utility over revenue justified such expenditures, but the state has only finite resources, and he recommended that it start by building the lines that require the least financial sacrifice. He proceeded to analyze four other lines in just the same way, concluding on strict accounting grounds that this one deserved to be second in priority, after a slightly more advantageous one from Lorient to Kernascléden.[41]

Baum was showing off his own skills in this report, yet in its essentials this was a well-established genre. Similar techniques, along with an analogy with Suez, were used to forecast volume and profits of the new Panama Canal that de Lesseps wanted to build.[42] Such predictions were known on occasion to fall short of perfect accuracy, but they were required by custom, if not by explicit rules. A three-page summary of a report on a proposed line from Falgueyrat to Villeneuve, prepared for the minister of public works, presented the crucial figures without explanation. They were standard quantities: maximum grades, minimum radii of curvature, construction costs per kilometer, annual revenues per kilometer (*produit kilométrique*). This last amounted "en nombre rond" to 10,000 francs, as compared to 8,000 for an alternative destination from Falgueyrat, which on this evidence was dismissed as inferior.[43] Another report of the same year, proposing two lines in the Sarthe, was sent back by the railroad section of the Ponts-et-Chaussées council because it lacked the necessary information for fixing the state subvention, as required by a law of July 17, 1865. The council demanded "an estimation of expenditures with a precise indication of how much will be paid by the département, and an estimate of the probable traffic on each of the two projected lines."[44]

This thoroughly administrative form of economic calculation was carried out in terms of cost and revenue, not costs and benefits. Still, behind it all stood utility, which, it was generally acknowledged, surpassed receipts to such an extent that local losses became social gains. The credibility of that assumption meant that state activity in transportation was not strenuously challenged most of the time. This reduced, but did not annul, the need to defend such interventions in quantitative terms. The measurement of utility became especially pressing toward the end of the 1870s, when a new minister of the interior, Charles de Freycinet, proposed a huge program of state subventions for new local lines all over France. The politics of local lines was so controversial that Sanford Elwitt has identified it as the main cause of the 1877 crisis of the republic. However astute politically it might be to provide rail service to thousands of small towns and villages, there were powerful financiers and big companies who preferred to see resources invested chiefly in main lines, as they had been under Louis Napoleon.[45]

Inevitably, the Freycinet Plan inspired a quantitative debate about the valuation of small lines. In terms of projected revenue (even if it materialized), they were dismal investments. Did they nonetheless pay for themselves by increasing traffic on the more profitable trunk lines? Might they, in this way, contribute to the "general interest," and not merely the local interest? Or could they at least save enough of the expenses incurred privately in hauling passengers and freight to make up for their cost to the taxpayer? Ever since the canal boom, various strategies had been employed to identify indirect benefits, such as increased property values in the vicinity of improved transport, to justify projects that could not be made to pay for themselves.[46]

Not too many cabinet ministers have performed arithmetic calculations of utility in speeches before a parliament. For Freycinet, though, such measurements provided the most powerful justification, or at least rationalization, for his plan. The "true revenue, the national revenue" of a railroad is "the economy it permits in transports." It costs 30 centimes to move a ton of freight a kilometer over the roads, whereas railroads charge only 6. "The community thus realizes a benefit of 24 centimes out of 30; in other words, the community realizes a profit equal to four times the tolls, four times the total of receipts." So, in a typical case, if revenue merely covers costs, making no contribution to interest on investments, a line will still yield a real profit of 14 percent even on the least optimistic assumptions.[47]

This was an inspiring vision. But was it true? Two engineers, Eugène Varroy and J. B. Krantz, criticized his computation (while supporting his program) before the Senate. Much of the traffic using the railroads would be new, because it couldn't pay the high costs of using the roads. Hence the presumed benefit of 24 centimes per ton-kilometer used by Freycinet, and before him by Navier, was no valid measure of utility. Krantz argued that, at a minimum, a line should cover the expenses of operation. Varroy made the easiest assumption consistent with Dupuit—that utilities to users are distributed evenly over the range from 6 to 30 centimes, and calculated the contribution of any line to public utility at 18 centimes per ton, hence three times the gross receipts rather than Freycinet's five. He was at a loss to say whether this was "local utility" or "general utility." He added that it was very tricky, requiring sagacity and experience, even to estimate levels of usage. Meanwhile, in the reports on Freycinet's various proposed lines, at least one engineer invoked an estimate of the utility represented by time and cost savings to cover a financial deficit.[48]

Etner points out that Dupuit's economics became well known to Ponts engineers at just this time, the late 1870s. Dupuit's idea of diminishing marginal utility was cited against Freycinet by cabinet ministers as well as engineers. Albert Christophle, Freycinet's predecessor as minister of public works, wrote a bitter preface to accompany a collection of his public speeches from 1876 and 1877. Freycinet, he explained, was not inspired by rational economics, but by craven politics. The only assurance of the utility of local railroads is the willingness of local units of government to contribute to their construction in proportion to the benefit and to their wealth. Freycinet's idea that the utility of a line exceeds its revenues by four or five times had been refuted in advance by Dupuit, whose calculations showed "irrefutably" that revenues of 6,800 francs correspond to indirect benefits of no more than 3,000 or 4,000 francs. He added that Freycinet also seriously miscalculated costs and revenues, as subsequent experience had shown. A writer in the *Revue des deux mondes* wondered if the ideal of rational planning demanded by declarations of public utility hadn't been supplanted by an indiscriminate drive to satisfy all appetites.[49]

One might expect that Ponts engineers would happily endorse the most generous calculations of the benefits of railroads. In fact they did so rather rarely, and many didn't support vast building programs at all. An argument by Félix de Labry in 1875 gives some idea of the possible convolutions. He appealed to the readers' intuition, and little more, to support his claim that of French national production of 26 billion francs, at least 5 billion could be credited to railroads. He also argued, however, that the state should invest in railroads only if the investment were paid back, not to society in the currency of general utility, but to the state treasury, in tax revenue. Since the state makes up 10 percent of the economy, any public money invested in railroads must generate economies at least ten times greater in private production and transport. The joint effect of his refusal to involve the state in promoting public utility and his extravagantly generous estimates of the economic consequences of railroads was indeterminate. Certainly his papers were not propaganda in favor of Freycinet's initiative. His fellow engineers disagreed with him mainly on the matter of principle, arguing that the interest of the state is identical to that of society, so that public utility is very much the state's business. But this, argued Antoine Doussot, does not mean building every railroad that might promote the public utility; it means spending the available funds on those projects that contribute most effectively to it.[50]

The most prominent economic spokesman for the Corps des Ponts around the fin de siècle was Clément-Léon Colson, a notable economic liberal. He did not oppose the whole Freycinet railroad initiative, but he considered it excessive and indiscriminate. Colson's entire career was a battle against indiscriminacy, in favor of nuanced judgment based on careful scrutiny of particular facts. At least this seemed the only way to run a railroad.

On the contribution of railroads to public utility, Colson defined his position against another engineer, Armand Considère. Considère was chief engineer of the Finistère, in westernmost Brittany. He published two long papers in 1892 and 1894 to demonstrate the vast benefits of state-supported local lines. He admitted that they rarely produce an adequate financial return, considered as separate enterprises. But they produce some important direct and indirect benefits, which, with care, can be approximately quantified. To begin, they increase the volume of traffic on the trunk lines. Considère estimated from charts of traffic volume over time that 50 percent of the freight carried by these local lines is new traffic. On average it will travel four times as far on the main lines as on the branch line where it originated. This effect is less decisive for passengers, but still, every franc received by the local lines means 140 centimes of increased revenues for the main lines. Next, Considère translated this added revenue into increased utility. He did not use the generous formulas of Navier and Freycinet, but incorporated the principles of Dupuit, and assumed that demand (and hence also utility) was a declining linear function of price. On a graph with price as one axis and demand as the other, this line could be drawn through the two points defined by current traffic at current prices (on the roads) and projected future traffic at railroad prices. The excess of utility above revenue would then be represented by a triangle, whose area could easily be found.

So much for direct benefits. Cheap transport also spurs economic development. A mine that was not worth exploiting for export until the rail lines came through might soon become a center of population and industry. Widely diffused railroad stations have a valuable advertising function, alerting peasants and craftsmen to the possibilities of exchange with a larger world. They help to overcome local inertia. These effects are of course not easily quantified. Considère believed in them enough to estimate benefits from the statistics of the whole French economy rather than from a model of their direct effects. In the last thirty years production had grown by 15 billion francs. Of this some 3.6 billion might be due to earnings on capital, and another billion to population growth, leaving more than 10 billion otherwise unexplained. He "conservatively" attributed only one-third of this to indirect effects of improved transport by railroads. Adding this to the direct benefits, he found that the advantages brought by local lines exceeded their receipts by at least six times. This meant, for example, that a line nominally losing 250 francs per kilometer actually returns 20 percent on capital in *utilité totale*. Of course one builds the best lines first, not every line that anyone happens to propose, but the whole Freycinet initiative looked very productive from Considère's perspective.[51]

Colson was unconvinced. Considère's formulas would lead to far too much construction. The calculations of indirect benefits were particularly vulnerable to criticism, but Colson also doubted the measures of direct benefits. He considered that Considère had generalized from unrepresentative cases, lines that were not typical. And how could he assume that new traffic generated by a branch line makes the mean journey? Long-distance traffic would not have been much discouraged by a few extra kilometers on the roads, so the new traffic induced by local lines would probably travel relatively short distances. Considère replied two years later with another long paper, including still more detailed attention to the statistics of several different lines, as well as attempts to measure the contested quantities in new ways. Colson returned that these statistical inquiries conceded his principal point: "The question cannot receive a general solution by the route of statistical studies." He did not mean that Considère should dispense with statistics, but that he should abandon the vain hope of a general solution. In the field of railroad planning, there was no substitute for judgment applied to a detailed consideration of each particular case.[52] The theoretician can help decide what quantities deserve to be measured or predicted, but there can be no rigorous mathematical formulas, only general guidelines. Colson emphasized the sense of tact that comes with long experience.[53]

This was Colson's perspective also on the vexed issues of prices. The French state, like every other, took a considerable interest in these matters. State guarantees of a fixed return on investment became systematic under the Second Empire, and the state's influence over prices was correspondingly strong.[54] In 1883 the railway companies were consolidated, and competition was replaced by state regulation. This did not quite mean that competition became inconsequential. The companies explained repeatedly that they needed the right to lower fares connecting points that were served by barges or ships.[55] In France, as elsewhere, the higher fares charged shipments between other points were loudly denounced, especially if a port city at the end of a line, hauling goods for import or export, received better rates than did intermediate destinations along the same line.[56]

Dupuit had sought to solve the problem of fares in terms of utility and demand. He distinguished between *frais* and *péages*, or variable costs and tolls. Those expenses directly associated with the transportation of persons and goods (*frais*), which go up with volume, should without fail be charged to users. The purpose of tolls (*péages*), on the other hand, is to recoup capital investment and to make a profit. They should be treated quite separately from costs, and set, so far as possible, in proportion to the utility derived by the user.[57] By the 1880s, all Ponts engineers agreed that his logic was impeccable. "However," warned Alfred Picard in 1918, echoing both Colson and Considère, "it contains also certain theoretical deductions whose application would be impossible."[58]

Most attempts to formulate a rational basis for rates, in fact, owed less to Dupuit than to the work of Jullien and Belpaire discussed in chapter 3. Or rather, the idea of charging to each passenger and unit of freight the costs attributable to it seemed so plausible, and indeed moral, that it was generally taken for granted, still more by the general public than by engineers. The social philosopher Proudhon argued vehemently that social justice requires a strict proportionality between transport rates and the cost to the company.[59] Baum thought so too. He called the cost per kilometer of transporting a passenger or a ton of merchandise the *prix de revient.* This is the minimum that railroads should charge in order to cover their expenses. Since any lower rates will bring a loss to society, the companies must avoid them even in response to the stiffest competition. Of course this price would vary in different locations according to circumstances. But it could be calculated using railroad statistics. Baum published a series of papers to show how.[60]

His solution, like Belpaire's, was basically a matter of allocating costs fairly to all users. It had all the defects of the genre. Other engineers did not fail to point this out. The sharpest critic, René Tavernier, remarked that a commission of American experts had tried in this way to settle the rate fights among Boston, New York, Philadelphia, and Baltimore, and had concluded that the *prix de revient* is impossible to determine. It is no constant, but varies by line, by season, and by traffic level. It will often be advantageous for a rail line to charge less than this value, at least for selected merchandise, since the variable costs associated with any particular load are much lower than the *prix de revient.* The best solution is to use flexible pricing, which the big lines are too bureaucratic to wield effectively. Hence it would be an improvement to split them up into smaller companies.[61] Baum responded, feebly, that Tavernier showed a lack of understanding of mean values; that variability no more undermines a calculation of *prix de revient* than of *vie moyenne.* Tavernier replied that the wide influence of Baum's useless quantity is testimony to the malign effects of the bureaucratic spirit of simplification.[62]

If not the last word, then certainly some very influential ones—and a very great many of them—were spoken by Colson. Himself a Ponts engineer, he taught political economy at the Ecole des Ponts from 1892 to 1926, and at Polytechnique from 1914 to 1929. There is little mystery about the content of his teaching, since he published one of the courses in six volumes totaling well over two thousand pages. Ponts engineers had always been taught economics by free-market liberals, whose doctrines they managed somehow to believe without compromising their faith in centralization and in beneficial state intervention.[63] In ideological terms, Colson's was a conventional, liberal introduction to political economy. He called attention to his use of mathematics, since the course was designed for engineers rather than merchants and lawyers. But, he argued, there are far too many unknowns for a mathematical strategy to be pursued consistently, or worked out in detail. Mathematics can suggest useful analogies and comparisons, and can help the economist to recognize when a problem has a well-defined solution. Those "rare authors" who are content to reason deductively and mathematically "have often deviated completely from real facts in their most ingenious theories."[64] Colson took pride in staying close to statistical facts, so that his economics would be useful in practice.[65]

Not surprisingly, his discussion of "public works and transports" made up the most important and original part of his course. Here he admitted a variety of influences, but he followed most closely the principles of Dupuit. Fundamental for him was the idea of a demand curve that declined as price rose, or equivalently of diminishing utilities. Colson knew of the new marginal economics. He spoke favorably of the nonmathematical Austrians, and also mentioned William Stanley Jevons, but ignored Walras, whose work was much more abstract and mathematical.[66] He did not develop their methods in the abstract; he only applied them to the very particular problem of setting prices for transportation. That is, he felt little need to take theory further than Dupuit had.

His mission, rather, was always to temper theory with practice. On the matter of prices, there is no way to recover the whole utility of transport, or even a given fraction of it, from every user. Differential tolls will sometimes be unworkable, in which case tolls should be kept low, so as not to discourage users who can benefit, even slightly, from a railroad or bridge. Here Colson stood firmly with the French tradition of state activity, against what he saw as the Anglo-American view that everything really useful must pay an entrepreneur to build it. This role of the state was the only reason he saw his calculations as having more than theoretical interest.[67] That is, even the liberal Colson used economic quantification mainly as an alternative to market mechanisms, often in opposition to market principles.[68] Still, he believed that the costs of transport, including capital costs, should be recovered from users to the maximum extent possible. He was suspicious of grand claims for the indirect benefits of new railroads or canals, and believed with Dupuit that a high proportion of the utility contributed by a canal or rail line ought to be recoverable (up to costs) in tolls.

Colson used Baum's language of apportioning costs fairly. Following Considère, however, he added an adjective, *partiel*, to Baum's phrase. Only variable costs entered into this calculation.[69] Even the *prix de revient partiel* contained ambiguities; it required ingenuity, and some tolerance for mere convention, to separate variable expenses from capital costs. But this quantity was at least consistent with Dupuit's theories, since it left capital costs to be allocated through tolls in proportion to utility derived. Ever a realist, however, Colson noted that Dupuit's was an impossible ideal. It would require some functionary to inquire into the value of every shipment, and, always somewhat arbitrarily, to fix the value of transport. So much discretion at this level of intervention was neither legally nor morally acceptable. It was necessary to charge according to categories of goods, following rules that are "fixed and not arbitrary, . . . rational and explicable."[70]

To deny this discretion to the employee on the line was not of course to eliminate judgment, but to concentrate it at a higher level of management. For the administrative elite, nothing could ever be reduced to simple formulas. Even the rules of tarification, whose fixity seemed so crucial, had to be "supple enough to adapt to commercial necessities." These considerations were so complex that they might well be judged diversely, "even by enlightened and impartial people."

Colson's acute sense of complexity mirrored the actual regulation of freight charges, which generated the most arcane disputes about classification of goods, and which was scarcely at all informed by any economic theorizing. François Caron suggests that rate-setting at the level of bureaucratic practice in France exhibited few scientific pretensions, and that the reigning theory was simply to charge *ad valorem*.[71] Colson, more ambitiously, favored calculation for this and other purposes, but he taught more than a generation of Ponts engineers that it could never be made rigorous. "There are many ingenious formulas to calculate the traffic volume on a planned route as a function of the population served; but to apply them with discernment requires taking account of the social, economic, and moral state of the population, and that is the great difficulty."[72] This is how engineers most often represented their methods to the larger public. It is also how they thought of themselves. The rhetoric of inflexible laws, followed self-effacingly by men whose expertise is purely technical, was not theirs. They were a self-conscious elite. Their uses of quantification can be understood in no other terms.

ENGINEERS AS ELITES

Given the almost unrivaled prestige of the Ecole Polytechnique, and the success of its graduates in industry and administration, one might not anticipate such skepticism about the possibility of rigorous quantification. It ought to have been especially appreciated by the eternally unstable Third Republic, which enshrined science as a basis for social consensus and an alternative to the conservatism of the Church.[73] The location and pricing of canals and especially railroads were enormously controversial throughout the century. Under the July monarchy, for example, départements not scheduled to get railroad lines actively campaigned against all state support for them.[74] State planners could of course compromise with their more powerful opponents, but in the end some départements had to be favored over others. Certainly it was expedient to be able to certify the fairness and objectivity of such decisions by numbers. Should we not expect an elite of mathematically educated engineers to take advantage of this veneration of science, and to make decisions in the classic way of engineers, by simplifying and quantifying?

The answer is no. They were quite capable of acting effectively in an informal way. Moreover, there was as yet no classic quantitative way to make engineering decisions. Ponts-et-chaussées engineers were themselves prototypes of the quantifying engineer. Their ambit extended beyond the domain of structures and machines, and, as we have seen, they made extensive use of numbers and calculations for economics, planning, and administration. But we cannot regard their inclination to quantify as a reflex, born of a faith intrinsic to modern engineering that every problem has a mathematical solution. These engineers believed that economic numbers, at least, became useful only when expertly interpreted.

Engineers of the Corps des Ponts were often accused of relying on numbers out of habit, or for lack of ability to understand social matters any other way. After the failure of a dam in 1895, they were mocked by one critic in the following terms: "The savant engineers of the sacrosanct *Ecole*, knowing the danger for having ascertained it in memoirs filled with numbers, knowing nearly to the penny how much the damages for the destruction of entire villages would amount to, and what the loss of human life would cost to the state, still filled the menacing reservoir up to the brim, until the definitive crack came to confirm the mathematical exactitude of their previsions."[75] This, however, is misleading, and not simply because it implies an absence of moral concern. French engineers used mathematics to plan bridges and railroads, but they rarely entrusted decisions to the numbers. Their prestige rested mainly on their background, education, and relation to the state. The authority of calculation and objectivity were secondary. Numbers were not powerful in themselves, and counted for little when deployed by outsiders. They could only provide a modest supplement to institutional power.

From this perspective, the modesty of their efforts to mechanize decision-making becomes less mysterious. They controlled, virtually unchallenged, the power of calculation, when they chose to use it. But polytechnicians were part of an elite so secure that they rarely needed to negate or conceal their own discretion. Mathematical prowess was not their principal claim to authority, and they preferred to rest their decisions on long experience and general culture. Thoroughgoing quantification involved costs, particularly its rigidity and its requirement that the weighing of factors be made explicit. Ponts engineers chose to manage their affairs in a different mode.

The intense mathematical study required to gain entry to the Ecole Polytechnique promoted its reputation as a purely technical institution. Thus Balzac, who had a brother-in-law in the Corps des Ponts, portrayed the state engineer in *Le Curé de village* as a beautiful orange blossom nipped by frost before it could bear fruit. The chill was presumed to be mathematical. In the realm of nonfiction, Joseph Bertrand reminisced that he was "prodigiously ignorant" when he entered the

Ecole Polytechnique, knowing absolutely nothing but mathematics.[76] Significantly, he became in maturity a paragon of broad learning and culture. This was an ideal to which engineers aspired. In fact, the entrance standards for Polytechnique made so narrow a preparation as Bertrand claimed (if indeed he spoke truthfully) almost impossible, except for extraordinarily gifted mathematicians.

After all, the democratically elitist Ecole Polytechnique created by the revolution, whose only concern was technical competence, lasted less than a decade. These were the years when a youth like Arago could discover that a Polytechnique education was the key to rapid military advancement, then immediately abandon his beloved "Corneille, Racine, la Fontaine, Molière" to devote all his attention to mathematics.[77] Napoleon tried to minimize radicalism by admitting a more elite student population. Since he could not return to the explicit requirement in the ancien régime that Ponts engineers come from at least a good bourgeois family, he instituted steep fees and reformed the entrance examination to require Latin. The restored monarchy in 1816 added literary study to the curriculum, and the more abstract mathematics championed by Laplace gained in importance a few years later. The result of these reforms, as Terry Shinn has shown, was to make a classical lycée education almost indispensable, and thus to screen out most students from the lower and middle classes. They did not, however, succeed in stamping out subversive politics, and from the 1820s until late in the nineteenth century Polytechnique was noted for its Saint-Simonian tendencies. Interestingly, Saint-Simonianism was far more influential among students from the wealthiest and most elite backgrounds, who tended to favor the Corps des Mines, than among the (often) proudly unpolitical Ponts engineers.[78]

Already in 1819, the council of the Ecole Polytechnique seemed to conceive it less as a place to recruit a new kind of elite than as an institution to educate and certify old ones. In a society that had become suspicious of privilege, meritocracy was a safely elitist form of democracy.

> We live in a time when the tranquility of the state can only be assured through the instruction of the superior classes. It permits them to obtain, through personal superiority of virtue and enlightenment, the influence they must exercise over others for the security of all. It is a happy necessity, if one thinks of it with an elevated spirit, that requires rank to be justified by merit, and wealth by talent and virtue.[79]

There remained some possibility of social mobility at Polytechnique, but, as André-Jean Tudesq observes, the family backgrounds of its graduates were never forgotten. Under the July Monarchy, those from privileged families often assumed high positions very quickly, and the very highest offices usually went to children of notables.[80]

Under the Second Empire, admissions began to favor applicants with a classical *baccalauréat ès lettres* from a lycée by adding points to their entrance examinations. Partly in consequence, three in four entering students between 1860 and 1880 had an education in dead languages as well as an intense preparation in mathematics. This preference made it still more difficult for students to gain entrance unless their parents could afford an expensive secondary education as well as two or three years of special preparation for the entrance examination. It remained a cause of much dispute, though it survived in one form or another until the First World War. John Weiss argues that the increasing prominence of the classical baccalauréat in preparation for many professions during the early nineteenth century reflected a deliberate policy to restore hierarchy to French society.[81]

This is convincing. And the effect was not due simply to patterns of recruitment, though these did tend to consolidate elites of birth and merit. Equally important, perhaps more so, is the sense of themselves as cultivated men with which graduates left Polytechnique. They were not mere specialists, whose standing in society would depend on their ability to calculate. In Paris, as in Cambridge, mathematics was regarded as anything but a technical skill. In 1812, its role in the curriculum was defended as training for the mind, indispensable in part because there wasn't enough time to give adequate instruction in engineering practice.[82] During the revolution of 1848 it was praised as the opposite of a mere practical training, a way of producing broadly capable men rather than mere technicians and specialists.[83]

This presumed inculcation of generalized abilities provided grounds for excluding *conducteurs*, who did much of the (narrowly defined) engineering work of the Corps. The *conducteurs* endeavored to exploit the democratic sentiments of 1848 and press their case that they should be allowed to rise through the ranks. A committee set up to consider these claims concluded that they lacked "the generality of knowledge—theoretical, practical, and administrative—of which an engineer cannot be ignorant." It added, however, that some of the education of engineers is useless, or worse, since it promotes an excessive confidence in theories over facts. In just this context, Dupuit referred favorably to a Polytechnique education as an uncrossable barrier protecting the Corps from *les incapacités ambitieuses.* This was not a matter of specific technical knowledge, but of a mind elevated above rote learning and capable of dealing with the unfamiliar.[84] Colson, who became a prominent spokesman for the Ecole Polytechnique, argued in 1911 that technicians are not enough for society. It needs leaders, and this leadership requires not just mathematical and scientific knowledge, but also a certain "instinct," difficult to define but intimately linked to culture and its antique roots.[85]

By 1900 Polytechnique was taking in students from a somewhat wider social base, but it lost none of its elitist spirit on this account.[86]

Ezra Suleiman, who used questionnaires and interviews to study French elites in the postwar period, found among former students at Polytechnique and the other *grandes écoles* attitudes continuous with those that had prevailed in the Third Republic. At the heart of the success of the French elite, he writes, "lies its profound belief in generalized skills, which are the only kind of 'skills' that enable one to move from one sector to another without prior technical training for a particular post." It "believes very firmly, much as the British civil service has believed since its creation, that a general preparation for leadership positions is the most desirable."[87] The engineers were comparatively loyal to their corps, but still took pride in polyvalence, not mere technique. J. Mante remarked serenely in 1967: "Our role as engineers of the Ponts et Chaussées does not consist in making calculations (this is the task of the forecasting engineers and their collaborators), but to verify their legitimacy, to weigh the consequences of their eventual deviation from reality, to determine how much can be left to chance."[88] As in Britain, this administrative elite learned mainly on the job; their formal education, as Bourdieu would have it, was mainly a matter of credentialing.[89] Suleiman concludes that France is by no means ruled by technical experts, notwithstanding (perhaps even because of) the high standing of the Ecole Polytechnique, the Ecole Nationale d'Administration, and other educational institutions that appear superficially to be trade schools. Against the customary picture of French technocracy, he argues that one must be more impressed by the entrenched power of a narrow elite than by its "rational, scientific, precisely-calculated decisions."[90]

The careers available to early-nineteenth-century polytechnicians seem not quite commensurate with the education they had received and the hurdles they had leaped. Gérard, the Ponts engineer in *Le Curé de village*, complains of mediocre pay, limited prospects, and (especially) intellectual stultification in the provinces. The letters of youthful engineers, such as Dupuit, Comoy, and Jullien, support Balzac's fiction, especially on this last point. At least, however, they enjoyed the solidarity of a corps, revealed by the standard salutation: *mon cher camarade.*[91] Later in the century, and even more in the twentieth, careers began to be improved by *pantouflage*: state service came increasingly to be regarded as a stop on a career track that might lead soon to better-paying positions in private industry. Ponts engineers and other polytechnicians were the original industrial managers in France, and French industry formed the habit of recruiting its leaders from the civil service rather than from within the enterprise or from other comparable enterprises. In what remained a thoroughly hierarchical society, such men com-

much too sensitive for mere politicians to be allowed to interfere in its execution. France, he wrote, should "disgorge that mixture of politics and administration from which it has suffered so much, and recognize that in a democracy administrative power exists rationally beside democratic power. France needs permanent technical administrators, responsible to the nation, to assure the technical direction of public services."[96]

French administration idealized hierarchy. Each official should be responsible only to his superior. Henri Fayol, a determined administrative innovator, still wanted no changes that would supplant these clean lines of authority. "[C]entralization belongs to the natural order; this turns on the fact that in every organism, animal or social, sensations converge toward the brain or directive part, and from the brain or directive part orders are sent out which set all parts of the organism in movement."[97] This ideal, of course, did not always work in practice, but at least it justified the insulation of officials from authorities other than their immediate superiors. Meritocracy, honored sometimes only in the breach, was also a part of this ideal. Beginning with the Third Republic, positions were customarily filled by a *concours*, or competition, following in a general way the pattern of recruitment into Polytechnique itself. The relative formality of the *concours* system was an answer to a pervasive fear of favoritism.[98]

Suspicion and careerism, the spirit of this bureaucracy, live on in historical studies of it. There is very little on what it actually did, but a great deal about the frustrations of civil servants as they attempted to ascend the career ladder. French bureaucracy has often been criticized for its rigidity. Courcelle-Seneuil argued in 1872 that the movement toward meritocracy, the increasing use of the *concours* and the prestige of the *grandes écoles*, tended to isolate the administration and encouraged that unfortunate esprit de corps that rendered bureaucrats indifferent to the public interest.[99] Hippolyte Taine argued plausibly in 1863 that the reason for this rigid system was not promotion of the best candidates, but rather removal of suspicions of injustice.[100] Fayol thought that mathematics was emphasized in the entrance examination for Polytechnique mainly because it permitted easy assessment. Certainly it bothered the council of Polytechnique when it appeared that different examiners in letters applied discrepant standards.[101] But polytechnicians could do without such rigid forms once they finished their education. Colson, as we might expect, believed the best system was to let those at the top of the hierarchy choose the most meritorious as their subordinates. "The *concours* has no grounds when we are judging men who have proved themselves by their work, but rather tends to elevate theoretical studies over experience and practice."[102] This, then, was a meritocracy sufficiently elevated and sufficiently homogeneous that informed judgment had ceased to be suspect. Members of the Corps did not care to subject themselves to oversight by a larger public.

At lower levels, the French bureaucracy became famous for its rigid adherence to a byzantine set of rules, largely unpublished. Since outsiders could never master them, the rules permitted officials to act with almost complete discretion. At higher levels, even the appearance of impersonal rule-following was often unnecessary. Balzac spoke in *Les Employés* of a France that since the Revolution had idealized the state, and thus come to be ruled by an army of bureaucrats. Especially in the Third Republic, the administration had much more staying power than political leaders.[103] Stanley Hoffmann has argued that "to a large extent the Republic was a facade behind which the bureaucracy made decisions." Ezra Suleiman makes a similar point about the more contemporary period. Nowhere have bureaucrats been more deeply involved in policy formulation than in France.[104] The process, moreover, has offered a wide latitude for administrative discretion. After the Second World War, in Herbert Luethy's words, France acquired a planned economy but no plan. The separate ministries retained a large measure of autonomy. It has been argued that high French officials continued to view their offices as property more than a century after the Revolution abolished venal offices. Family connections were so crucial that there were virtual dynasties in the French administration. These, as Pierre Legendre notes, could survive the formalization of the *concours* system under the Third Republic partly because many offices were outside it, but also because the knowledge upon which candidates were examined was sponsored mainly by the elite lycées. Moreover, the *concours* were locally controlled by each branch of the Administration, and since they had oral as well as written components, they tested style, culture, and poise as much as knowledge.[105]

The French Administration, then, operated with considerable autonomy, and was almost closed to public scrutiny. Roger Grégoire argued in 1954 for setting up committees, formally powerless, to which the bureaucracies would have to explain their decisions. This was strongly resisted on the grounds that it would lead to a complication of lines of power, and to delays.[106] The officials were defending a right to privacy in public administration. Suleiman notes that even in the 1970s the Corps des Ponts continued to protect itself by withholding information, in contrast to the U.S. Army Corps of Engineers, which was compelled to adopt the less desirable expedient of supplying too much.[107] This same freedom from outside scrutiny was manifested in a long-standing disinclination to keep, much less release, reliable statistics, which has frustrated numerous researchers. Walter Sharp, in 1931, found "in many government offices a disconcerting reluctance to divulge facts which the files doubtless contain. This attitude of secrecy is apparently a vestige of the aristocratic inheritance from monarchical and imperial regimes, when official posts were in the main the private patrimonies of the occupants." He added that "officialdom has not as yet been greatly impressed by the value of keeping accurate, comparable statistics on personnel practices, let alone publishing them promptly."[108]

An unwillingness to collect and make available statistics and a lack of enthusiasm for quantitative decision criteria reflect a similar set of attitudes and conditions. Statistics were withheld because the affairs of an office were regarded as its own business, and not something into which elected officials or the public ought to pry. If this private domain could be preserved, then there was little point in trying to quantify and mechanize the decision process. Ponts engineers were no different from other administrators in this respect. Like others in the French elite, they believe "that the growing complexity of society's problems requires, above all, men whose breadth of view, and understanding of a vast set of interdependent problems that involve the entire society, enable them to transcend the limitations of technicians."[109]

It might even seem that the education of polytechnicians was out of harmony with their ideals. Roger Martin, president of a large industrial firm, told an audience of polytechnicians that his education there, and most particularly his training in mathematics, was quite useless to him.[110] Fayol argued that engineers and industrial managers needed far less mathematical education than they habitually received. He preferred to see training in finance and accounting, but he wanted also to emphasize literature, history, and philosophy. "[I]ndustrial heads and engineers . . . need to know how to speak and write, but they do not need higher mathematics. It is not sufficiently well known that the simple rule of three has always been enough for business men as it has for military leaders." To attribute the success of polytechnicians to mathematics, he added, is to mistake effect for cause: "Mathematics count for nothing, or almost nothing, in the renown attaching to the Ecole Polytechnique."[111]

TECHNOCRACY

All this should help us to understand why French engineers did not take their emphasis on quantification to the point of seeking impersonal decision rules. Still less did the interwar pioneers of technocracy attempt to mechanize economic or social decisions. French technocrats were highly interested in management—there was considerable enthusiasm for F. W. Taylor, and even more for Saint-Simon and for Walther Rathenau.

This reflected their characteristic preference for administration over politics.[112] But theirs was an ideal of expert judgment and general managerial skills, not of specialized or technical routines.

"Technocracy" is a term used with notorious looseness, but its most common meaning reflects an impulse that in one important respect is quite opposed to the spirit of quantitative rigor. Richard Kuisel provides an instructive definition: technocracy, he writes, supposes

> that human problems, like technical ones, have a solution that experts, given sufficient data and authority, can discover and execute. Applied to politics this reasoning finds interference from vested interests, ideologies, and party politics intolerable. Its antithesis is decision making through the weighing of forces and compromise. Technocrats thus tend to suspect parliamentary democracy and prefer the "rule of the fittest" and a managed polity.[113]

The opposition to the give and take of politics is shared by technocracy and practical quantification. But the reference to "a solution" bespeaks the emphasis on impersonality of militant quantifiers. Technocrats in the French tradition have insisted that a cultivated judgment is required to solve social problems, and would be hard put to explain why different experts should not on occasion reach somewhat different decisions.[114]

The suspicion of parliamentary democracy does not mean the same thing to technocrats as to quantifiers. Technocrats wanted the authority to manage without being subjected to the constant scrutiny that parliamentary government entails. Quantifiers too may suspect that the legislative process will produce less than ideal results, but they have at least accommodated themselves to it by concealing, even denying, their own authority as men of culture and discernment. Technocracy means elitism tending to authoritarianism, in the interest of productivity and efficiency. The pursuit of quantitative rigor flourishes mainly in conjunction with democracy, though perhaps not a vigorous participatory democracy. Technocracy implies experts in authority. The technocrat Hubert Lagardelle even called for "the reintroduction into social life of the aristocratic element . . . , the rehabilitation of government by elites."[115] The regime of calculation involves a bid to empower experts who have at most a limited ability to subvert democratic control. Technocracy presupposes relatively secure elites. Quantitative decision rules are more likely to support a bid for power by outsiders or the effort of insiders to fend off powerful challengers.

The pursuit of quantitative objectivity did not become widespread in France until after the Second World War, and then largely under the influence of Americans. As is clear from Bertrand de Jouvenel's discussion of economic forecasting in his account of the Futuribles group, this was heavily dependent on American sources.[116] François Fourquet's study of national accounting and cost-benefit analysis in postwar France makes this indebtedness equally plain.[117]

From the standpoint of knowledge alone, this priority of Americans should be surprising. Until the 1930s, American science was distinguished by its weakness wherever sophisticated mathematics was required.[118] Practical quantification, then, was no simple result of elite technical education, but must be understood in terms of social structures and political cultures. The French, through institutions like Polytechnique, maintained a mathematical tradition second to none, and regularly employed calculation as an aid to management. But the systematic use of IQ tests to classify students, opinion polls to quantify the public mood, elaborate statistical methodologies for licensing drugs, and even cost-benefit and risk analyses to assess public works—all in the name of impersonal objectivity—are distinctive products of American science and American culture.

CHAPTER SEVEN

U.S. Army Engineers and the Rise of Cost-Benefit Analysis

What modern Pythagoras, what Einstein of our own age, can determine with unquestioned accuracy the proportionate share of the benefits to be derived from the construction of reservoirs in distant lands?
(Theodore Bilbo, senator from Mississippi, 1936)

THE ARMY CORPS of Engineers was permanently established in 1802, on the model of the Corps des Ponts et Chaussées. Its officers were recruited from among the top graduates of the military academy at West Point, the American Ecole Polytechnique. The French emigré L'Enfant, designer of the great geometric capital of Washington, had a hand also in its planning. At its creation, much of its technical library was in French. Like its predecessor, the Corps of Engineers stood for administrative unification. This, and the proud elitism of its officers, made them politically suspect in nineteenth-century America.[1] Its enemies sustained this critique into the twentieth century. Harold Ickes, Franklin Roosevelt's secretary of the interior, could have forgiven their centralizing ambitions had they not blocked his, but he was happy to play the populist against them. He called them "the most powerful and ambitious lobby in Washington. The aristocrats who constitute it are our highest ruling class. They are not only the political elite of the army, they are the perfect flower of bureaucracy."[2]

This is engaging hyperbole, but nobody ever quite believed it. Perhaps the Corps of Engineers has been a kind of elite, but its pretensions as a ruling class have never extended beyond the bounds of its administrative domain. The same could not be said of the Corps des Ponts et Chaussées, which for two centuries has been intertwined with a real, relatively unified elite. The history of Polytechnique has been most interesting to the French as an exemplar of an educational system that has perpetuated hierarchy in their society since the Revolution. That of the Corps des Ponts is, in addition, a story of bureaucratic autonomy, the triumph of administration over politics. The Army Corps of Engineers, to American historians, has less to do with social hierarchies than natural ones—the control of nature. In political terms, it is synonymous with interest groups, lobbying, "logrolling," and above all "pork barrel." Finally, and most revealingly, the historian of bureaucracy does not portray the Army Corps at the center of an administrative ruling class, but in a scene of utter disunity and savage infighting. This, I argue, is the appropriate context for understanding the pursuit of uniform cost-benefit methods. That form of economic quantification grew up not as the natural language of a technical elite, but as an attempt to create a basis for mutual accommodation in a context of suspicion and disagreement. The regime of calculation was imposed not by all-powerful experts, but by relatively weak and divided ones.

This chapter gives a history of cost-benefit analysis in the United States bureaucracy from the 1920s until about 1960. It is not a story of academic research, but of political pressure and administrative conflict. Cost-benefit methods were introduced to promote procedural regularity and to give public evidence of fairness in the selection of water projects. Early in the century, numbers produced by the Corps of Engineers were usually accepted on its authority alone, and there was correspondingly little need for standardization of methods. About 1940, however, economic numbers became objects of bitter controversy, as the Corps was challenged by such powerful interests as utility companies and railroads. The really crucial development in this story was the outbreak of intense bureaucratic conflict between the Corps and other government agencies, especially the Department of Agriculture and the Bureau of Reclamation. The agencies tried to settle their feuds by harmonizing their economic analyses. When negotiation failed as a strategy for achieving uniformity, they were compelled to try to ground their makeshift techniques in economic rationality. On this account, cost-benefit analysis had to be transformed from a collection of local bureaucratic practices into a set of rationalized economic principles. In the American political context of systematic distrust, though, its weakness became strength. Since the 1960s, its champions have claimed for it almost universal validity.

THE BEGINNINGS OF ECONOMIC QUANTIFICATION IN AMERICAN ENGINEERING

As in France, so in America, academic training for engineers was not the spontaneous creation of the marketplace—of entrepreneurs seizing every opportunity for competitive advantage. Peter Lundgreen shows that "school culture" in engineering had more to do with bureaucracy than with industrialization. Formal engineering study first arose in countries where state engineers provided the model for the profession.

In Sweden and several German states, mining academies defined the role of the educated engineer as rational bureaucrat. The French Corps des Mines was modeled mainly on the Saxon Mining Academy in Freiberg, while its Corps des Ponts et Chaussées was itself at the forefront of scientific civil engineering. The Army Corps of Engineers was never powerful enough to shape a national profession, as was the Corps des Ponts in France. Still, it was from the outset an important presence on the American scene.[3]

None of the engineers on the Erie Canal had formal training before the project was undertaken. When the Corps of Engineers surveyed the route of the Chesapeake and Ohio Canal in the 1820s and estimated the cost at $22 million—three times that of the Erie Canal—the Congress rebelled and brought in some practical men, who duly reduced the figure by half. The project then failed utterly. The Corps was limited mainly to river and harbor work after 1838.[4] Although it surveyed a number of routes to the Pacific, it lacked administrative authority over the vast net of railroads that spread across the North American continent in the nineteenth century. Military engineers were nevertheless mainly responsible for the forms of accounting and administration through which railroad companies became prototypes of the modern, managed corporation in America.[5]

Military engineering also had something to do with the application of mathematics to such problems as bridge design. But the sources were more French than American. Charles Ellet Jr. worked his way up the ranks on the Erie and the Chesapeake and Ohio canals, then traveled to Paris in 1830 to study as an external student at the Ecole des Ponts. The calculation of stresses on suspension bridges unfortunately proved a bit more complicated than he had imagined, and he suffered some disastrous failures. Ellet introduced a new variety of economic thinking about public works to the United States, advocating that monopolistic canal charges be based on utility rather than on allocation of costs. Here the stresses that defeated him were of a more political character.[6] Railway rate experts, who tried to settle disputes among cities or between farmers and companies, did not depend on any tradition so organized as in France. Expertise was fashioned as needed by engineers and lawyers in response to political and judicial pressures.[7]

Still, the growing legal and regulatory apparatus in the United States did give some continuity to their efforts. In contrast, American efforts to provide economic evaluations of public investments before the Corps of Engineers entered this domain were almost completely ad hoc.[8] To be effective, cost-benefit analysis had to be institutionalized and routinized. This was the distinctive achievement of twentieth-century army engineers.

Samuel Hays has argued that the growth of governmental expertise and rationality in America depended on the breakdown of small communities in the face of increasingly centralized power. The penetration of the local by the national now provides a major theme of the political and intellectual history of the United States in the period called progressive.[9] The Corps of Engineers, famously, worked on both sides of this divide, exploiting its ability to mobilize intense local interest to gain support for a nationwide program of projects. The idiom of cost-benefit analysis, though, was clearly adapted for the audience in the capital, and not, for example, in Oologah, Oklahoma. When Herb McSpadden came to town to complain that a proposed reservoir on the Verdigris River would cover up the birthplace of his late relative Will Rogers, he ventured to convert its tourist value to money terms. In total the project would cause damages of $70 million, he claimed, "so it is, to use your words, not 'economically feasible.' That is a mighty big word for us out there, but I have got to use your words back here." To which the Mississippi chairman of the Flood Control Committee, Will Whittington, replied: "If when you boys come to Washington, you don't get some big words to take back, it is a loss of time."[10]

In Europe, technical agencies like the Corps des Ponts were often at the forefront of bureaucratic rationalization. In the United States, decisions about public works began to be systematized only near the end of the nineteenth century, as Congress moved away from particularistic legislation toward some conception of its role in terms of enacting general policies. This required, in turn, a stable bureaucracy, and provided room for increased influence by experts. The professionalization of the civil service was advanced by the ending of the spoils system in 1883. Americans were inspired in part by the British model. But the British created space for Oxbridge-educated generalists atop their civil service, whereas in America only politics and money were superior to specialized expertise. And good politics sometimes required deferring to experts. Theodore Roosevelt even appointed a Committee on Scientific Methods.

Expertise did not stand naked. The discipline of science—of facts—was to be a forge of morality and character. Carroll Wright, head of the very influential and effective Massachusetts Bureau of Statistics, said of government statisticians in 1904:

> No matter for what reasons they were appointed, no matter how inexperienced in the work of investigation and of compilation and presentation of statistical material, no matter from what party they came and whether in sympathy with capital or labor, and even if holding fairly radical socialistic views; the men have, almost without exception, at once comprehended the

> sacredness of the duty assigned to them, and served the public faithfully and honestly, being content to collect and publish facts without regard to individual bias or individual political sentiments.[11]

For practical and moral reasons alike, efficient democratic government seemed to require improved methods of accounting, statistics, and other forms of quantification.[12]

Could quantification settle important issues of public policy? Experience was often disappointing, but hope sprang eternal. The best American engineers, like their French counterparts, understood by the 1880s that railroad pricing could never be fully rationalized by economic calculation. Yet in 1913 Congress required the Interstate Commerce Commission (ICC) to fix the value of all railroad, telegraph, and telephone property, including that of franchises and goodwill. The ICC, though capable of heroic feats of accounting standardization in the interest of systematic regulation, argued that this was impossible. It identified an insurmountable problem of circularity: property, and especially "goodwill," didn't even have a fixed value until after the prices of service were known.[13] The Supreme Court refused to let it off the hook. This appraisal was necessary, the court held, in order to calculate just rates based on cost of service. The result was 50,000 pages of hearings, which still did not suffice to reach a conclusion. Morton Keller calls this "an emblematic Progressive attempt to find fixed grounds for regulating an enterprise whose prime reality was flux," and refers to the investigation of public utilities as "the same black hole into which the railroads had plunged." Courts and Congress learned nothing from the experience. In the 1920s, they were at it again.[14]

Behind this frenzy of quantification, inevitably, was a lack of trust in bureaucratic elites. Another possible strategy for regulating railroads and public utilities was bruited at various times. The Interstate Commerce Commission, as conceived in the 1880s, was to be made up of experts who would be allowed to exercise judgment in the settling of disputes. This conception was even put into law. Soon, "five wise men" on the ICC moved aggressively to change the railroad rate structure. The Supreme Court promptly struck their initiatives down, with the revealing exception of their drive to gather better statistics. In most years Congress was similarly disposed. For legal and political reasons alike, administrative discretion was highly suspect, so the regulators had little alternative but to search relentlessly for facts and to reduce them, if at all possible, to a few decisive numbers.[15]

Such constraints applied less forcefully to the navigation projects that, until early in this century, constituted virtually the entire mission of the Corps of Engineers. Congress could be persuaded to systematize the regulation of railroads but was not at all inclined to give up its power to choose federal water projects. There was no great demand for efficiency; protective tariffs brought in more revenue than the government knew how to invest usefully. It was expended instead on pensions for civil war veterans and river and harbor work. Opponents of this spending worked with modest success to reduce opportunities for purely political choices. After 1902, a Board of Engineers for Rivers and Harbors, within the Corps, had to certify projects as beneficial before they could be recommended to Congress. One secretary of war, Henry L. Stimson, tried in the early 1910s to require the Board to rank projects in order of merit. The Corps resisted this, recognizing, it seems, that congressional choice was the key to congressional favor.[16]

Still, the Corps was anything but a rubber stamp for every proposal that reached it. Since any project could at least bring construction money into a community, and since navigation was a nonreimbursable federal service, there was no shortage of local requests to study the feasibility of waterway improvements. More than half were turned down. Economics was the usual basis for decisions, or at least for their explanation. For example, in 1910 the Board of Engineers recommended a narrower channel than originally proposed near Corpus Christi, Texas, on the ground "that resulting benefits to general commerce and navigation would not at this time be sufficient to justify the cost" of the larger one.[17]

In the 1920s, something more nearly approaching an economic routine began to appear even in favorable reports. It involved estimating project cost, and then itemizing benefits until they exceeded this cost, or fixing potential benefits as a cap on expenditures. In 1925 the Board of Engineers adopted an unfavorable report on Port Angeles Harbor, Washington, "on account of the large expense involved in proportion to the possible benefits."[18] A preliminary report on flood control on the Skagit River, Washington, set mean annual flood damages at $125,000 or $150,000, and added: "These figures will give an approximate basis for considering the feasibility of plans for flood control."[19] None other than U. S. Grant 3d, then district engineer in Sacramento, explained how a $2,670,998 dam and locks on the Sacramento River would save $25,000 in maintenance per year on existing projects, $45,000 per year on costs that would otherwise be needed to maintain a uniform flow on one part of the river, $260,000 capital expenditure plus $80,000 annual maintenance that would otherwise be required to insure a six-foot river depth on another part, and so on. Assuming an interest rate of 4 percent, these converted into capital values of $625,000, $1,125,000, and $2,260,000, respectively. One more benefit of $1,828,000 on the Feather River made the economic justification of the project crystal clear, or so it seemed to the former president's grandson. Unfortunately, powerful shipping interests disagreed, fearing that a lock on the lower river would retard traffic—and so did Grant's immediate superior, the division engineer in San Francisco. The Board of Engineers concurred with the opponents, and recommended instead some channel work.[20]

A district engineer in West Virginia had better luck with a report of 1933 recommending navigation improvements on the Kanawha River. The annual cost of $173,000 exceeded his estimate of annual benefits of $150,000, though it was far surpassed by the $1 million of annual benefits claimed by local navigation interests. Still, an increase of only 300,000 tons per year in coal transport would justify the improvement, and such a prospect in fact did, at least to the relevant authorities in the Corps.[21] One last example is a report by district engineer M. C. Tyler on three sections of channel proposed for Bayou Lafourche, Louisiana. Tyler in every case recommended the largest possible channel dimensions whose cost would not exceed the potential benefits. The Board of Engineers approved smaller ones, thereby increasing the surplus of estimated benefits over costs, though it did not explain the decision in terms of any policy of maximizing net benefits. It simply noted that the smaller channels would be adequate for handling anticipated traffic.[22]

There was not much pretense of rigor in these reports. Still, they show that by sometime in the 1920s, the Board of Engineers expected its recommended projects to promise benefits in excess of costs. Economic calculation was encouraged by legislation in the early 1920s,[23] including new standards for cost allocation. But a strict cost-benefit hurdle was not written into law until 1936. It has sometimes been supposed that the Corps took up cost-benefit analysis only in response to the 1936 act. This clearly is wrong, and indeed it is difficult (though not quite impossible) to imagine that Congress would have required the Corps to base project planning on a form of analysis that scarcely existed, or was entirely foreign to the Corps.

The growth of cost-benefit quantification at the Corps of Engineers was not simply a response to legal mandates. The Hoover era, even before the Hoover presidency, was an exceptionally favorable one for economists. They argued for the neutralization of partisan influence on public works spending.[24] Growing budgets due to flood control acts of 1917 and 1928, the latter in response to the exceptional Mississippi River floods of 1927, created pressure for greater accountability. In 1927, Congress directed the Corps to study all the major river basins of the United States with an eye to improved navigation, water power, flood control, and irrigation. In response, over the next decade the

Corps produced a mass of documents and proposals, called "308 reports" after a House of Representatives document that listed them. As the Corps began to acquire a huge civilian labor force, it relied increasingly on quantification to impose discipline. Hence it was not caught unprepared by the Flood Control Act of 1936, with its famous requirement that no flood control project could receive federal funds unless its benefits, "to whomsoever they may accrue," were projected to exceed its costs.

NUMBERS JUSTIFIED BY AGENCY AUTHORITY

The cost-benefit provision of the 1936 Flood Control Act was one of the heroic efforts of the United States Congress to control its own bad habits. The act was precipitated, as usual, by floods, but also by the continuing depression, for which public works seemed an appropriate remedy. Edward Markham, chief of engineers, explained that the House Flood Control Committee had put together its bill in 1935 by going over the 1,600 projects contained in the "308" reports and choosing those with the best ratio of benefits to cost. We can be sure that regional balance was also a consideration.[25] The bill made it all the way to the full House and Senate, but then, in a last-minute display of animal spirits, it got loaded by floor amendments with a huge collection of projects that the Corps had viewed unfavorably, or even had never studied. The display was so unwholesome that it defeated the bill. No major flood-control legislation was passed in 1935. The language requiring benefits to exceed costs was part of an effort to avoid such an unsavory spectacle in 1936.[26]

The particular hurdle was probably less important than the institutional regularity it implied. Hereafter Congress could only authorize works that had been studied and approved by the Corps. A preliminary examination and then a full survey, each running through several levels of Corps bureaucracy, required months or years, and could not be completed to satisfy the sudden whim of a legislator. When, now more rarely, really disgraceful projects were authorized, a modest standard of decorum was maintained. Official economic analyses helped to cut off debate and bargaining in Congress.[27] Flood control chairmen in the House and Senate routinely invoked the cost-benefit rule in floor debate to block amendments proposing new projects. The rule was construed as a dam, holding back a flood of legislation. The Senate, explained John H. Overton in 1944, cannot make exceptions. "If we did, we would soon be at sea." Whittington's metaphor in the House warned against this and other disasters: "Mr. Chairman, if we propose to make an exception in one case, you let the bars down and you crucify the sound, fundamental principles of flood control."[28]

The Corps did not thereby become all-powerful. After authorization came appropriation, which left ample room for Congress to make political choices. Still, this regularization of the planning process could not but enhance the standing of the Corps. Except when it was challenged by powerful opponents, its numbers were generally accepted on no more authority than its own reputation. That authority was enough. As Overton of Louisiana told the Senate in 1938: "In order to determine whether a project is of value as a flood-control measure, it should be submitted first to the judgment of experts, and the chosen and recognized experts upon this question are the Army engineers."[29]

The expression of this judgment in quantitative form invited Congress to advertise its rationality and objectivity. The cost-benefit standard was an instant cliché. "All of these projects have been studied by my department and on all of them favorable reports have been made and their construction recommended," reported chief of engineers Julian Schley at the beginning of the 1940 House Flood Committee hearings. "We never report a project to Congress," announced Whittington in 1943, "until it has been recommended by the Board of Engineers and the Chief of Engineers stating that . . . the benefits of the project will exceed the cost." He added that "the ability of this committee to secure annual flood-control authorizations up to the war and the invasion of Poland by Hitler, we believe, is due largely to the fact that this yardstick has been adhered to."[30]

Especially on quantitative matters, the responsible congressional committees could be dazzlingly uninquisitive. They asked many factual questions, but it rarely mattered what the answer was. Often the record was left blank for a time, and the response to a statistical query would be inserted afterwards. If a benefit-cost ratio proved to be 1.03, this never provoked comment or alarm, unless perchance there had been recent flooding on the endangered river, when committee members might wonder aloud what miscalculations had generated a number so low. In 1948, local interests in Texas proposed to modify a project on the Neches-Angelina River system to stabilize the local water supply. The Corps didn't mind, though it would, as Colonel Wayne S. Moore explained, "slightly reduce the theoretical ratio of general benefits to costs but not materially." Someone asked how much. "The cost benefit ratio is estimated in the report as 1.08, and as modified by the proposed legislation it will be 1.035, or possibly somewhat greater, a difference which is within the limit of error in the estimates." Nowhere else have I seen margins of error mentioned in these hearings. Nobody noticed or cared that a probable error of .05 might not redound to the credit of the proposed project. The numbers were almost never questioned. In 1954, Prescott Bush of the Senate Flood Control Subcommittee learned that the local contribution for a project in California was "estimated at $22,500, sir. It is calculated according to a rather complex formula. I won't worry you with the details of that formula." "All right," replied the senator.[31]

On what basis did Congress place such implicit faith in these economic numbers? Perhaps its members were frightened by talk of complex formulas. But fear itself was superfluous. In these cozy committees, inquisitiveness was a deadly vice. The congressmen did not leave their faith in the Corps unspoken. It was the better part of valor not to challenge this powerful agency except privately, where factual claims didn't matter so much, and always to praise it publicly. Senator Royal Copeland of New York, who inserted the cost-benefit provision into the 1936 Flood Control Act, told the Senate that Corps engineers are incorruptible, calling them "honorable, straightforward, patriotic men." Whittington proclaimed to the House that "the chief of engineers is impartial and represents Congress and the country." Vandenberg of Michigan explained in 1936 that the new system requires "an independent, nonpolitical, unprejudiced decision as to priorities," adding disingenuously that "no one has ever heard a suspicion or a remote challenge" regarding the integrity and competence of the Board of Engineers."[32]

If anyone did, it was in their interest to suppress it. Senator Robert S. Kerr of Oklahoma, who received not only the customary political benefit, but also a good deal more than the usual measure of personal economic benefit from the projects he sponsored as chairman of the Senate Committee on Rivers and Harbors, reacted with righteous indignation when some Corps numbers were criticized in 1962. These are the finest graduates of West Point, he thundered, and it would be "presumptuous" to challenge their calculations.[33] The Corps studiously avoided any involvement with politics, in public. The record is clear that it was possible to be chief of engineers without even knowing what politics is. When the distinctly friendly Homer Angell of Oregon asked General Lewis Pick, during an uncharacteristically unfriendly congressional investigation in 1952, if his bureaucratic enemies in the Bureau of the Budget might sometimes "put in a dash of politics," Pick was at a loss. "Sir?" he replied. Angell explained: "Sometimes they give it a dash of politics, too, do they not, in determining what projects should go along?" Pick remained shamelessly disingenuous: "I do not know, sir. If they do I do not see it." Such blindness was politically farsighted. In the same hearings, George A. Dondero of Michigan remarked that "I can only recall one or two occasions in 20 years where the committee ever doubted the wisdom of the Corps of Army Engineers in sending a project to us."[34]

Sometimes a conspiracy theory is tempting—that the whole enterprise of congressional hearings was a masquerade, to disguise a system of mutual patronage. But this is certainly inadequate. Patronage alone did not make the Corps. It derived prestige from its military connection in a century of frequent war. It had the advantage of military discipline. It was the government's most effective emergency relief agency. It built up considerable expertise on dikes and levees, and whatever the economic justification of its dams, at least they didn't fall down. Its engineers earned a reputation for technical competence. Still, politics seems the best explanation for the failure of Congress to require the Corps to follow rigid rules in its economic analyses. "Do you think there is any agency in this Government anywhere that operates in a more scientific way in arriving at their conclusions than the Army Engineers?" asked Orville Zimmerman of Missouri, thereby disabling a critic. William M. Corry mobilized the advertising expertise of the Zanesville, Ohio, Chamber of Commerce to make the same point:

> I want to say in the beginning that I am no engineer. If I had a stomach ache, I would go to the doctor. If something were wrong with my automobile I would take it to a garage mechanic. By the same token, when I want flood control, I go to the best source possible, to the group of people trained and who through the years have earned the distinction of being the most capable exponents of proper flood control in the world, namely, the Corps of Engineers, United States Army.[35]

Dependence on the Corps is nowhere more complete than along the lower Mississippi. There it has struggled mightily to satisfy the contradictory interests of chemical plants and crayfish packers, barge companies and flood-plain residents, New Orleans and Morgan City. Many suppose that the Corps is already too optimistic about the possibilities of managing this huge river, but the interests always demand still more. Even while protesting, though, they remain assiduous in their deference to an agency whose discretionary power they understand all too clearly.[36]

One of the few real issues engaged by the House Flood Control Committee in the 1930s and 1940s concerned the plan for a Mississippi floodway, to carry off excess flow in times when it is next to impossible to contain the river with levees. The Corps proposed to buy up rights to send this water down what would become the Eudora Floodway, across a corner of Arkansas and a good deal of Louisiana. Louisiana representatives like Leonard Allen complained bitterly, but were almost unfailingly gracious to the agency that drew up the plans. "I know there is not a group of men in Washington I have more confidence in than I have in the Corps of Engineers," he proclaimed in 1938. Whittington, the committee chairman, identified their decisions with engineering and economic necessity. "We have given them the yardstick, have we not, when we say that we ask the Chief of Engineers in the most economical way and at the most advantageous place to provide for 3,000,000 cubic feet per second."[37]

By chance, Whittington's district was in Mississippi. Some of the testimonies from his state were even more deferential to the Corps. Of course they could afford to be. W. T. Wynn, representing a Mississippi flood control district, explained: "The problem, I think, has gotten beyond our local engineers. It is a national problem, and we are the patients, and we think it should be turned over to the Army engineers. Now, how we can tell those engineers how to operate or what kind of medicine to give us, I do not know." But would you just let the army engineers run the water over your state, if the situation were reversed? Allen asked another witness. "Yes, sir," replied Mr. Rhea Blake, "we are saying that right now. We are saying we should turn the whole matter over to the Corps of Engineers."[38]

Three years later, with nothing solved, the rhetoric had reached full flower. Leonard Allen asked again why "every plan that has been proposed and that Mississippi has endorsed has been a plan to run the water over Louisiana?" J. S. Allen, chief engineer of the Mississippi Levee commissioners, replied that "God almighty fixed that; we didn't." He continued: "We recognized the geography of this situation and we respected the opinion of the Army engineers." And finally, God descended to earth: "The ranking engineers in the United States decided on that point."[39] In a subsequent exchange during the same hearings, a witness referred to political pressure on the Corps, and Representative Norrell of Arkansas reacted in horror: "Do you mean to tell this committee that the Army engineers are susceptible to political pressure and influence." The witness denied implying any such thing. Norrell continued: "I just want to get it clear in the record that you didn't mean that public or political influence could be brought to bear upon the Army engineers and further that they are always guided solely by the technical field in which they operate."[40]

Of course the congressmen knew that technical considerations did not abolish choice. In more relaxed moments they happily admitted as much in public discussion. A 1948 proposal to improve the harbor in Half Moon Bay, California, showing a benefit-cost ratio of 1.83, included a dazzling assortment of benefits: "[I]ncreased catch of fish and savings in production and transportation costs, elimination of lost fishing time, decrease in damage to fishing craft and in loss of gear, reduction in marine insurance premiums, availability of local marine repair facilities, increased recreational activities and associated business, and from change in land use attributable to harbor improvements." To this list developed by the district engineer, an inspired division engineer added the benefits to a local rock quarry. Congressman Jack Anderson could not contain his enthusiasm: "Mr. Chairman, I think that the Army engineers should be highly commended for having exhausted every possible public benefit and for having surveyed every one that might accrue in the event this is constructed."[41]

A more striking case involved the Savage River, a tributary of the Potomac, in western Maryland. A dam was begun during the late 1930s by the Works Progress Administration, after the Corps of Engineers, in 1935, declared the project "unjustified economically" since benefits, even "at their most liberal evaluation," were only 0.37 of costs. The work was interrupted by war, and in 1945 this embarrassing, half-finished dam was thrown back into the lap of the Corps. By adding hydroelectric facilities to the project, it managed, barely, to rationalize the economics of completing it. Unfortunately, this power generation was so vigorously contested in a public hearing before the Board of Engineers that it was dropped. Now, as General Crawford of the Corps explained with revealing redundancy, "[T]he over-all economic justification of the project was not sufficient to justify the project." The local congressman, J. Glenn Beall of Maryland, professed alarm that his constituents would remain vulnerable to floods. Pressure came also from Senator Jennings Randolph of West Virginia. And indeed the Corps hated to leave standing such a monument to waste and futility.

A few days later, Crawford returned to the hearings, to vent his eloquence on the results of a "further investigation." "We have asked the district engineer to consider the Savage River Dam again, as an individual project separate from the main report. In doing that he has developed other benefits that he did not find it necessary to develop when he wrote his main report. The result is that he finds greater benefit, on further investigation, than he had in his report." Now, it turned out, annual flood-control benefits of a mere $2,700 were augmented by $5,000 for power benefits downstream owing to a better-regulated stream flow, $45,000 for pollution abatement, and $130,000 for improved water supply. The benefit-cost ratio for completing the dam was now 1.5, so "it would be perfectly proper to add this Savage River Dam project" to the report.[42]

This multiplication of benefits provided a helpful general strategy for getting projects over the cost-benefit hurdle. Some classes of benefits were long recognized by the Corps as important, but considered unquantifiable. Occasionally in the 1940s the Corps cited such intangibles to justify projects whose tangible benefits could not be made to exceed costs. A river channel in Michigan with a calculated ratio of 0.82 was "considered meritorious and necessary for the general welfare of the communities affected," on account of pervasive local anxiety. Improvement of the port of Skagway, in the territory of Alaska, was justified despite its ratio of 0.53 "in view of the importance of the port in encouragement of future development in the area." Flood control on the Lackawaxen River in Pennsylvania showed a ratio of only 0.8. But a 1942 flood had cost twenty-four lives, and the intangible benefit of avoiding such loss of life in the future was sufficient for the Corps to recommend the project.[43]

The Corps, however, never relied much on exceptions to the regime of calculation. It was better to systematize them. As the best harbors were developed, levees erected, and dam sites used up, more and more of these so-called intangible benefits were made tangible, and quantified. In consequence, many projects that were turned down, some decisively, in the 1940s or 1950s were eventually approved and built. Boosters recognized this general shift and urged it forward, though the Corps was often hesitant. A private report calling for development of the Red River for the benefit of Arkansas, Oklahoma, Texas, and Louisiana noted that while various individual dams and waterways had failed the cost-benefit test, an integrated project could easily pass it. The Corps of Engineers, it optimistically supposed, recognizes "the pernicious effects of trying to measure national concerns in terms of dime-store economics." How were such losses on each project to be made up in volume? "Present-day procedure was used in computing the ratio of costs to benefits." "Present-day procedure" turned out to permit multiplying unit recreation and water supply benefits by the entire population of the area, and unit irrigation and drainage benefits by all potential agricultural acres, among other extravagances.[44] This was too much for the Corps, even in its most expansive moments, and it refused to endorse this report.

CORPS OPPONENTS AND THE PUSH TO STANDARDIZE

The examples given above demonstrate that Corps economic methods could not, by themselves, determine the outcome of an investigation. This will come as no surprise to most readers. But it is important to understand that these are not typical instances of the quantification of costs and benefits. The Corps transgressed its customary standards most egregiously when the political forces were overwhelming, and when they were all arrayed on one side. In routine matters its prestige sufficed to contain the politics. Generally, congressional investigation was so perfunctory that the Corps was not bound to observe any particular rules of quantification. To the extent there were checks on discretion, they were mainly internal. As will appear later, the top officers of the Corps made real efforts to impose some uniformity on the economic analyses reaching them from the districts and divisions, but this never amounted to a campaign to neutralize personal judgment.

The most powerful force for standardized methods, and in this sense for objectivity, was supplied by opponents of the Corps. Of course there was unhappiness whenever a hoped-for navigation or flood-control project was turned down. The Board of Engineers might be obliged to travel from Washington and conduct a special hearing.[45] Disappointed local interests might complain to a congressional committee.[46] But local interests were generally weak, and were rarely in a position to contest official numbers. Only powerful interests, interests that systematically opposed a whole class of Corps projects, could exert much pressure toward the rigorous standardization of its cost-benefit methods. The most effective of these opponents were the utilities, the railroads, and two rival agencies within the federal government: the Soil Conservation Service of the Department of Agriculture, and the Bureau of Reclamation, in the Department of Interior.

Electric Utilities

Although electric power generation was not part of the Corps' official mission, it was routinely considered as a possible secondary benefit, and occasionally it far outweighed the nominal primary benefit. The Corps was more open to multiple-use river development than the prevailing historiography allows.[47] Bureau of Reclamation dams, especially on the Columbia River, were even more important as sources of power. Private utility companies objected to this government-sponsored competition. Their spokesmen hinted that the Corps was an agency of creeping socialism, and that big dams were in any case unwise.[48] They also pursued the more mundane strategy of scrutinizing economic analyses, often cogently, though it seems they won few victories.

In 1946, both the House Flood Control Committee and the Senate Commerce Committee heard testimony on the Rappahannock River, which runs through and occasionally floods Fredricksburg, Virginia. The opposition was led by the Virginia Electric and Power Company, represented by Frederick W. Scheidenhelm, a hydraulic engineer from New York City. He told the committees that the power benefits for the main dam in question, at Salem Church, had been exaggerated because the Corps had not considered such essential technicalities as load factors. He argued that their cost estimates were outdated, since they had been made before the war. He also pointed out that power production was not an authorized Corps mission, so that here they were building a dog to wag the tail. Only 9 percent of the claimed benefits were for flood control. But even this 9 percent was an exaggeration, for about a third of them pertained to land that would be protected from floods only because it would be underwater, in the reservoir basin. "I think the bottom of the barrel was scraped a little hard . . . in this case."

Scheidenhelm's point commands our respect. On the other hand, the project provides modest evidence that the Corps' economic standards were not infinitely flexible. Its engineers wanted to distribute projects over every region of the country. The people of Fredricksburg didn't want levees, which would reduce property values. Colonel P. A. Feringa of the Corps explained that its engineers had been unable to make a single-purpose pure flood-control dam meet the cost-benefit standard. So they tried various options until they found something whose economics could at least be defended, even at the cost of arousing the ire of electric utilities. It no doubt helped that this opposition generally failed, as it did here. The flood-control committees disagreed with Scheidenhelm, preferring the testimony of one D. C. Moomaw. He pointed out that the Corps had found many projects not to be economically feasible, so "when they state they are, I think we are entirely justified in accepting their statements."[49]

Railroads

The railroad companies had no objection to flood control, but were bitterly opposed to the government-subsidized competition created by expensive canals and channel dredging. Objections on principle got them nowhere, especially since many in Congress regarded them as greedy monopolists. So they argued instead that canal projects were economically unjustifiable. Here, too, the obstacles were very great.

The railroads opposed for decades that most famous of Corps boondoggles, on the Arkansas River, which made Oklahoma a maritime state. The Corps, under great pressure, planned a "truly multiple-purpose project," because only with many different kinds of benefits was there any prospect of getting them up to the level of costs. Colonel Feringa proudly reported to the House Committee on Rivers and Harbors in 1946 a benefit-cost ratio of 1.08, without relying on intangible benefits. "The Corps of Engineers once presented a project to this committee which we thought was very good, but in which we tried to evaluate benefits which are not readily evaluated in dollars and cents." R. P. Hart of the Association of American Railroads objected that for the cost of $435 million the government could build a good double railroad line and haul everything for free. Perhaps he was found to be convincing. The project was not approved in 1946, or indeed for the next fifteen years. But in 1946 Robert S. Kerr was merely governor of Oklahoma. By 1962 he was a senator and chairman of the Senate Committee on Rivers and Harbors. Kerr-McGee Oil Industries had a huge financial stake in the waterway. In 1946 he had testified: "Let us not confine this hearing to the minor subject of comparative water-rail freight costs. Rather let us think about building a greater nation." The line evokes speeches written by Theodore Sorenson. At least, it resonated with the Kennedy administration, which wanted to get the country moving again. In the interim, Congress had decreed that increased employment in undeveloped areas should be recognized as a social benefit of water projects. Kerr sponsored legislation to solidify and increase the valuation of recreation benefits. Such procedures made it easier for projects like this one with powerful political support to clear the formal economic hurdle. Freight to Tulsa now passes through the Robert S. Kerr lock and dam and over the Robert S. Kerr reservoir.[50]

On a few rare occasions the railroads were able to disturb the tranquility of congressional hearings on public works and force the legislators to consider in detail the economic merits of a water project. A nice discussion was generated in some Senate hearings on rivers and harbors in 1946 over a canal in Louisiana and Arkansas. It was proposed by local boosters as part of their ceaseless efforts to develop the Arkansas-White-Red river system after the fashion of the Tennessee Valley Authority, but with the Corps of Engineers firmly in charge. The Association of American Railroads, represented by Henry M. Roberts, found itself at a disadvantage. The project description, sent over by the House of Representatives, hinted at doom: "Red River below Fulton, Arkansas, in accordance with the report of the Chief of Engineers dated April 19, 1946; *Provided*, That the improvement herein authorized between Shreveport and the mouth shall, when completed, be named the 'Overton-Red River Waterway' in honor of Senator John H. Overton, of Louisiana." Overton was chairman of the subcommittee conducting the hearings.

He set their tone by challenging Roberts's credentials. He thought he had settled with the railroads privately, and was not happy to be confronted with their opposition. But the informal settlement hadn't held, as he at last understood. Roberts argued against spending public money to favor one kind of transportation over another. Without massive subsidies, he explained, inland waterway transportation is not cheap. We know, and "we are not amateurs" in these matters. The benefits of this proposed canal, he continued, had been greatly exaggerated. In estimating freight volume by sampling, the Corps seemed to have forgotten that railroad offices are closed on Sundays and holidays. Their tonnage estimates were high in relation to comparable projects. So also was the figure for savings per ton-mile. Besides, the Corps had ignored the cost of getting freight to the river or canal from wherever it originated. Faced with all these defects, the railroads had hired their own experts to recalculate. Roberts proposed a much lower estimate of benefits.

Overton tried to discredit these attacks on the official numbers, and on the experts who made them believable.

> OVERTON: Let me ask you, is it true, or not, that the Board of Engineers have rate experts in their employ?
>
> ROBERTS: Well, that word 'expert,' sir, takes in a lot of territory. I have met two or three men up there I thought were pretty good rate men. Whether they had anything to do with this, or not, I don't know.
>
> OVERTON: They do have them in their employ. Does the Interstate Commerce Commission have some pretty good rate experts?
>
> ROBERTS: They are supposed to be.

It seems the "government experts" were parties to an oxymoron. The difference between railroad rate men and the Corps is that "we are realistic in getting facts. Two and two is four with us." "And not so with the board?" interjected Overton. "Well, we don't reach up in the air and get figures and say they represent actual facts." That is, he disapproved of sampling. "Private enterprise could not survive under such a system. It is like taking your street number and dividing it by your telephone number and getting your age."[51]

Feringa countered Roberts in the most general terms possible: the Corps pursues a just mean in its economic analyses, and it must be succeeding, since it generates antagonism on both sides. "We steer a middle course. We try to be neither proponents nor opponents, but merely the consultants of Congress with no axes to grind, trying to give you the figures as best we know how." The state of Louisiana, in the spirit of boosterism, calculated a benefit-cost ratio of 1.92; Roberts's figures implied one of 0.80. The Corps reported 1.28. Overton inserted that the Corps was too conservative; that real benefits are almost always higher than their estimates. "Still it is a commendable conservatism because it creates confidence by the public and the Congress in the recommendations of the Board of Engineers." This was the favorite posture of the

Corps: beset by enthusiasts on both sides, they had learned to take the claims of boosters and opponents alike with a grain of salt. Feringa explained that rate work demands a special kind of expertise. "It is a science in itself, and a man has to be trained for it." He confirmed the special nature of the expertise by presenting it incoherently to the Senate committee.[52]

There followed a revealing exchange. Roberts requested that one of these so-called rate experts be brought in to testify before the committee. Overton demurred: "Oh we have got too much to do now." Roberts: "I thought so." Overton read from the report of the chief of engineers, which he proclaimed to be both thorough and fair. Roberts said he knew the Interstate Commerce Commission had not put correct rates on the bills, because his own crew of seven real experts had gone over them. Overton reacted with horror: "That is quite a reflection on the Interstate Commerce Commission, and on the Board of Engineers." If his committee had to call in witnesses and compare rates in detail, they would require two or three weeks for every project. They had no choice but to trust the report of the Board of Engineers. But then, at last, somebody supported Roberts. Guy Cordon of Oregon inserted that "this is the first time I have had experience with opponents coming in and controverting facts and making their allegations specific." If the committee refused to call in the experts and set the record straight, he couldn't understand why they have hearings at all. It's all wasted time if they just get embalmed in the record.

At last, Overton relented. In came Eric E. Bottoms of the Economics Division, under the Board of Engineers. Roberts was not allowed to challenge Bottoms item by item, so the Corps was given the benefit of the doubt. But Bottoms made clear that economic analysis was a serious business. It involved an immense amount of filing and counting by people who had in fact thought through their methods. The Corps researchers had censused freight movement on the rail lines, then sorted all bills for one day each month according to their judgment of whether the load in question would be more advantageously transported by water. Either for internal reasons, or as a defense against external challenges, the Corps was careful in observing formalities. It had consulted other agencies, such as the Interstate Commerce Commission, for judgments falling within their special competence. If its numbers were too generous, this was accomplished mainly at the level of minute details, which of course could not easily be challenged.[53]

Another waterway that the railroads vigorously challenged, again unsuccessfully, was the Tennessee-Tombigbee. This was a huge project, arousing political forces far too powerful for the cost-benefit analysis to remain innocent. Still, the numbers were not simply fabricated. In 1939, the Board of Engineers managed to raise the benefit-cost ratio above 1.0 only by attaching numbers to certain benefits that had always been regarded as intangible, including $600,000 for national defense and $100,000 for recreation. The chief of engineers, Julian Schley, doubted the propriety of these and some other values, and refused to make an official recommendation. He concluded that the economic analysis was not straightforwardly valid, but fell "within the realm of statesmanship to which the Congress can best assign the proper values." The railroad spokesman, J. Carter Fort, argued that the waterway was merely a huge subsidy for a few special interests, and complained that it depended on the most extreme economic inventiveness. In particular: "That figure for national defense is a figure that must, in its very nature, have been pulled out of the air. No one could possibly put a value on it in money."[54]

After the war, inevitably, the project came up again. Perhaps it was also inevitable that it would now appear economically justified without the intangibles. But it was by the slimmest of margins, a ratio of 1.05. Unquantifiable items make the project rather better, explained chief of engineers R. A. Wheeler; "some day we are going to have to have some sort of a formula" to evaluate them. For now, the Corps had relied on 2,500 questionnaires sent to shippers, of which 1,338 were returned, to estimate potential traffic volume and savings. The railroads again doubted the analysis. But they could do almost nothing. The forms contained privileged business information, which could not be released to private parties.[55]

Six years later, though, this project fell afoul of the powerful House Appropriations Committee. It had been authorized in 1946, and the Corps promptly began preparation of a detailed "definite project report." But before this was finished, it requested a relatively small appropriation to build the first leg of the project. This, its enemies charged, was a scheme to commit Congress to the whole thing. John J. Donnelly of the committee staff subjected chief of engineers Lewis Pick to a withering interrogation. As usual, what mattered most was in the details. Could operators haul eight barges in one tow across Mississippi Sound from Mobile to New Orleans? The Corps assumed they could, but the committee staff had been told they could not, in which case a whole class of purported advantages of the waterway would vanish. Should the costs include the added expense of rebuilding locks on the Mobile River for long barges? The Corps argued that they needed to be rebuilt anyway. There were also doubts about the true time savings for waterway traffic in comparison to Mississippi River traffic? The committee concluded that the most recent benefit-cost ratio of 1.13 was based on serious mistakes for both costs and benefits, and offered its own ratio of 0.27.

Pick obviously suffered some moments of acute discomfort, but in the end he was undaunted. He did not attempt to refute the committee staff in detail, but simply claimed greater expertise.

> Without doubt some of the opinions gleaned by the investigative staff from informed sources as to the feasibility of the project have been in sharp conflict with observations and testimony from similar sources found acceptable by the Corps of Engineers' analysts as determinative. In such a situation, the comparative competence and familiarity of the respective staffs with the practical problems of water transportation and their respective experience in canvassing the field, and weighing the information offered by those with special interests to serve, would seem to afford the most reliable test of credibility. The ability to make sound appraisals of the sometimes overenthusiastic claims of waterway advocates is highly important, but it is equally essential to discount the natural hostility of intrenched carrier enterprises which want to forestall troublesome competition and which are dependent upon the good will of existing regulatory agencies. . . . Experience of the Corps of Engineers in the development of successful waterways would seem to furnish the most reliable guide in estimating the future performance of such projects as the Tennessee-Tombigbee improvement.[56]

The experience of the Corps carried the day. It was evidently impossible for private interests opposing particular projects to discredit its officially sanctioned numbers.

Upstream-Downstream: The Agriculture Department

Industries and interest groups were able to enforce some standards of care in the preparation of cost-benefit analyses by the Corps of Engineers. But effective pressure to spell out, and sometimes even to change, cost-benefit practices came mainly from other branches of the federal government. There were dozens of agencies involved in what was called water resources development. Many had a well-defined role that did not threaten the Corps. A few did not. The most bitter rivalry in this field, by all odds, involved the Corps and the Bureau of Reclamation. Second place among the antagonists was held by the Department of Agriculture, and particularly its Soil Conservation Service.

The missions of the Corps and Agriculture were not obviously in conflict. The 1936 Flood Control Act divided their jobs between downstream and upstream. Downstream meant bigger dams. Almost immediately after the Corps began building dams regularly for flood control, it was faced by opposition tinged with populism. This was not simply a matter of ideology. No matter where a dam was to be located, those upstream from it faced the double indignity of being deprived of flood control themselves and, for some, of being flooded out of their homes and off their farms by the reservoir. Many came to believe that big dams were unnecessary for flood control—that floods were artifacts of poor land management and could be avoided through reforestation, contour plowing, and small dams near the headwaters of streams. For such reasons, opposition to the Corps was very often attended by a strong preference for the policies of the Soil Conservation Service.[57] The Corps complained of this, and Congress tried to neutralize it, but with scant success.[58]

Congress considered that upstream people, like everyone else, had something approaching a constitutional right to a cost-benefit analysis for their proposed flood-control measures.[59] The Department of Agriculture had its own approved cost-benefit methods. These treated big downstream structures less generously than the Corps, but they could often justify a network of small, cheap dams as part of a systematic program of soil conservation and small-scale irrigation. The Corps viewed many of these as uneconomical. Such divergent outcomes of economic analysis, of course, fed the controversy. Still worse, from the standpoint of the Corps, a network of small dams protected towns and cities downstream only from small floods. This might be sufficient to tip the benefit-cost ratio against a large dam on the main river without reducing at all the impact of catastrophic major floods.[60] So far as the Corps was concerned, suspect economic practices sanctioned by the Department of Agriculture were undermining its effort to provide real flood protection. This was one of its main incentives in seeking a single, standardized method of cost-benefit analysis throughout the federal government.

The Bureau of Reclamation and the Kings River Controversy

"Hitler could not have selected better people to sabotage the American interests than those who have done that in the San Joaquin Valley," complained Congressman Alfred Elliott of California in 1944.[61] What had these traitors done? They had performed a cost-benefit analysis showing irrigation benefits in excess of flood control benefits for a proposed reservoir on the Kings River. In this case, evidently, the politics of quantification had gotten out of hand. But perhaps this should not have been unexpected, in a time of war. In the files of the Bureau of Reclamation for the Kings River we find the following:

> Repeatedly since 1939 I have written and spoken to Commissioner [John] Page and later to you about my growing apprehension concerning the 'institutional ambitions of the Corps of Engineers.' The battles over the Missouri River and Kings River are the present highlights in a campaign, long-

> planned and thoroughly planned by the Corps, which is intended to cover the entire West. On the Missouri the Corps is using its navigation divisions; on Kings River, its flood-control battalions. It is trying to carry out a huge pincers movement. . . . We are making the fight on an unfavorable terrain, from the standpoint of irrigation possibilities. The Corps, not the Bureau, picked the battlefield. If the Corps wins the battle decisively, the whole war may be lost to us; there may remain no secure and important sphere of action for the Bureau on other western rivers. A defeat for the Corps could not be similarly crucial, not with flood control pork barrels in every valley.[62]

The inevitability of war was often denied in early 1939. The Commissioner of Reclamation thought he could negotiate with the Corps. Appeasement seemed at first to work, for on March 28 he issued a triumphant memorandum announcing that "the California district is an outstanding example of cooperation between representatives of the Army and the Bureau of Reclamation."[63] But, as Ickes recognized in a communication to the president, it is dangerous to negotiate from weakness. "It has been alleged that a contest is developing between the Corps of Engineers and the Bureau of Reclamation to see which agency should build the big dams in the west. . . . If such a contest should develop, obviously the Bureau would lose, because it operates under the Reclamation Law which requires that all or most of the money expended be repaid to the Federal Government."[64]

These differences in law governing the Corps of Engineers and the Bureau of Reclamation were poisonous. The Bureau of Reclamation was created in 1902 to provide irrigation water for (what soon became) the seventeen states west of the 97th meridian. It was required to charge farmers for the cost of supplying water, though without interest. This might seem generous, but the Corps required no local contributions at all for navigation projects, and rather little for flood control. The really crucial advantage of the Corps of Engineers in California, though, was that the Bureau was governed by an ethic of homesteading, and was not allowed to provide water to holdings over 160 acres. By 1940 it had found ways of compromising this standard, but not to the extent that it could provide satisfaction in the agricultural plutocracy that was California's Central Valley. In the Kings River valley, big farmers were already pumping great quantities of water out of the river and the ground for irrigation. They were in fact pumping too much, and needed governmental help to be saved from themselves. But they were not about to invite in some federal agency to build expensive water works if it would require them to divest of all but 160 acres each.[65]

By 1939, the Bureau of Reclamation had been involved in California's ambitious water program, the Central Valley Project, for about a decade. So it was natural that when planning began to dam up the Kings River, the Bureau was contacted first. In February 1939, the local congressman introduced legislation for the Bureau to construct a dam at Pine Flat, as compensation to his constituents for a bill to create Kings Canyon National Park.[66] Evidently he was out of touch. In March, under pressure, he withdrew both bills. Not only did local people oppose the park, but many wanted the Corps of Engineers to build this dam. Local water interests had already been negotiating with the Corps for some months, and had even proposed to finance privately the project survey in hopes of getting congressional approval within a year. The district engineer replied that the complexity of procedures made this unlikely. The Corps began drawing up plans early in 1939.[67]

Urged from higher up, local officers of the Corps and the Bureau worked almost from the beginning to harmonize their projects. This would help keep the politics under control. S. P. McCasland, who designed the dam for the Bureau, reported back to headquarters that "the interests" were being kept in the dark about negotiations. Initial plans called for a reservoir at Pine Flat with a capacity of about 800,000 acre-feet, and this was recommended in McCasland's report, dated June 1939.[68] A capacity of 780,000 was proposed to the Corps by L. B. Chambers, district engineer in Sacramento. As justification he sent along to Washington a graph of benefits and costs, and of their ratio, as a function of reservoir capacity. This was sufficiently standard, at least in Sacramento, that such information could be communicated easily by telephone.[69] But the office of the chief of engineers was worried that this would not control the "maximum possible flood," and wondered if a one million acre-foot reservoir might not be better, especially if flood control were to lead to further development in the valley below. The principal engineer on the project, B. W. Steele, explained that the smaller reservoir was based on the flood of 1906, and that a larger one might be justified if the huge 1884 flood was considered as a basis for planning. The district engineer, in contrast, argued simply that the last 200,000 acre-feet were "not economic," and that the downstream valley was already fully developed. But in Washington they wanted to control the maximum possible flood. So Chambers's boss at the division office in San Francisco, Warren T. Hannum, redrew the plans. While the increased capacity was not economically justified (at the margin), there was a sufficient surplus of benefits over costs in the smaller dam to cover the deficit in the last 220,000 acre-feet and still leave a benefit-cost ratio of 1.4.[70] This became the Corps plan. The Bureau of Reclamation immediately acceded to the larger reservoir, not wanting to contest the judgment of the Corps about the requirements of proper flood control.[71]

But Page extracted a quid pro quo. The Corps had estimated a much lower cost for their dam. Page thought they were too optimistic about foundation conditions, for which they lacked detailed data. Bureau engineers also pointed out that the army added only 10.1 percent for contingencies, in contrast to their 16 percent. Page persuaded the Board of Engineers to raise its estimate by $1 million to $19 million, and tried to get them up to $20 million.[72] Finally the agencies compromised on $19.5 million. They also compromised on the all-important allocation of benefits. The Corps considered that nearly 75 percent of the reservoir was needed for flood control, so only about a quarter of the costs should be charged to irrigation.[73] This percentage, as everybody recognized, was in fact almost arbitrary, because flood levels could be predicted rather well from the winter snow pack, making it possible to have most of the reservoir available for flood control without much compromising its capacity to hold irrigation water. Normally the Bureau was happy to assign the maximum of costs to flood control, thereby reducing charges to water users for irrigation. In this case, though, Page did not want to see the storage capacity of the reservoir allocated overwhelmingly to the business of the Corps of Engineers. A Solomonic compromise was reached. Half the costs would be allocated to flood control, and half to irrigation.[74]

This is how replication worked in politically charged economic analyses. The two agencies, under pressure from higher authorities, negotiated a settlement. The project reports, both issued as House documents in February 1940, proposed the same structure with the same costs allocated in equal proportions to the same functions. They also agreed on the annual benefits of flood control ($1,185,000) and water "conservation" ($995,000). The reports were never fully harmonized, though, because the Corps released its report to Congress before the negotiations were complete, to the dismay of the Bureau. The Corps claimed that 54 percent of benefits were for flood control, making the dam predominantly a flood-control project. The Bureau of Reclamation included a hydroelectric power facility, with a cost of $2.6 million and annual benefits of $260,000. This power would be used to pump water, and hence counted as conservation. So the Bureau report claimed a total benefit of $1,255,000 for irrigation, giving it the advantage in both costs and benefits. Its report contained a few additional quantitative tricks, subtle but admirable. By counting reduced evaporation from Tulare Lake, into which the Kings River flows, the Bureau managed to increase the available irrigation water annually by 277,000 acre feet, as opposed to only 195,000 acre feet according to the army engineers. It calculated an annual cost for flood control, based on repayment in forty years of capital plus 3½ percent interest, at $486,000. Irrigation, although allocated the same costs, was exempt by law from interest charges, so the annual cost was figured at $263,750. Accordingly, the benefit-cost ratio for the flood control portion of the project was a mere 2.4, while the functions of the Bureau could boast a superb ratio of 4.8.[75]

There were a few other differences. The report by the Bureau of Reclamation had an attachment, written five months later. It was a letter signed by the president, Franklin D. Roosevelt, who momentarily forgot that the Corps had any business but navigation. Ickes exploited the lapse: "Again we find the armed forces of the United States are massing to protect farming communities from the floods of an unruly river in the interior of California." On the basis of the Bureau's numbers, the president concluded "that the project is dominantly an irrigation undertaking, and is suited to operation and maintenance under reclamation law. It follows, therefore, that it should be constructed by the Bureau of Reclamation."[76] He elevated the point to a general principle: these jurisdictional conflicts should hereafter be settled by the numbers.

Since both Corps and Bureau were agencies under the executive, the president's letter could scarcely be ignored. Ultimately, though, it proved less important than the other difference between the two reports. The Corps proposed to collect the present value of all future water payments in a lump sum at the outset, and to let local interests take charge of the distribution of the water. The Bureau considered water distribution as part of its mission, and refused to turn it over to local interests. It promised to respect existing water rights, but it also set about renegotiating water contracts. Secretary of Interior Ickes later gave a speech vowing to use the conserved water to create small California farms for returning soldiers. In 1943 a new commissioner of reclamation, Harry Bashore, announced his intention to enforce reclamation law, and not to supply water from Bureau projects to big holdings.[77] It was not clear the Kings River would be exempt.

Already in 1939, water interests near the Kings River had been inclined to favor the Corps. In the 1940s they refused to negotiate water contracts with the Bureau of Reclamation, holding out for the almost complete autonomy offered them by the Corps. In some 1941 Congressional hearings on the Kings River, engineers representing the main water companies—and hence the biggest landholders—testified in favor of construction by the Corps of Engineers. In 1943 this was still more decisive, and by 1944 had become almost hysterical: Hitler himself couldn't have subverted American agriculture more effectively than the economic analysts who assigned a majority of benefits from this project to irrigation. A parade of farmers and engineers traveled to Washington to testify solemnly that they were not much interested in irrigation, but desperately needed flood control. Engineers and lawyers for the water companies recalculated the benefits, and determined that at least three-fourths pertained to flood control. Their testimony was convincing to the congressional flood control committees.[78]

Although the Pine Flat Dam was eventually built by the Corps, the disposition of the water had to be negotiated between users and the Bureau. Predictably, the negotiations were bitter. They lasted from 1953 until 1963. The Kings River Water Association now insisted that irrigation should have priority over all other uses; it even rejected a draft contract giving priority to irrigation "subject only to flood control requirements."[79] Bashore had repeatedly pointed out that all agriculture in the Central Valley of California depended on irrigation. So it might seem that the testimonies of the big water interests and the Corps itself to Congress in the early 1940s about the paramount need for flood control were simply mendacious. This is not quite right. Without irrigation the land was worth practically nothing for agriculture, but given the existing irrigation development in 1940, most of the measured benefits of a dam at Pine Flat could reasonably be attributed to flood control. Tulare Lake, into which the Kings River flows, had no outlet in most years; it expanded and receded with the seasons. It was the scene of extremely large-scale agriculture. The big growers planted in the late spring as the lake receded, and hastened the retreat of the waters by pumping from the lake to irrigate higher ground. This worked adequately in normal years, but in flood years much of the land remained underwater too long to grow crops. Ignoring, as all analyses did, the value of this huge expanse of wetlands to migrating waterfowl, the chief benefit of the proposed works was indeed to contain and stabilize the lake. It accrued almost exclusively to big investors, the only ones who commanded the resources to manage the fluctuations of water level and earn a profit on the intermittent lake bed.

These were the growers whose interests were most effectively represented at the hearings in faraway Washington. Some small farmers, especially members of the Pomona Grange, sent eloquently ungrammatical letters, and even gathered a petition, in favor of the Bureau of Reclamation. They did not call for the breakup of the big holdings, though a few did testify in 1944 at some hearings in Sacramento in favor of enforcing the acreage restrictions of Reclamation law in the Central Valley. The main appeal to them of the Bureau's plan for the Kings River was the cheap power it would provide to help them move water onto their crops. When representatives of the big water interests, especially Charles Kaupke of the Kings River Water Association, told the House committee that his people would prefer no project to one subject to the rules governing the Bureau of Reclamation, the *Fresno Bee* editorialized against him. It anticipated, wrongly as it turned out, that Roosevelt's opposition to the Corps on this issue would prove decisive. It warned that selfish interests might block a federally supported dam promising great local benefit.[80]

There is no indication that any of this mattered much in Washington. The local congressional delegation, and the relevant committees, were firmly allied to the big owners in favor of construction by the Corps. The White House was no less firmly behind the Bureau of Reclamation. Quantitative analysis, to which both sides looked for a resolution, was too loose to provide one. If anything, it obstructed negotiation, as both sides claimed to be conclusively vindicated by a preponderance of benefits. Discrepancies in forms of calculation had contributed to an embarrassing standoff and a political quagmire. Bureaucratic battles like the one over Kings River seemed to reveal a compelling need for the standardization of cost-benefit analysis throughout the federal government.

DISCREPANT ECONOMIC PRACTICES IN FEDERAL AGENCIES

Any effort to reconcile diverse cost-benefit practices faced the most severe obstacles. Cost-benefit analysis was not merely a strategy for choosing projects. It structured relations within bureaucracies and helped define the form of their interactions with clients and competitors. The Bureau of Reclamation, the odd man out in most interagency discussions of cost-benefit analysis, could least afford to give up its distinctive procedures for measuring benefits. Not just in retrospect, but even at the time, some of its practices were regarded as indefensible, bordering on ludicrous. Be that as it may, they were explicitly codified. The Bureau accepted the methods of other agencies, including the Corps and the Federal Power Commission, for evaluating benefits such as flood control and power generation that were collateral to its primary mission. But it was the specialist on irrigation, and in deference to an economic test written into the Reclamation Act of 1939 it created a set of distinctive, Depression-era methods for quantifying that class of benefits.

The Bureau's analysis of direct irrigation benefits began with the agricultural production made possible by a new supply of water. These products were assumed to provide an irreplaceable livelihood for a set of farmers. To the revenues they receive must be added the "extended benefits radiating outward." First, the new production provided raw materials for processing and sale by others. This embraced five classes of activities: merchandising, direct processing, other stages of processing, wholesale trade, and retail trade. Economic analysts for the Bureau assigned percentages to each of these activities for each of ten crop groups. For grain, these were, respectively, 8, 12, 23, 10, 30, a total of 83 percent. Increased production imputed to irrigation was multiplied by .83 to measure this class of indirect benefits. Different factors applied to other crop groups. This was not the only multiplier. Farmers who benefit from irrigation water spend most of their income in the local community. The Bureau defined nineteen classes of enterprises to which farmers extend their custom, and, once again, assigned percentage factors to each. These factors were then multiplied by the increased revenue of the enterprises in question. Some 12 percent of increased retail trade purchases, for example, were credited to new irrigation works. So also were 29 percent of increased expenditures on auto repair and, most famously, 39 percent of the new revenues of motion picture theaters. Finally, at least in principle, the grand total was to be reduced by applying a "federal cost-adjustment factor," the ratio of net to gross farm income.[81]

This was by no means the outer limit of accounting inventiveness at the Bureau. It was required to charge farmers for expenses allocated to irrigation. It considered itself at a disadvantage in this respect compared to flood control and navigation, which required no reimbursement at all. As a task force of the first Hoover Commission observed in 1949: "Interagency rivalry has fostered a sort of Gresham's law with respect to Federal financial policies, the tendency being for higher standards of repayment by State, local, and private beneficiaries to be replaced by lower."[82] The Bureau undertook, with dazzling effectiveness, to minimize this disadvantage. By law the farmers were exempted from paying any interest. The period of amortization stretched gradually from ten to forty and then fifty years. By 1952 it had reached the lesser of one hundred years or "the life of the project." A hundred years without interest was a nice subsidy. But irrigators were never charged even this much. The Bureau calculated benefits for flood control, hydroelectric power, pollution abatement, recreation, and fish and wildlife, among others. Its announced policy was to allocate costs to nonreimbursable functions first, up to the entire cost of the project.

During the debates about the Kings River, Harry Bashore happily explained that "the larger the flood-control benefits are the better it suits us, in a way, for the reason that the burden then becomes lighter on the irrigators who have to pay for irrigation benefits." If not all costs could be allocated to nonreimbursable functions, what remained were assigned preferentially to power production. If this power might be used to pump irrigation water, it could (like the irrigation water itself) be exempted from interest on the cost side, but credited with interest on the payment side, so that even less of the initial investment would remain for irrigators to pay off. And still the farmers often defaulted, partly because they became aware that the Bureau had no teeth, but partly also because these "conservation" projects generally brought them such modest increases of income. The Bureau of Reclamation needed its extravagant measures of the benefits of irrigation to keep its mission from drying up.[83] It was much criticized, and in 1952 a panel of academic consultants was enlisted by Commissioner Michael W. Straus to "make an objective appraisal" of its disagreements with other federal agencies. The panel was sympathetic to the concept of secondary benefits, but still concluded that "the applications actually made by the Bureau go far beyond what can be soundly identified as quantitatively measurable secondary benefits . . . attributable to public water-use projects."[84]

Apart from disappointed interest groups, few have thought the Corps of Engineers too strict in its economic analysis. But it always had more requests for projects than there was any hope of building in the immediate future. In the mid-1940s it recommended and received authorization for a great backlog of works, which in the early 1950s became a source of embarrassment on account of delays and the inevitable cost overruns. The backlog would have been worse had the Corps not rejected more than half the requests that reached it.[85] Critics have generally cited those cases in which the Corps departed most flagrantly from its own economic standards as evidence of the political pressures it faced. On this account, many have supposed that its economic analyses were just for show, honored only in the breach. But conspicuous creativity was not the norm. The engineers were embarrassed whenever they had to put a money value on "intangibles"—which in practice meant any act of quantification (even involving uncontroversial values like the saving of life) not yet formulated in terms of rules. In the ordinary run of business, the Corps had to decide about a host of small and intermediate projects, all with some political support. The credibility required to approve some and reject others depended on its reputation for following the rules. Huge and exceptional cases sometimes overwhelmed them. For ordinary decisions it was politically expedient not to play tricks, but to establish and maintain routines.

This was not easy. Since World War II, the Corps has had about forty-six district offices grouped into eleven divisions. After 1936 the number of civilian engineers grew hugely. In 1949 the Corps was made up of 200 army engineers, 9,000 civilian engineers, and 41,000 other civilian employees. The top officers in Washington tried to use cost-benefit analysis to impose some coherence on planning within this unwieldy bureaucracy. District engineers used the economic results to defend their decisions against disappointed supplicants, who might even be supported by higher authorities within the Corps. Boosters were endlessly imaginative in finding economic arguments for projects. They might, for example, calculate the number of seagulls that would reside on a new reservoir, then multiply by their rate of grasshopper consumption and by the value of grain eaten by each grasshopper.[86] If such extravagances were permitted, project planning would be reduced to naked politics, and flood control would lose credibility.

The office of chief of engineers sent out a series of circular letters in the late 1930s and early 1940s specifying the appropriate categories of benefits, and how they were to be quantified. The rules were restated in the army's "Orders and Regulations," section 283.18. By the late 1950s, the Corps was printing, revising, and reprinting whole volumes on the quantification of various classes of benefits. The tone, as befitted a military bureaucracy, was always strict and serious. The first of the circular letters, dated June 9, 1936, urged that the economic analysis should discount "the natural optimism of the engineer" as well as the exaggerations of industrial companies.[87] "Orders and Regulations" declared it appropriate to consider as a benefit the "higher use" of land protected by flood control, but if the flood plain was being developed anyway, the correct measure of benefit was the anticipated reduction in flood damages. The engineer must never use both measures: this is double counting, the cardinal sin. "Indirect damage" estimates must in each case be confirmed on their own merits. Simply adding a percentage to direct damages is not permissible, "except in cases when such relations have been established for certain selected areas and are applied where comparable conditions exist."[88]

To be sure, the rules were not overwhelmingly restrictive. It was permissible to measure "collateral" benefits, such as pollution abatement, as the alternative cost of achieving the same effect, even if nobody intended to do so. Some navigation projects showed as the principal benefit a saving of time, which fishermen and shippers were notably unwilling to subsidize. A Mr. McCoach, who took economic questions for the chief of engineers in some hearings on flood control in 1937, participated in the following exchange:

> MR. [CHARLES R.] CLASON [MASS.]: Is not the important factor the increase in property value, does not that go up into the millions, while others are floating around in the thousands?
>
> MR. MCCOACH: That is correct, but, of course, that is one of the most-controversial items of benefit that you can find.
>
> MR. CLASON: In the absence of an increase in the value of the lands, no dike or levy would ever be considered beneficial?
>
> MR. MCCOACH: That is correct.

McCoach went on to explain that Corps measures were actually conservative, because "there are so many indirect and intangible things that you cannot evaluate by what I call the invoice methods." He also acknowledged that "no two men in this room" would agree on how to value property, and that while assessed valuation is taken into account, it is not decisive. Clason was troubled: "If voters write in to me to ask me on what basis they build a dike, I would like to be able to tell them something more definite than guess at the increase in the value of the property." "I would not say it is a guess," McCoach replied, "it is an estimate."[89]

After 1940, though, the Corps moved away from such heavy reliance on changes in property value to justify flood control works. In some districts, they began to be called intangible. Increased property values should, after all, reflect potential or historical flood damages. This was the most formalized of all benefit categories. It was still rather tricky. Even if flood records were good, and went back for many decades, an average of recorded damages was an inadequate measure. Since there had almost always been population growth, an equivalent flood could be expected to cause greater damages in the future than in the past. Moreover, the mean annual level of flood damages was extremely sensitive to the size of the largest probable flood, the project flood, which remained hypothetical. To estimate this, the engineers used probability techniques as well as weather records to plot a flood frequency curve, then drew maps showing the expected extent of the water, depth contours, and the duration of flooding. Average damages calculated from historical records might be only a third of those estimated when a hypothetical maximum flood was factored in. It is only fair to add that on a number of embarrassing occasions the hypothetical flood was promptly exceeded by a real one.[90]

Clearly there was much room for judgment in economic estimation. There was also a perpetual effort to define the terms within which it would operate, and to mark its permissible bounds. Engineers were warned not to take on faith the damage claims of those actually flooded, since they were prone to exaggerate. Quantification of "intangibles" was strongly discouraged. A 1940 report by a district engineer that relied too much on intangibles to get benefit-cost ratios of 1.01 and 1.06 was rejected at the divisional level. The division engineer had no doubt that these benefits were real, but since they were "not susceptible of exact evaluation," they should only have been relied upon to justify projects that at least appeared as marginal based on proper, tangible costs.[91]

There is ample evidence that the Corps took its cost-benefit calculations seriously, even in the face of political pressure. A big flood near the headwaters of the Republican River, in Colorado and Nebraska, caused considerable damage to small towns and farms, and even killed 105 people. A study was promptly requested. The Corps found that a big dam could be justified, but only because of its contribution to flood control on the main stem of the Missouri and Mississippi—that is, it had to be downstream, meaning that it wouldn't help those who had suffered in the 1935 flood. All potential upstream reservoirs showed very low benefit-cost ratios, averaging 0.46. The downstream reservoir, though, had a comfortable surplus of benefits over costs—a ratio of 2.35. When this became known, the state engineers of Colorado, Kansas, and Nebraska joined in a petition demanding that this excess be used to cover the cost-benefit deficits of several upstream dams. "We believe it to be the intent of Congress that either the plan should be a complete plan for flood protection on that stream system, or that it should provide as much protection within the Basin as can be provided, and keep the benefits, to whomsoever they may accrue, in excess of the estimated cost of the project." They prepared an integrated package of flood control on the Republican River showing a collective benefit-cost ratio of 1.6. The district engineer resisted this. Such a policy "would conceivably lead to demands by local interests for construction of the maximum number of economically unjustified reservoirs . . . that could be included in a multiple reservoir project." He had to concede that there was precedent for this kind of packaging, though.[92]

An engaging memorandum from the division engineer in Kansas City to the chief of engineers remarks that Senator Norris of Nebraska had come into his office to inquire about plans for the Republican River and to explain the "distress of farmers." It turns out that they wanted to use the flood damage as an excuse to build a big irrigation project, and in fact that they were likely to refuse flood control without it. The division engineer told the senator that the best reservoirs showed benefit-cost ratios of about 0.16 for flood control alone, so they were working on dual-purpose reservoirs.

> It was explained to Senator Norris . . . that the cost of these reservoirs would be between 40 and 60 million dollars; and that the cost-benefit ratio will not be better than 2:1, and will probably be nearer to 3:1 even with very liberal assumptions as to benefits. He was told that we were making every effort to improve the showing of the project, . . . that we have not yet found a justifiable project for him, have scant hope of doing so, but are exhausting our ingenuity to make the report convincing to all concerned.[93]

The balance of politics and objectivity in this letter seems about right. Norris, evidently, accepted the division engineer's reasoning. Perhaps the rejection would have stood for a while, had not the Bureau of Reclamation and Corps of Engineers decided in 1944 to divide up the whole Missouri River watershed, to avoid another bloody war and to block plans for an independent Missouri Valley Authority. Survey reports in the next few years managed to justify a number of projects on the Republican River, most with benefit-cost ratios in the range of 1.0 to 1.2. A huge flood in the lower basin in 1951 settled the matter, and a current map shows reservoirs everywhere.[94]

But this perhaps only exemplified the unfortunate effects of competition from an agency whose cost-benefit standards permitted almost anything. On every possible occasion, the Corps of Engineers favored the establishment of firm standards for cost-benefit analysis, made uniform for all agencies of government. This did not mean discouragingly stringent standards. When an investigative committee asked chief of engineers Lewis Pick whether the number of projects might be reduced by requiring a benefit-cost ratio of at least 1.5, or by demanding large local contributions, he responded with characteristic affability and eloquence: "That is true. I think it is very easy to stop it, sir. If you wanted to stop conservation programs in the United States it would be very easy to stop."[95]

Rather, the Corps was engaged in a perpetual effort to push back the frontiers of cost-benefit analysis so that there would always be a manageable supply of economically approved projects. Occasionally, Corps officers complained of the excessive narrowness of sanctioned benefits, and spoke of "the need for new methods of economic analysis which, by improving the benefit-to-cost ratios, would justify the construction of projects currently judged unfeasible."[96] Such talk suggests that cost-benefit analysis was, to a degree, constraining, at least at any given time. But new methods were indeed forthcoming. The flood-control construction boom of the 1960s, for example, was promoted by new, liberal methods for assessing recreation benefits.

Remarkably, Congress sometimes displayed more commitment to fixed standards than the Corps itself. Through the 1950s, recreation was only made "tangible" by treating it as a source of profits for tourist establishments on or near reservoirs and waterways. But the benefits to tourists themselves are important, announced Pick's successor, R. A. Wheeler, in 1954, and "some day we are going to have to have some sort of a formula" to evaluate them. The National Park Service provided aggressively generous measures in its efforts to justify extensive recreational facilities at reservoirs, after the decision to build them had already been made. The Isabella reservoir on the Kern River was assigned recreation benefits in 1948 by summing travel costs by anticipated tourists, per diem living costs for overnight visitors, a "recreational value" of 12½ cents per visitor-day, benefits to local businesses, and summer-home tract values. This was a Chinese encyclopedia, and Corps engineers would have readily perceived the double counting. The National Park Service, wanting to do better, consulted ten expert economists in the late 1940s, in hopes that they would agree on a correct formula. They did not.[97]

Finally the Congress itself took the bull by the horns. It did not give the Corps a blank check, but attempted to create a rigid, though not especially parsimonious, yardstick. An act was considered in 1957 to credit all projects with $1 per visitor-day as recreational benefit. The Corps considered this rather foolish. The value per day must depend on what use people are making of a reservoir, and on whether there are other equally attractive bodies of water in the immediate vicinity. It would be better, testified assistant chief of engineers John Person in a Senate hearing, to substitute "reasonable value" for this inflexible measure.

Immediately Senator Roman Hruska of Nebraska expressed alarm that "the word 'reasonable' might mean a different thing to different people at different times." Francis Case of South Dakota replied: "Of course that wouldn't be any greater discretion than is accorded the Engineers of the Bureau of Reclamation in evaluating other criteria. We don't spell out the measure of flood damages, nor do we spell out the measure of irrigation values." "Yes, yes, I think we do," interjected Robert Kerr, the chairman, who of all people should have known better. "Not in terms of precise dollars," said Case. "I think we do," repeated Kerr. "We don't tell them what the specifications are, but we tell them to advise us and fix the value in terms of dollars as to what the benefits will be from flood control or flood prevention and they arrive at it." After some more discussion, somebody thought to ask their expert engineer.

KERR: How do you fix it, General?

PERSON: Well, the flood damages prevented we determine by a study of the flood frequency curve, records of each flood, the actual damages experienced, and other related matters.

KERR: It is a fixed specification that guides you then rather than a reasonable estimate?

PERSON: It is fixed to the extent that we have to have something concrete on which to base it, yes.[98]

The railroads and the Bureau of the Budget opposed this whole initiative to put values on recreation as a loosening of requirements. Congress didn't mind such a relaxation, of course, but it did insist on "objectivity."[99]

THE PUSH FOR UNIFORMITY

In 1943, an interdepartmental dinner group of officials from federal water agencies formed in Washington. The organizer was R. C. Price, from the Bureau of Reclamation, who presented a memoir, complete with graphs, showing how "incremental analysis" could be used to design dams of optimum size.[100] The dinner group was soon formalized as the Federal Inter-Agency River Basin Committee. The records of early gatherings show little sign of personal hostility. At the first formal meeting, "it was stressed by all members that the most fundamental reasons for variations in reports and divergence of views originated in the field" and that a "spirit of cooperation" prevailed among the top officials in Washington. There followed a discussion "at considerable length" of "the status of the Bureau of Reclamation and War Department proposals for the construction of Pine Flat Reservoir on Kings River, California." This, they agreed, was already beyond rescue, but they hoped that other conflicts might somehow be forestalled.[101]

The next meeting was about economic analysis. "The discussion . . . centered around the possibility of setting up principles for determining cost and benefit factors and the necessity of freely admitting that certain items cannot be solved by standardization of the method of approach between the different agencies." Someone suggested a subcommittee on cost allocation. In June one was appointed, under Frank L. Weaver of the Federal Power Commission. Its members chose to work through a case study, a project on the Rogue River in Oregon. In October, they announced their intention to report back at the next meeting. But in November, Weaver had to concede that the subcommittee "had not as yet prepared in final form their memorandum report," though G. L. Beard, of the Corps of Engineers, had written up a draft report. When the report did come in, there was substantial disagreement about its recommendations. This could not easily be resolved. After a year of further meetings, a consensus was reached that the mandate of the subcommittee should be broadened to embrace in full generality the measurement of benefits. It was not only to review existing practices, but also "to consider the possibilities of formulating entirely new principles and methods based on a purely rational approach and unencumbered by present practices and administrative limitations." For this it needed a staff. In April 1946 a new "subcommittee on costs and benefits" was appointed.[102]

That subcommittee's members were high-level administrators from each of four central agencies: the Corps of Engineers, Bureau of Reclamation, Department of Agriculture, and Federal Power Commission. "Also present" were some staff people, who attended far more meetings than their superiors, and who did most of the work. Immediately their assignment was subdivided. A working group from the Department of Agriculture was charged with the modest task of preparing "an objective analysis of the problem, including what constitutes a benefit and what constitutes a cost. . . . [T]he analysis must be purely rational and not influenced by present practices or administrative limitations." Meanwhile, members from each of the main agencies would report on their current practices, and the subcommittee would seek to identify the most important similarities and differences.[103] Both jobs proved more difficult than anybody had expected, but the objective analysis took longest.

In April 1947 and December 1948 the subcommittee printed "progress reports" for use of the larger committee, which aimed to describe existing agency practices. Summaries of those reports were eventually published as appendices in the subcommittee's 1950 publication.[104] They did not contribute much to the main report. Having clarified their points of difference, there was no way to resolve them. Neither the interagency committee nor its subcommittee had any authority to bargain away customary procedures. The subcommittee didn't even try. After completing the descriptive sections, it almost stopped meeting. The only hope for agreement was the objective analysis. Its authors had a relevant academic identity as well as their bureaucratic one: they were economists with the Bureau of Agricultural Economics. But there was no precedent for the work that had been requested of them, and since nobody was assigned full-time to the subcommittee, it took them three years to complete a draft. Finally it was distributed in mimeograph form on June 13, 1949, under the title "Objective Analysis."[105]

This formed the heart of the eventual report. The changes made by the full subcommittee were more than trivial, but they did not depart from the basic form of the original statement. It called for maximizing the excess of benefits over costs, meaning that each separable portion of a project should show a surplus of benefits. It mentioned the possibility of discounting according to "social time preference" rather than government interest rates, which in the published version was discarded as a needless complication. Neither the mimeograph nor the published report worked through the problems of quantifying benefits of flood control, navigation, irrigation, recreation, or habitat for fish and wildlife in sufficient detail to serve as a manual, but both offered advice about difficult points and warned against neglecting various classes of side effects. They acknowledged that flooding a wild river valley might well involve scenic or recreational losses as well as gains, a possibility the Corps had generally ignored. On issues of controversy between the Bureau of Reclamation and other agencies, the reports took gentle but unmistakable stands against the Bureau. The assumption of a fifty-year project life seemed quite long enough for an economic analysis. "Secondary benefits"—grinding the grain and baking the bread from wheat grown on newly irrigated land—should only be considered in unusual circumstances.[106]

The printed report, especially, was more ambitious than customary bureaucratic practice in calling for the quantification of intangibles.

Since there is no available framework but market values for evaluating project effects in common terms, it held, these should be assigned whenever possible. It argued firmly that recreation benefits must reflect value to the user, not revenue to concessionaires, and should be assigned a price, though it didn't explain how. Improved health should also be given a price. The published volume, but not the draft, considered that it might be useful to assign some "generally accepted judgment value" to human life, based on consideration of the economic factors involved." It added that lives saved or lost should also be listed as a separate entry on the accounts.[107] Among the most ambitious, and least explicable, moves in the draft and published report was to call for projections of future relative prices. Nobody seems to have had any idea how to accomplish this.

The completed volume came to be known affectionately among water analysts and cost-benefit economists as the Green Book, on account of its cover. Its influence was considerable. But it failed utterly to reconcile the cost-benefit practices of the participating agencies. It was most seriously considered by some interagency water development committees, particularly those concerned with the Columbia River and the Arkansas-White-Red river system. But requests from the former came in too early for the subcommittee to be of much use, and the latter found much of the advice too abstract. It was particularly vexed by the insoluble problem of projecting prices, and the subcommittee was not able to provide much help.[108]

The early 1950s was a difficult time for the Corps of Engineers. Its battles with the Bureau of Reclamation had alienated it from the executive branch. It was accused, especially by friends of the Bureau, of being a pork-barrel agency, more interested in transient political advantage than in systematic water management. The huge cost overruns resulting from the construction in 1950 of projects that had been planned in 1940 and authorized in 1946 led to some severe scrutiny by its customary ally, Congress. The chief of engineers, Lewis Pick, displayed signs of paranoia, though his enemies were real enough. He told one committee that all these inquiries "reflect adversely on the wisdom and ability of Congress."[109] For whatever reason, there was a rash of efforts in the late 1940s and early 1950s to rein in spending on water projects by imposing stricter standards of quantification.

The agency best situated to watch over government spending was the Bureau of the Budget. Beginning in 1943, all project authorizations were sent to the Bureau before going on to Congress. Congress, almost without fail, ignored its advice. In 1952 it attempted to strengthen its hand by issuing cost-benefit instructions in a budget circular, A-47. These were in many respects similar to the recommendations of the

Green Book, though they placed greater emphasis on local cost-sharing. Still, the Bureau of the Budget lacked the personnel to enforce these standards, except superficially. This generally meant refusing to recognize new classes of benefits, and opposing projects that depended for their justification on unquantified intangibles.[110]

The failures of the Budget Bureau, and the evident weakness of the Federal Inter-Agency River Basin Committee, inspired various champions of rationality in water planning to propose that projects be submitted to a panel of independent experts. The first Hoover Commission, in 1949, called for a "Board of Impartial Analysis." It proposed to eliminate interagency rivalry and duplication of effort by consolidating all water planning within the Interior Department. Lewis Pick, always gracious, responded for the Corps: "Those in government who would in all likelihood be charged with centralized authority and responsibility are presently engaged in spearheading the movement to set up in this country, through unbridled exploitation of our natural resources, a totalitarian form of government by regions."[111] Congress was not about to approve the annihilation of its favorite agency, and the Corps argued successfully that a new board of impartial analysis would be redundant in its case. The second Hoover Commission did not try again to eliminate agencies, but recommended that project beneficiaries be required to pay almost all costs. It also called, with no more success, for "objective review" by a special panel, and issued its own "principles to be applied in determining economic justification of water resources and power projects and programs." While admitting that cost-benefit analysis is "easily corrupted," it considered this the fault of incompetent or biased practitioners, not a weakness of the method. People make mistakes in arithmetic too, it noted.[112]

TAKEOVER BY THE ECONOMISTS

In the early 1950s, the Corps of Engineers began what was to become a huge expansion in its employment of economists and other social scientists. Soon, every district office had a section devoted to economic analysis. Some of the early economic specialists were failed engineers, shunted off to a domain where they were likely to do less harm. But the combination of criticism from disgruntled interests, pressure from other agencies of government, and an expanding range of authorized benefits upon which numbers had to be fixed created a need for economic expertise that could not be ignored. The environmental legislation of the 1960s and thereafter, and the increasing likelihood of being subjected to judicial review, further intensified this pressure.[113]

Economic expertise relevant to cost-benefit analysis, however, was almost nonexistent in the early 1950s, outside the bureaucracy itself. When professional economists wrote on the benefits of public works, this was likely to be more closely related to a bureaucratic discourse than an academic one.[114] In the 1950s, there was a convergence. The bureaucracy was looking to quantify an ever more diverse and recalcitrant array of benefits. The new welfare economics presupposed that all pleasures and pains in life were commensurable under a single, coherent, quantifiable utility function. It seemed both intellectually serious and practically useful to try to work this out for such difficult issues as recreation, health, and the saving or loss of life.

Richard J. Hammond, whose early critiques of cost-benefit analysis have never been surpassed, considered that the entry of fancy economics brought its downfall. As a handy bureaucratic convention, the comparison of readily quantifiable benefits with investment costs was perhaps not to be sneered at, but now, he believed, this form of analysis had become a license to concoct imaginary data. Hammond was aware, though, that Adam and Eve felt temptation even before the economic serpent presented them with this apple. Implicit already in its bureaucratic uses, especially in the United States, were pressures to reify its terms, to deny the validity of human judgment, to lust after the impersonality of purely mechanical objectivity. To some economists, this sounded like a definition of science. Cost-benefit analysis first became a respectable economic specialty in the late 1950s.[115]

The analysis of water projects was not its only inspiration, although I believe it was the most important. Transportation studies, especially of highways, provided a largely independent source, though a readily commensurable one.[116] There is also a more distant connection to military uses of operations research, where a form of cost-benefit analysis was developed by the RAND corporation as a strategy of optimization. Operations research itself, in turn, was continuous with Taylorism.[117] Words like optimization and Taylorism, though, should warn us that we are dealing with broad trends of twentieth-century American bureaucratic history and history of science. RAND's form of cost-benefit analysis points toward the wider context of militant quantification. It was also of decisive importance for efforts by Robert McNamara and Charles Hitch during the Johnson administration to reformulate government accounts in a way that would permit comparison of the costs and benefits of various government programs. But economic analysis of defense was accomplished informally, not as public knowledge. Military economics never became a research specialty, and it was not a crucial point of reference for the economists who around 1960 began measuring the benefits and costs of almost every form of government activity.[118] The analysis of water projects was.[119]

From this standpoint, the expansion of terms and the importation of the language of welfare economics in the Green Book appears particularly significant. This was mainly the work of economists, though bureaucratic rather than academic ones. The role of economists at the Bureau of Agricultural Economics deserves closer investigation, but before about 1950 they seem to have preferred a language of rational, systematic planning to the evaluation of projects, one at a time. When they finally took up cost-benefit analysis, they did so with specific reference to water projects.[120] This does not explain where they learned to apply welfare economics to public investment analysis. Citations by Mark M. Regan, the most important author of the "Objective Analysis" that provided a template for the Green Book, do not suggest a direct translation from high theory.[121]

The effort to redefine cost-benefit research according to the standards of economists began in earnest in the mid-1950s. Most authors of the first generation wrote on water projects, often in the guise of a case study.[122] In general, economists agreed with budget officials and with the champions of private industry who dominated the Hoover commissions that the cost-benefit test for water projects had not been strict enough. The most favored vehicle for eliminating marginal projects was the imposition of a uniform discount rate, higher than the rate of interest on government bonds. At the same time, economists did not recoil at the idea of placing money values on the previous generation's intangibles, and in this way they may even have contributed to the construction boom of the 1960s. Only in the 1980s was the quantification of intangibles mobilized as a strategy for discouraging the development of wild places, as researchers began using surveys of citizen preferences to place monetary values on scenic landscapes.[123]

A still more important consequence of this pursuit of unbounded quantification was the spread of cost-benefit techniques to all kinds of government expenditures, and later even to regulatory activities. An early, seemingly unpromising, topic was the economics of public health, which required placing a value on days of sickness and even on lives saved and lost. The economist Burton Weisbrod did not flinch, but used lost productivity as the measure of both, and concluded that even polio vaccination was of doubtful net benefit. Education was another. Gross returns from the labor market permitted an endorsement of high school and college, and, inevitably, of MBA programs, but not of graduate study in science or engineering. The authors duly recommended a shift of educational resources to where salaries were highest.[124] By 1965, economists had used cost-benefit methods to evaluate research, recreation, highways, aviation, and urban renewal. Perhaps the available data were less than ideal for some measures. But as Fritz Machlup commented: "The economic valuation of benefits and costs of an institution, plan, or activity must attempt to take account of values of any sort and to apply reasoned argument and rational weighting to problems commonly approached only by visceral emoting."[125]

Cost-benefit analysis is often criticized for preferring easy answers based on what can be measured to complex, balanced investigations.[126] Economists have by no means been immune to this. Although they routinely concede by way of preface that calculation can never replace political judgment, cost-benefit and risk analysts clearly want to rein it in as much as possible. So, typically, they insist that a decision can never be left to the judicious consideration of complex details, but must always be reduced to a sensible, unbiased, decision rule. An effective method should not be a mere language, focusing discussion on central issues, but must be constraining. The great danger, announced the authors of a major study of risk, is that "combatants may learn to conduct their debates in, say, the nomenclature of cost-benefit analysis, transforming the technique into a rhetorical device and voiding its impact."[127]

Cost-benefit analysis was intended from the beginning as a strategy for limiting the play of politics in public investment decisions. In 1936, though, army engineers did not envision that this method would have to be grounded in economic principles, or that it would require volumes of regulations to establish how to do it, or that such regulations might have to be standardized throughout the government and applied to almost every category of public action. The transformation of cost-benefit analysis into a universal standard of rationality, backed up by thousands of pages of rules, cannot be attributed to the megalomania of experts, but rather to bureaucratic conflict in a context of overwhelming public distrust. Though tools like this one can scarcely provide more than a guide to analysis and a language of debate, there has been strong pressure to make them into something more. The ideal of mechanical objectivity has by now been internalized by many practitioners of the method, who would like to see decisions made according to "a routine that, once set in motion by appropriate value judgments on the part of those politically responsible and accountable, would—like the universe of the deists—run its course without further interference from the top."[128] This, the ideal of economists, originated as a form of political and bureaucratic culture. That culture has helped to shape other sciences as well.

PART III

POLITICAL AND SCIENTIFIC COMMUNITIES

Therefore I insist that good sense is the principal foundation of good manners; but because the former is a gift which very few among mankind are possessed of, therefore all the civilized nations of the world have agreed upon fixing some rules for common behaviour, best suited to their general customs, or fancies, as a kind of artificial good sense, to supply the defects of reason. Without which the gentlemanly part of dunces would be perpetually at cuffs.

(Jonathan Swift, "A Treatise on Good Manners and Good Breeding," 1754)

CHAPTER EIGHT

Objectivity and the Politics of Disciplines

> Statistical science is one of the precision instruments available to the experimenter, who, if he is to make proper use of the knowledge at his disposal, must either learn to handle it himself or find someone else to do so for him.
> *(Donald J. Finney, 1952)*

RESEARCHERS in accounting, insurance, applied economics, and quantitative social science have generally been so keen to model their work on the better-established disciplines that they may be a little nonplussed to find their specialties represented as the prototypes of quantification in science. Others who have learned some history of physics or biology from science textbooks, or even from the standard historical literature, may reasonably wonder if my chapters on quantification in the context of bureaucracy could have much to do with its uses in the more academically respectable sciences. In the rest of the book I will take up these issues directly. This chapter is about the pressures, some of them deriving from politics and governmental regulation, that have pushed certain disciplines in the direction of standardization and mechanical objectivity. Chapter 9 will be more concerned with contrasts, with the cultural and political circumstances that lend or deny credibility to judgment and personal authority in science.

I do not aspire here to a grand unified theory of science, but to a sharper appreciation of disunity. Certainly I would not claim that, in making knowledge, a scientific collective behaves just like any bureaucracy. The similarity to bureaucracy applies only in certain respects. Also, perhaps more crucially, there isn't much sense in the phrase "just like any bureaucracy." Recent scholarship reaffirms what readers of, say, Balzac, Dickens, and Gogol already know—that bureaucracy is a heterogeneous category. This book is concerned with only one dimension of that heterogeneity, but even that is enough to make the point. Neither the French administration nor the British civil service have conformed very well to Max Weber's precepts:

> Bureaucratic administration means fundamentally domination through knowledge. This is the feature of it that makes it specifically rational. . . . The dominant norms are concepts of straightforward duty without regard

> to personal considerations. . . . Everywhere bureaucratization foreshadows mass democracy . . . The 'spirit' of rational formalism, which bureaucracy [embraces] . . . is promoted by all the interests. . . . Otherwise the door would be open to arbitrariness.[1]

Weber's formulation combines the rigidities and suspicions of Prussia and America. Even allowing for the exaggerations of the ideal type, this is much too restrictive. Nevertheless, his is a valuable formulation. For this is the form of bureaucracy that has been most receptive to science, and hence that has worked hardest to remake science in its own image.

BUREAUCRACY, AMERICAN STYLE

At the top, American bureaucracy is made up of political appointees. Hugh Heclo, who subtitled his study (with Aaron Wildavsky) of the British bureaucratic elite *Community and Polity inside British Politics*, wrote a comparable book on American administration called *A Government of Strangers*. The Americans, no less than the British, need "relationships of confidence and trust" to organize and administer effectively, but such relationships "require time and experience, both of which are in short supply in the political layers."[2] Partly for this reason, it is very difficult to keep secrets, or even to negotiate in confidence, in American government. Perpetual exposure to political demands made it impossible for the Corps of Engineers and the Bureau of Reclamation to negotiate settlements of some of their fiercest controversies. Cabinet secretaries and agency heads in America tend to be preoccupied with politics, and have few defenses against its pressures. These reach all the way to the middle levels of bureaucracy, and sometimes still lower.[3] Often the agencies are faced with incompatible political demands, for American government lacks clean lines of authority.

James Q. Wilson argues that while bureaucracies are not so aggressive in expanding their turf as is generally supposed, the concern for autonomy is universal. This means not being preempted or overruled by outsiders. Faced with the contradictory expectations of the executive, a myriad of congressional committees, and the courts, it is little wonder that they should seek to minimize responsibility by adhering whenever possible to rules. This preoccupation with rules, calculation, and fact-finding is not the essence of a bureaucratic-legal mode, as some would have it, but a defense against meddlesome outsiders and a strategy for controlling far-flung or untrustworthy subordinates. Wilson writes, "The United States relies on rules to control the exercise of official judgment to a greater extent than any other industrialized democracy."[4] A similar lack of trust has inspired Congress to impose rules on every agency, dictating how to award contracts or hire and fire employees, as well as how to carry out its central mission. It sometimes even imposes such standards on itself. Cost-benefit analysis, for example, is a monument to the halfhearted desire of Congress to bind itself in red tape. As currently practiced, it is a distinctive achievement of American political culture.

Under such a system, there is more need for objectivity than in Britain or France. "In a country where mistrust of government is rife, the temptation to substitute supposedly impersonal calculation for personal, responsible decisions and to rely on the expert rather than size up the situation by oneself, cannot but be exceedingly strong," observed Richard Hammond.[5] In the United States, mere experience or know-how is not sufficient to ground public expertise. At first this seems surprising, even contradictory. The country prides itself on democracy, and continues to nourish a distinguished tradition of anti-intellectualism. Americans fear expertise, writes Sheila Jasanoff, yet insist that administrative decisions be depoliticized; they oscillate "between deference and skepticism toward experts."[6] There is no contradiction, however—merely a paradox. The current forms of expertise often come close to meeting the strenuous demands of anti-intellectualism. Where experts are elites, they are trusted to exercise judgment wisely and fairly. In the United States, they are expected to follow rules. This disposition to place faith in regulations, not to look for persons of character to fill offices, was already a powerful one by 1830. The vast twentieth-century demand for experts has not eliminated it. "The truth is," wrote Richard Hofstadter disapprovingly, "that much of American education aims, simply and brazenly, to turn out experts who are not intellectuals or men of culture at all."[7]

But the know-nothings should not get all the credit. American courts, which in recent years have spread their dominion over more and more of public life, have worked also to limit everybody's discretion except, perhaps, their own. Since they do not generally make it their business to assess complex and alien practices from the inside, they prefer instead to see those practices subordinated to explicit rules.[8] Hence, too, they prefer professional experts over seasoned participants, and theoretical over practical knowledge.[9] Although Anglo-American courts authorize experts to give opinions, and not merely to testify to matters of fact, courts have been particularly stubborn in believing that science should mean the straightforward application of general laws to particular circumstances. On this account, the testimony of real living scientists often holds up rather badly in the adversarial courtroom situation. Legal questioning, as Brian Wynne points out, is much more searching and critical than peer review. Lawyers ask questions from outside the "socially filtered climate of considerable trust, credulity, common background, common assumptions, understandings, values, and interests" that characterizes scientific communities. The courts themselves maintain a pretense of merely ascertaining the facts and applying the law, divorced from any social or economic context. They look to science to emphasize the separation, and hence to support their own claims to objectivity as impartiality.[10]

The courts are by now preeminent among those meddlesome outsiders who exert unremitting pressure to turn private knowledge into public knowledge, and in this sense to expand the domain of objectivity. The ideal is a withdrawal of human agency, to avoid the responsibility created by active intervention. Subjectivity creates responsibility. Impersonal rules can be almost as innocent as nature itself. John McPhee gives an eloquent example:

> [L]ava deflected from one route could wipe out houses on another. And this is not Iceland, the home of the fair; this is the United States, the home of the lawyer. When Mauna Loa erupted in 1984, the state was asked if, in dire emergency, an attempt would be made to save Hilo. The answer was no. The Department of Land and Natural Resources regarded such a struggle as futile in the first place, and, moreover, could not imagine any way to deal with the legal consequences of lava diversion.[11]

If an individual or firm suffers harm as the result of discretionary human actions, this is at least suspect. Courts will often undertake to reverse it. They are less inclined to overturn the nature of things, even if this is an artificial nature sustained by rules and conventions.

Research done according to the standards of scientists is often not impersonal and lawlike enough to stand up to political and judicial scrutiny. Even the standards of mathematical proof have now been critically examined by a British court of law.[12] In America, as Jasanoff argues, administrative decisions have come to be patterned after judicial ones, relying on a form of open and adversarial argument that is scarcely distinguishable from litigation. The regulatory context thus demands an especially rigorous and objective form of knowledge, to the point that new research specialties have been created to provide it in a form that courts and regulatory agencies can accommodate. The most influential of the new methods is risk analysis, a close relative of cost-benefit analysis, which must often be undertaken as proof against arbitrariness in technological decisions. The Supreme Court ruled against the Occupational Safety and Health Administration (OSHA) in 1980 for relying on expert judgment, when it should have calculated risk levels using mathematical models. Such calculations were presumed to provide some guarantee against abuse of discretion, where discretion was defined to mean lack of support by "substantial evidence." Procedures have become as important as outcomes, and rules may be maintained even though they are unable to accommodate new kinds of relevant scientific information.[13]

It is now a commonplace in the literature on regulation that European agencies behave differently from American ones. The Europeans of course also vary among themselves, but all are capable in some measure of formulating policies and determining how to apply them through negotiation with the interested parties, behind closed doors. Americans, on the whole, are denied this: "Unable to strike bargains in private, American regulatory agencies are forced to seek refuge in 'objectivity,' adopting formal methodologies for rationalizing their every action."[14] The point should not be exaggerated. Nobody goes so far as to claim that value judgments can be excluded entirely from the regulatory process. But there is a strong incentive to systematize them, so they will be applied uniformly, and to isolate them so they do not corrupt the process of establishing scientific facts.

Such a stance is, of course, sanctioned by a hoary philosophical doctrine about the impossibility of deducing values from facts. Scientists, especially those without administrative responsibilities in the regulatory agencies, are generally sympathetic. The alternative, it seems, is to politicize the laboratory and to invite public debate about scientific conclusions. So they call for a clear separation between the scientific phase of objective, quantitative risk determinations and the political one of subjective management decisions. They inveigh against the surreptitious mixing of "sociopolitical judgments" with "technical findings," with the expectation that the latter will provide a firm factual basis for the former.[15] As the scientists are well aware, though, that basis is subject to massive uncertainties in regard to such crucial problems as regulating suspected carcinogens. Environmental agencies often have to make choices before there is time for scientific consensus to emerge. Scientists always negotiate the content as well as the meaning of what are presented to the world as facts, and opponents of regulation routinely contest their findings, using research by their own scientists. Regulatory agencies must decide whom to believe as well as what to do.[16] Hence the model of applying general principles to the hard facts of each case is rather implausible, especially in the context of science done for regulatory purposes. In countries like Sweden where negotiation occupies the place of the formal, open hearings in the United States, the pretense of a rigid separation between facts and policies is unnecessary.

Even in the United States, the quest for formal decision methodologies in a context of bureaucratic weakness and exposure is a relatively recent phenomenon. To be sure, the bureaucracy was never weaker or more suspect than in the half-century following Andrew Jackson's presidency; and while it drew on the authority of rules, it was too politicized and transitory to put in place an elaborate apparatus of calculation. Progressivism meant a considerable enhancement of the status of professionals. Those, like doctors, who dealt personally with their clients could increasingly count on their trust, without giving reasons.[17] A similar conception of professional expertise animated the Progressive effort to staff regulatory agencies with economists, accountants, and scientists, and to authorize them to exercise judgment. The New Deal strongly affirmed expertise, and the regulatory bureaucracies it created were given considerable autonomy. They were closer to the European model of bureaucracy than to the American one that emerged clearly in the 1960s.[18]

But there were always pressures tending to undermine this bureaucratic autonomy. The 1936 Flood Control Act offers testimony to these forces. As the preceding chapter shows, its demand for an analysis of costs and benefits became more rigid as a result of opposition to the Corps of Engineers from certain American industries and especially from other federal agencies. A more wide-ranging attack on the expert agencies was initiated by free-market economists, who complained that they had become captive to the industries they professed to regulate. The economists' solution was to deregulate, and this was tried in a few areas, but their opposition to private negotiation between regulators and regulated was built into a new round of regulatory legislation in the 1960s. It resounded widely in calls for openness as an antidote to self-interest and to corruption masquerading as expertise.

This bureaucratic impulse has by now invaded every domain. A front-page article in the *Los Angeles Times*, for example, reports that American college accreditation agencies are working on "a plan to measure how well students are educated. . . . There is a very significant body of opinion in higher education that says to the public, 'Trust us. And don't require us to produce any evidence [of results].' What we're saying is that those days are over." Like every institution, the university must be refashioned as a panopticon to open it to surveillance by law courts and regulatory bureaucracies. A "pervasive expectation" of inquiry into "what constitutes learning and what promotes it" should move universities to subsume their activities within a "culture of evidence."[19]

The massive effort to introduce quantitative criteria for public decisions in the 1960s and 1970s was not simply an unmediated response to a new political climate. It reflected also the overwhelming success of quantification in the social, behavioral, and medical sciences during the postwar period. I suggest in the remainder of the chapter that this was not a chance confluence of independent lines of cultural and intellectual development, but in some way a single phenomenon. It is no accident that the move toward the almost universal quantification of social and applied disciplines was led by the United States, and succeeded most fully there. The push for rigor in the disciplines derived in part from the same distrust of unarticulated expert knowledge and the same suspicion of arbitrariness and discretion that shaped political culture so profoundly in the same period. Some of this suspicion came from within the disciplines it affected, but in every case it was at least reinforced by vulnerability to the suspicions of outsiders, often expressed in an explicitly political arena. It was felt most intensely in fields treating matters of public interest, and in many cases quantitative methods were initially worked out by applied subdisciplines, migrating only later to the more "basic" ones.

INFERENCE RULES

A multitude of examples might be given of applied quantitative tools and methods that were subsequently taken up or formalized within university disciplines. Cost-benefit analysis is an obvious and important one. So is accounting. The social survey, a genre invented by public-spirited citizens who wanted to understand and manage the poor, became part of academic sociology. The study of voting behavior entered political science from market research by way of electoral polling. By now, almost every form of quantification used in business or the professions is taught and researched in the academy.

The movement of a genre to a new site often promotes important changes of method and rhetoric. When academics take up a branch of practical quantification, they commonly complain that their predecessors were moralists and lacked objectivity. This, for example, is the way sociologists have customarily interpreted the early history of the social survey, arguing that disciplinary autonomy is needed to attain a proper state of objectivity. The converse may be more nearly true: the Weberian language of objectivity was adopted in part as a defense of the incipient discipline against political interventions.[20] Moving away from a descriptive, empirical style and using ever more recondite quantitative techniques brings similar advantages.

For many disciplines, the most important source of quantitative sophistication has been mathematical statistics. Although even its basic ideas derive in part from social reform and administration, the statistical methods of twentieth-century sciences were not taken over directly from philanthropic institutions or public bureaucracies in the way that the most basic methods of cost-benefit analysis were. They flourished among academic researchers and other skilled professionals with advanced degrees. Users of statistics, not implausibly, regard their techniques as deriving from mathematics, which, somewhat reluctantly, nurtured a statistical subdiscipline. Still, the extraordinary modern success of inferential statistics must be understood partly as a response to conditions of mistrust and exposure to outsiders similar to those that have been so important in the history of accounting and cost-benefit analysis. On the whole, statistical inference has not made its way down the hierarchy of science, from mathematics and physics to the biological and at last the social sciences. Rather, it was seized most readily by weaker disciplines, such as psychology and medical research, and indeed by their relatively applied subfields.

The point here is not to argue that science, or quantification, is merely a tool of politics and public administration. It is to recognize how strategies of quantification work in an economy of personal and public knowledge, of trust and suspicion. In recent decades, especially, democratic politics has been decisive in forming a context of overwhelming distrust, or at least distrust of personal judgment. But lack of trust is also characteristic of new or weak disciplines. It might almost be taken as the defining feature of weak disciplines. Standard statistical methods promote confidence where personal knowledge is lacking. They are also used to train and discipline outsiders, such as students and uncredentialed assistants.

The problem of recruiting, training, and supervising unprofessional labor was central to the early history of error theory, the first reasonably routinized form of inferential statistics. It is not, to be sure, the whole story. Like every kind of statistical analysis, this one was impossible until researchers had some basis for confidence that their measures were homogeneous. For that reason, as Stephen Stigler points out, many of the earliest uses of what we would call statistical data reduction were carried out by astronomers on measurements they had made for themselves, and knew to be of good quality.[21] This was rather a personal matter, and was evidently not in response to any interesting problem of social or intellectual distance. It was also not very strictly disciplined. Eighteenth-century astronomers felt few compunctions about discarding measurements that seemed wrong somehow—if, for example, the sky was not perfectly clear, or the telescope was unsteady. And who could know better when conditions were optimal than a master astronomer, his skills honed by spending endless nights with his eye at the eyepiece?

In the nineteenth century, greater rigor came to be demanded. This was promoted by new instrumentation, which reduced the skill level necessary for various kinds of observations and promoted the standardization of data across observers. Still, the standardization of observers did not follow automatically, but had to be vigorously pursued. As observatories grew larger, the actual business of observation fell increasingly to subordinates. It was far from clear that assistants could be trusted to decide which were their best observations and which should be discarded. Error analysis was a protocol for running a centralized organization, rather like the inflexible rules that, according to Adam Smith, defined the natural state of large business organizations.[22] It was universalized as part of a routine of discipline, which involved averaging over measurements rather than picking out the best ones. To discard an observation for no clear reason became an offense against sound morality.

Soon the professional astronomers themselves internalized these values. A prominent astronomer was thought to have lost his sanity and then his life from guilt at having discarded his most discrepant data. Clear, quantitative rules governing the rejection of outliers, it was implied, could have saved him.[23] A similar regime of impersonal rules seemed necessary to regulate the presentation of photographic images made through telescopes. When Warren de la Rue proposed to print "an engraved representation" of solar prominences observed during a total eclipse, "compounded of my own drawing and the photographs," George Airy discouraged him. Any human interference would detract from the authority of the pictures, for the originals, not "touched-up photographs, contain the evidence on the case. The interpretation . . . *may* be fallible (I do not believe that it is), but the whole question about the prominences is strongly debated, and you must proceed with exactly the same caution as in a disputed case in a court of law."[24]

Error theory, like photography, was a strategy for eliminating interference by subjects. It became integral to the intense nineteenth-century pursuit of precision in the physical sciences, and in some areas of biology. Precise measurement was of course valuable where experimental or observational results were to be compared with mathematical theory. It was also appreciated for its own sake, as a mark of competence and of the moral character that gave assurance of honest and careful work. Its main role was to protect against false judgment or bias. The anthropometrist Paul Broca scorned the idea "that it suffices to study and measure a few individuals of each race, chosen *with discernment* as the representatives of the mean racial type."[25] Precision was especially emphasized in the classroom and seminar, as a form of discipline, and the laboratory course in physics became above all an exercise in exactitude.[26]

The distinctively moral value of mathematics was particularly emphasized by British actuaries. The author of an 1860 paper on the construction of mortality tables conceded that various simplifying approximations could give values "as near the *truth* as the values correctly deduced would be." But we should aspire to make our conclusions fully consistent with our premises. "To this very proper regard for logical consistency, which is the foundation of mathematical science, we owe the construction of tables of annuities certain to five or six decimal places; for it cannot be pretended that *any* assumed rate of interest represents the value of money so exactly as to render such extreme accuracy at all necessary to the abstract justice of the case."[27] The Institute of Actuaries defended its mathematical examination not mainly on account of its importance for calculating insurance rates, but rather because it helps to keep the profession free of "admixture with all 'baser matter.'" A spokesman explained that mathematics "has the effect of promoting and improving our powers of judgment, or creating in us care and caution, and of indirectly producing those very qualities for which, I believe, actuaries are noted."[28]

BIOLOGICAL ASSAYS AND THERAPEUTIC TRIALS

It is worth observing that the actuary just quoted, H. W. Porter, went on in his lecture to stress the value of Latin and Greek for members of his trade. Had these subjects been part of the examination, this would have been even more effective in keeping the trade free of baser matter. The actuaries aspired to the status of a liberal profession. Merely technical knowledge would never suffice. This was also recognized by practitioners of the classical profession of medicine. Science was central to the medical curriculum, but physicians strongly resisted the idea that scientific medicine could be applied by rote. Medical tact, the ability to recognize the most crucial symptoms and to propose an appropriate, individualized treatment, was defended as a matter of skill and experience, not just formal knowledge. Especially for the elite, as Christopher Lawrence observes of Victorian physicians, to believe otherwise would be "to place the superior claims of character and breeding on an equal footing with those of scientific merit when making appointments." These gentlemen practitioners opposed specialization, and even resisted the use of instruments. The stethoscope was acceptable, because it was audible only to them, but devices that could be read out in numbers or, still worse, left a written trace, were a threat to the intimate knowledge of the attending physician.[29] They stood for an ethic of impersonal facts rather than personal trust, and were long held suspect by most doctors in America as well as Europe.

This stress on private knowledge was challenged first by "quacks" and then by researchers, both of whom often criticized such claims as obscurantist. Medical research only began to form the basis for an identity separate from that of the practicing physician in the later nineteenth century. Even then, most researchers—especially clinical researchers—did not stop thinking of themselves also as physicians and teachers of physicians.[30] Still, as medical scientists they accepted an ideal of public knowledge, and they often took quite a different view of objectifying instruments from that of other elite practitioners. Instruments appealed for a variety of broadly consonant reasons, as Robert Frank shows in his historical study of tools to record the action of the heart. The first sphygmographs, made by Etienne-Jules Marey, were perhaps less sensitive than the expert finger. Marey stressed their advantage of providing a permanent record, immune to observer prejudices, and their ability to speak across boundaries of language, time, and place. The London clinician William Broadbent, author of an 1890 textbook, praised the sphygmograph as a teaching tool. No mere verbal description could convey so concrete a sense of the heartbeat to distant readers. Half a century later, in America, electrocardiograms enabled nonphysicians to diagnose acute coronary thrombosis without even seeing the patient. The several phases of the development of this instrument, Frank summarizes, were animated by a desire for "visible and permanent records of great precision" that would be accessible to all and "were *not* dependent upon the acuity of the cultivated sense of the physician."[31]

If the role of statistics in medicine depended on a trial of strength between statisticians and doctors, its failure was assured. And for most of its history, it fulfilled this reasonable expectation. The idea of a medical statistics was as old as statistics itself, but only in the domain of public health was the language of quantity very successful. Even there it was strongly challenged by some physician-administrators who preferred to follow the course of epidemics by examining in detail the transmission of disease from victim to victim, ship to port, and town to town.[32] The idea that a numerical method might provide the best test of medical therapies was urged repeatedly in the nineteenth century, and sometimes even put into practice by its champions. A few, such as Jules Gavarret, argued that probability methods ought to be used to decide if the differences in rates of cure between two populations of patients should in fact be attributed to the treatment regimes. Most physicians would have none of this. Gavarret's methods, argued one anonymous reviewer, would require physicians to "accept servilely all medical ideas, which would be imposed by the professors."[33] While physicians did not unanimously oppose quantification, they doubted that medical numbers could have meaning apart from clinical judgment.[34] Even the professors were divided. In the late nineteenth century, experimental physiology and then bacteriology seemed more promising bases for finding effective therapies than did medical statistics.

How, then, did controlled clinical trials and statistical analysis become standard, even obligatory, for the evaluation of new therapies? The remarkable flowering of mathematical statistics in the twentieth century provides part of the answer. Austin Bradford Hill, who supplied the statistical expertise for the first large-scale controlled clinical trials, was a student of the medical statistician Major Greenwood, himself a student of Karl Pearson. But statistics was, crucially, part of a regime of public knowledge. Greenwood worked mainly on public health statistics. Hill was deeply involved in the debates of the early 1950s on the relation of smoking to lung cancer. He argued that it was still possible to take advantage of R. A. Fisher's conception of experimental design, with its emphasis on comparison of similar treatment and control groups, by defining in advance observational controls.[35] Where strict experimentation was possible, as in the case of clinical trials, the methods Fisher worked out for agriculture could be applied more rigorously, and comparability assured through randomization. Still, the statisticians needed medical allies. Physicians in private practice rarely admitted a need for statistical methods to analyze their experience, but researchers in medicine, as in agriculture, psychology, ecology, economics, sociology, business, and most other biological and social disciplines, began to redefine their fields in statistical terms. The reasons for this are in every case complex. In the broadest terms, statistics supported a research ideal of openness and public demonstration.

Researchers, though, were often doctors too. Moreover, they had their own craft skills to defend. When Major Greenwood argued against the champions of tact that science must seek data and employ explicit methods of reasoning, medical researchers felt divided loyalties.[36] The statisticians had to win them over to an approach that most found too rigid, and many thought unethical. Hill spoke often of medical opposition to statistics in his invited lectures at medical schools during the 1940s and 1950s. He tried to appease his audiences, promising that statisticians didn't scorn clinical judgment, and wouldn't want to lose the benefits of individual medical experience until they were in a position to replace it with something more deliberate and objective.[37] But the ideal of objectivity, as the statisticians conceived it, was difficult to reconcile with clinical judgment. A tuberculosis researcher, quoted approvingly by Hill, observed that for clinicians to be "willing to merge their individuality sufficiently to take part in group investigations, to accept only patients approved by an independent team, to conform to an agreed plan of treatment, and to submit results for analysis by an outside investigator involves a considerable sacrifice."[38]

Hill also liked to quote Helmholtz: "All science is measurement." This, as he construed it, implied that the opinions of participating physicians about the success of the treatment and the health of the patient were poor substitutes for hemoglobin counts and sedimentation rates as an indication of outcomes. The clinical trial, he observed, "demands to the greatest possible extent objective measurement of results and the use of subjective assessments only under a strict and efficient control which will ensure an absence of bias." "Measurements," wrote another spokesman for the controlled clinical trial, "are potentially more acceptable than clinical assessments because they are less subjective, and so less likely to be influenced by either the clinician's or the patient's knowledge of the treatment." Hill graciously conceded that good numbers count for little if the patient promptly dies.[39]

Hill's extremely successful textbook of medical statistics, which began as a series of articles in *The Lancet*, discussed concepts of experimental design and emphasized that medical experiments and observations can easily go astray if the control group is not strictly comparable to the treatment group. Much of the book, though, consisted of basic statistical mathematics: means and variances, standard deviations, and some simple inferential tests. This is what the doctors called statistics, and the mathematical component provided the main intellectual basis for the authority of the statistician. In fact, as Harry Marks argues, the statistical mathematics was of mainly instrumental importance in the history of the clinical trial. The real problem for therapeutic research was the organization of labor. Doctors were firm individualists, little inclined to "merge their individualities" into a large-scale research program. They made up a professional elite, whose daily work was even more poorly standardized in 1945 than it has since become. Their characteristic faith in their own judgment, refined through long experience, made it exceptionally difficult to subject them to a shared discipline.[40]

Hill was acutely aware of this problem. He sometimes gestured at a conciliatory stance, suggesting that the experimental protocol might remain loose, and the particulars of each case be left to the discretion of each physician. The statistical analysis could be conceived as a test of therapy plus clinical judgment, rather than of a strict regimen. The disruptive effects of physician judgment could also be minimized by not telling the doctor (or the patient) who was in the treatment group, and who in the control. This was the main purpose of double-blind methodology in medicine: to neutralize the effects of expert discretion without disbarring it. Even with this modest refinement, the issue under investigation was muddied somewhat by physician judgment. And certainly no conclusion could be reached about the value of any unplanned additions to the treatment regime. If a therapy is altered in response to the peculiarities of a patient's conditions, wrote Hill, it can no longer be used to explain that condition, except by reasoning in circles.[41]

These tensions between statistical and clinical ideals suggest that the impetus for statistical study of medical treatments did not come mainly from doctors. In Britain, the first large-scale clinical trials were organized by the Public Health Service in the immediate aftermath of World War II, and hence backed by the power of the state. A shortage of streptomycin eased the moral decision to withhold this treatment from some tuberculosis patients, chosen at random, and made it much easier to control the physicians who were given access to the drug.[42]

The political and bureaucratic background to these British therapeutic trials has not been so well studied as has the encounter between statisticians and doctors. This background is clearer, indeed obtrudes conspicuously into the foreground, in the American case. Here, too, the expertise of statisticians was indispensable. A cultural and political account of statistics in medicine that denied the intellectual dimension would not only be false, but inconceivable. Medical schools paid tribute to this statistical knowledge by hiring statisticians in great numbers to help with the analysis of experimental data and, increasingly, with the design of experiments. But American doctors were no more eager than their British counterparts to abandon their own expert judgment and defer to the objectivity claimed by statistics. How did the controlled statistical trial gain its overwhelming public authority?

This kind of experimental rigor was not demanded by doctors. Neither did it arise naturally within any particular community of medical researchers. Doctors, of course, learned to accept it, but the impulse for uniform and rigorous standards came mainly from regulatory authorities. Therapeutic credibility was a valuable and dangerous commodity, in their view. They considered that the expertise of doctors provided an inadequate control on the bold claims of drug manufacturers. The alternative was a more centralized decision process, to be based mainly on written information. In medicine, as in accounting, standardized measurement was already a familiar way of dealing with distance and distrust. Life insurance companies required doctors to submit instrumental measurements as evidence of the good health of applicants in late-nineteenth-century America, thereby contributing to the advance of instrumentation and quantification in medicine. In the early twentieth century, industrial medicine became thoroughly quantitative in part because workers assumed that the exercise of judgment on the part of corporate doctors would work to their disadvantage.[43] The Food and Drug Administration, challenged on all sides, endeavored to reduce decisions about the licensing of new drugs to questions of uniform measurement.

As always, the task of quantification depended on an infrastructure of standardization, which it promoted still further. A pharmacopoeia of variable organic materials, mixed somewhat differently by every druggist, could not be tested with the conclusiveness of synthetic chemical treatments. The huge international push for biological standardization, discussed in chapter 1, was also a contribution to regulation. That standardization soon became a statistical problem, as researchers learned that the variability of drugs derived from living materials was matched or exceeded by the variability of laboratory animal response. On this account the "minimum lethal dose," learned from a single trial, was replaced by the median lethal dose, or LD 50, requiring the sacrifice of large numbers of animals. Biological assay became an important topic for statisticians interested in estimation.[44] The demands for expertise and for labor made biological standardization a full-time job for specialist laboratories, whose work was too expensive for any but large manufacturers.[45] In the same way, but to a still greater extent, the increasingly formalized and demanding protocol for gaining approval of new drugs made this impossible without a large centralized laboratory.

The Food and Drug Administration gained authority to regulate drugs mainly from two acts of Congress, in 1938 and 1962. Especially in 1938, Congress was worried mainly about new drugs that had proved dangerous, and this early legislation authorized the FDA to reject drugs only on account of their danger to health, not because they lacked efficacy. But this was the New Deal, and the agency used its discretion. As Harry Marks shows, it defined drugs to be safe if they could be expected to cause more health benefits than harm. Hence an inert drug might be rejected as harmful, since it could be prescribed in place of something really useful. Initially, the FDA considered the opinions of clinical specialists as well as research data in determining the balance of benefits and dangers.

The regulators, though, had no great faith in ordinary doctors. Regulating drugs was in part a substitute for the impossible task of overseeing medical practice. The FDA kept close watch on the literature sent to doctors by the drug companies. More than that, it felt obligated to consider the tendency of doctors to misprescribe drugs in assessing whether the net effect of a new drug would be beneficial. Once approved for one purpose, a drug could be prescribed for any, so a medication shown effective for a relatively uncommon disease might, if it was at all dangerous, cause still more harm to others to whom it was administered in error. The drug companies believed such issues lay outside the agency's legitimate regulatory authority, and challenged the whole policy of assessing safety against efficacy.[46]

The balance of power was tipped decisively in favor of the FDA by the Kefauver Bill of 1962. Like most such legislation it was provoked by a disaster, this time involving Thalidomide, but it gave a clear mandate that drugs must be shown to be effective as well as safe. The determination of safety has never been standardized, though both the companies and the regulators would prefer that it were. Arbitrary rules don't work here, as they have for certain aspects of cost-benefit analysis, because a failure to recognize potential hazards can have manifest consequences that, if fatal to users, are disabling to the agency. Against everybody's preferences, evaluating the safety of new drugs remains a matter of medical as well as political judgment, though subsidiary questions can sometimes be settled according to a routine protocol. The demonstration of efficacy is less of a problem, because for this it has been possible to define relatively strict and objective criteria. Those criteria were designed to withstand challenges in court.[47] A statistically correct experimental design, carried out by qualified practitioners and yielding a statistically significant difference between drug and placebo, constitutes an acceptable demonstration. This definition was sometimes followed to the letter, in defiance of good sense. In the early 1970s, the FDA approved amphetamines for weight loss, despite very modest benefits and serious drawbacks such as addictiveness, on the ground that the advantages of a drug must merely be shown to be statistically significant, not clinically significant.[48]

Since the 1960s, drug testing has defined the standard of rigor in clinical trials. It is a bureaucratic and political standard as well as a scientific one. Professional support has been at best uneven: doctors were and have remained suspicious of advances in the regime of objectivity.[49] Even statisticians often insist, very plausibly, on the importance of what has been called "statistical tact" in the design and interpretation of experiments. But they have worked effectively to impose standards on clinical judgment, especially in the context of large-scale therapeutic experiments, for which their discipline supplied the discipline. In the United States, as well as Britain, this was not always specifically dictated by regulatory authorities. The complex disciplinary situation of medicine, especially the difficulty of integrating research and practice, creates its own problems of cultural distance and distrust and encourages the drive for objectivity. Still, the pioneering uses of experimental design and statistical analysis in medicine seem always to have been organized by government agencies.[50]

Since the 1940s, the use of statistical tests has become obligatory in many if not yet most areas of medical research. It is by no means my intention to deny that there are good reasons for these developments, or to claim that they reflect nothing more than the effects of bureaucratic and disciplinary politics. I am arguing, though, that they work mainly as social technologies, not as guides to private thinking. The advances of statistics in medicine must be understood as responses to problems of trust, which have been most acute in the context of regulatory and disciplinary confrontations.[51] This, and not any inherently statistical character of clinical medicine, explains why inferential statistics entered medicine through therapeutics.

MENTAL TESTING AND EXPERIMENTAL PSYCHOLOGY

"Standardizing the mind is as futile as standardizing electricity," wrote a New York judge in 1916, in denying that performance on an intelligence test could provide adequate evidence of feeblemindedness.[52] It is far from clear that most psychologists would have disagreed, at least about the mind. In 1916, the intelligence test had not yet crossed the threshold from the psychologist's office to the classroom. Mass testing was developed the next year in a bid by psychologists to contribute to, or perhaps profit from, the American mobilization for war, and it was applied to nearly two million army recruits. Until then, the intelligence test was a diagnostic tool used by physicians and by psychologists acting like physicians. To be sure, it was designed by a firm opponent of medical tact, Alfred Binet. But the test was generally administered to one patient at a time by an expert professional, and its credibility was greatly enhanced by its ability to match intuitive judgments.[53]

The standardized test, which had industrial as well as military roots, eventually found its most secure base in public schools. It was, in many ways, distinctively American. National tests have been used in many countries in the twentieth century to separate university-bound students from technical ones. But most have not aimed at anything like mechanical objectivity; the multiple-choice test, a staple of American schooling, was almost unknown in Europe until very recently. European schools had little need for mental measurement, because their *lycées*, *Gymnasia*, and "public schools" remained secure in a reasonably settled class hierarchy. Elite education, meaning instruction in dead languages, mathematics, religion, and perhaps philosophy, history, literature, and science, was assumed to provide the best preparation for a university education. Students from cultured or wealthy backgrounds were expected to master these subjects. A few less privileged youths could prove their worth, not by excelling on an aptitude test, but by succeeding in the classical curriculum.[54]

American education was less differentiated. The public schools experienced a demographic explosion from 1880 to 1910, when the number of high schools increased from about 500 to 10,000, and the number of students at that level from 80,000 to 900,000. Only a tiny minority boarded at academies. The tracking of students took place mainly within schools, not between them. Standardized testing was used above all to sort students, who arrived in a rather undifferentiated mass and had to be segregated into vocational, commercial, and academic tracks. Tests were also used to identify the "feebleminded" and, sometimes, the "gifted." Intelligence tests did not create this tracking, but provided a scientific basis for it. Or rather, they rationalized it, in more than one sense.[55]

Indeed, as Kurt Danziger points out, the schools began using group tests decades before psychologists undertook to show them how to do it better. The new educational systems, with their age-grading and standard curricula, "actually created the kinds of statistical populations that Galtonian psychology took as its basis." Danziger also shows that reliance on tests designed by psychologists was anything but a logical necessity. There were other ways to sort out students. The most obvious was to leave it to the teachers. They were far from enthusiastic about the invasion of standardized tests, designed to provide an objective alternative to their judgment. But teachers, most of them unmarried women, made up no self-confident elite or secure profession. They were largely unable to resist this onslaught of objectivity. It was sponsored by a new generation of educational administrators, most of them men, who were looking for ways to distinguish themselves from the people in the classroom. They keenly appreciated the virtues of statistical analysis, which provided "a culturally acceptable rationale for the treatment of individuals by categories that bureaucratic structures demanded."[56]

This cultural acceptability owed partly to the scientific status, or pretensions, of educational psychology. Above all, though, it meant impersonal objectivity. That objectivity was of course not absolute. The differentials in performance among racial and ethnic groups has come in recent decades to be widely regarded as evidence of unfairness or bias in the construction of the tests. At the level of individuals, though, advocates of the tests could at least claim that nobody was the victim of a hostile or ignorant examiner. Whatever the validity of the test, the scoring was free of judgment. It was perfectly mechanical, and came very early to be done by machines—this is how objectivity is defined in the world of testing. Finally, it permitted complete standardization, so that, for example, college admissions officers could readily compare a score from Wichita with one from Boston, even if they had no experience with the school or region.

Statistics entered "pure" psychology from educational testing—that is, from "applied" psychology.[57] As with medicine, the push for quantitative rigor was in the first instance an adaptation to public exposure, not the achievement of a well-insulated community of researchers. In the 1930s and 1940s, up-to-date statistics became a mark of self-consciously scientific experimental psychology. Even then, it was scarcely a sign of a secure discipline. As Mitchell Ash remarks, psychology has been more self-consciously scientific than the natural sciences precisely because of its institutional weakness and intellectual disunity. Inflexible methods of quantification compensated for the lack of a secure community. Revealingly, statistical tests in experimental psychology were pioneered by parapsychology, its weakest and least trusted subdiscipline. They were part of a regime of replication and impersonality, necessary if the study of psychical phenomena was to win even a modest degree of scientific credibility.[58]

Still, statistics was important to the internal life of the discipline, and not only to its external relations. Using the right methods of inference became a mark of professionalism, and helped to create a research identity. Firm statistical rules for designing experiments and processing data promoted disciplinary consensus by ruling out at least some of the diverse meanings that could be attached to ambiguous psychological data. A broadly statistical orientation also protected research against a distinctly psychological kind of subjectivity, that of the experimental subject. Twentieth-century American experimenters wanted general laws, not remarkable phenomena involving special persons. Finally, an insistence on experiments, yielding certain types of data that could be analyzed in a conventional way, promoted a kind of operationalism that turned attention away from theory. Broad theoretical commitments were dangerously divisive, and shared statistical methods did much to hold the field together.[59]

Because the identity of their field was so closely bound up with statistics, psychologists were almost compelled to believe in the unity and coherence of statistical method. This seemed unproblematical, since statistical reasoning fell within the territory of mathematics, the paradigm of rigor and certainty. But in fact mathematical statisticians were not unified. From early in the century they were bitterly divided, both personally and intellectually. Karl Pearson and R. A. Fisher agreed on little; Jerzy Neyman and Egon Pearson developed an alternative experimental protocol to Fisher's, and insisted on their incompatibility. Later a new level of complexity was introduced by the revival, really creation, of a more subjective Bayesian statistics. Psychologists, like most experimental scientists, preferred to remain oblivious to rival statistical programs. Their textbooks presented a synthetic, mostly Fisherian, version of experimental design and analysis, and called it simply statistics. A calculation of statistical significance, which was idealized as purely mechanical, often determined whether a researcher had a publishable result. The "null hypothesis" of no effect had simply to be "rejected" at the level of 0.05.[60]

This particular number, an indication of the probability that the results might have occurred by chance, is clearly no more than a convention. The argument of this section suggests that the methods, too, have an element of conventionality, particularly when one kind of test was deployed by almost every author—in defiance of the statisticians, who increasingly called for nuance. But in the disciplines that standardized it most severely, statistics was regarded as anything but conventional. Researchers were urged to follow statistical rules as a matter of scientific probity, and to feel guilt if, for example, they reformulated the hypothesis after the data came in.[61] Perhaps the most compelling testimony to faith in statistics as a mode of reasoning comes from a new field of psychology, the study of judgment under uncertainty. As Gerd Gigerenzer argues, psychologists became so accustomed to statistical testing that they naturalized it into a theory of thinking. In the 1950s, the mind began to be represented as an intuitive statistician, spontaneously applying analysis of variance to assign causality and signal detection theory to separate an object of interest from noise.[62]

By the 1970s, the "intuitive statistician" had failed too often to duplicate the results of calculation, and this program of research shifted to the "heuristics and biases" affecting subjective assignments of probabilities. Experimental subjects seemed not merely to calculate incorrectly, but to ignore certain kinds of data and even to make simple mistakes of logic. The errors were not limited to college students, the favorite subjects of psychological experiments, but were committed also by doctors, engineers, and graduate business students.

These findings, in turn, contributed to an ongoing debate about the relative merits of expert judgment and quantitative rules in making various kinds of practical decisions. As usual, there were political and moral as well as scientific arguments on both sides. An important source of this literature was a study of the actions of Illinois parole boards. Decisions to grant or deny parole were widely assumed to be crooked, or at least to depend heavily on family connections and clever lawyers. The Chicago sociologist Ernest Burgess prepared in 1928 a study of the factors determining whether parolees committed new crimes. He hoped to provide a superior alternative to parole board discretion—which, was not easily distinguished from corruption. In this way, he wrote, the administration of parole might "be raised above the level of guess work and placed upon a scientific basis."[63] He found the statistical rules to be superior both in the quality of their predictions and in their fairness and consistency.

More recent advocates of expert systems, too, have emphasized the moral superiority of explicit methods over unarticulated judgments. They insist also that computerized rules can predict outcomes better than human experts. There have been heated debates, with expert judgment most often defended by professionals who do not care to see their refined intelligence replaced by the artificial variety. An impressive body of evidence favors the rules. But a closer inspection reveals that the numbers work most effectively in a world they have collaborated in creating. A favorite topic of debate has been medical, especially psychiatric, diagnoses. The computer, obviously, cannot confront a real patient. In many studies, neither does the expert. Instead, both are given the numerical results of tests, along with statistical data about the rates of false positives, false negatives, and base rates. Benjamin Kleinmuntz, for example, tested the relative diagnostic abilities of psychologists and computers when given outputs from the Minnesota Multiphasic Personality Inventory (MMPI). Only a few psychologists beat the machine. There are, he conceded, situations where expert judgment is far superior to decision rules. But much of the time, physicians and other clinicians merely "process hard data from laboratory tests and other exams." In such cases they can do little more than apply decision rules, the same as machines, except that people make more mistakes.[64]

They are particularly bad at probability calculations, as the new studies of judgment under uncertainty have shown. It is not clear why professionals with graduate degrees including training in statistics should have so much difficulty solving elementary Bayesian problems. But it is no mystery why such problems do not succumb to the abilities of the "intuitive statistician" once thought to be within us all. Apart from a few games of chance—and even these are arguable—no human ever confronted a stable, quantified probability value, or even the data to construct one, before the seventeenth century. Probabilities are in every case artifacts, created (but not arbitrarily) by instruments and by well-disciplined human labor. By now, an economist, doctor, or psychologist who cannot comprehend statistical arguments involving variances and probability values will work less effectively on that account. This is not because the world is inherently statistical. It is because quantifiers have made it statistical, the better to manage it.

CAN OBJECTIVITY REPLACE EXPERTISE?

The question of whether mechanical objectivity can replace expert judgment has generally been framed as a scientific one. It raises also a political and cultural question. Is mechanical objectivity capable of replacing expert knowledge in human societies and polities?

There is no simple answer. One must begin by recognizing that the ideal of mechanical objectivity is just an ideal. Sociologists of knowledge have shown that there is an element of unarticulated expertise built into every attempt to solve problems according to explicit rules, not excluding computer analyses of quantitative data.[65] Moreover, the problem of trust can never be eliminated, nor can it be fully separated from hierarchies and institutions. Credible numbers are produced by agencies of government, university researchers, foundations, and research institutes. Coming from lobbying organizations and business corporations, they may yet be accepted, but are more likely to be scrutinized. Lay people, and even other specialists, can rarely repeat the entire operation; numbers can at best be checked for internal consistency, and compared with related numbers from other sources.

In short, it requires institutional or personal credibility even to produce impersonal numbers. If experimental reports or the numbers fed into calculations cannot be replicated at will, their authors will only be believed if they can impress readers somehow with their skill and probity. The demands on personal credibility are greatly reduced, though, if it appears that other competent people are in a position to check or recalculate some numbers, and especially if some of these people have contrary interests. In practice, objectivity and factuality rarely mean self-evident truth. Instead, they imply openness to possible refutation by other experts. Trust is inseparable from objectivity, rather like a Doppelgänger. But the form of trust supporting objectivity is anonymous and institutional rather than personal and face to face.

For most purposes, credentialed scientists are not automatically suspected of deceit or incompetence when they announce findings within their own field of expertise. If, by deferring to statistical routines, academic psychologists and medical researchers were able to convince one another, outsiders were unlikely to disrupt these knowledge claims. But even the most insecure disciplines do not act forever like societies of strangers. Through the experience of working and talking together, researchers gain confidence in one another, and acquire a more nuanced sensibility of what and whom to believe. It is possible that the imposition of uniform standards may even promote the formation of a more secure community. As this is achieved, though, the narrow definition of an acceptable analysis tends to be relaxed. Mechanical objectivity, when it succeeds, becomes less urgent. Shared knowledge can have the effect of alleviating distrust, thereby loosening the straitjacket of impersonal rules.

In the wider public domain, a different dynamic applies. When quantitative rules are supported by institutional power or credibility, as was cost-benefit analysis during its early history within the Corps of Engineers, they may be sufficient to keep a process running smoothly and even to settle minor controversies. The need for institutional support does not imply that the act of quantification itself is ineffectual; the status of the Corps was not the prime mover, the uncaused cause, but was enhanced by a reputation for impersonality and even rigor in its economic analyses. This, however, depended on an absence of powerful rivals, which is much more difficult to maintain in a contentious political culture than in a largely academic discipline. Within the Corps of Engineers, cost-benefit analysis meant something reasonably unambiguous, and helped to settle controversies. But outside a particular institutional framework, economic quantification could and did have quite different meanings. The idea of measuring costs and benefits, by itself, was much too indefinite, or at least flexible, to lead to consensus. In postwar disputes involving the Corps of Engineers, it proved impossible even to negotiate a single authorized method, though in a less contentious setting a compromise might have been reached.

Recent history suggests that the pursuit of mechanical objectivity cannot suffice to settle public issues in conditions of pervasive distrust. Brian Balogh's study of American nuclear power helps to explain why. For some years after the Second World War, commercial nuclear power remained firmly within the control of the Atomic Energy Commission, because its military ties permitted secrecy. But the construction of nuclear power plants in large numbers required the collaboration of a variety of interests and specialties. Soon the monopoly of knowledge by nuclear physicists and engineers within one agency broke down. A decade before the eruption of a vigorous public argument over nuclear power, there were debates among the various kinds of experts who considered some aspect of nuclear power as falling within their own specialty. All of them controlled reasonably credible forms of knowledge; the problem was that they couldn't agree. Their conflicts opened this highly technical field up to the larger public, thereby eliminating the possibility that agreement might be reached through quiet negotiation. No matter how rigorous its methods, a discipline cannot make convincing objectivity claims when it has strong rivals. The proliferation of specialists within the disjointed American political system, at least, means that there will almost always be rivals.[66] It will often be impossible to reach agreement without judgment and negotiation. There can be no consensus in a world of specialists, all attempting to follow strictly the rules of their own discipline—all in this sense forms of local knowledge.

Partly in response to the current political stalemate, the recent sociological critique of objectivity has been remarkably favorable to expert judgment, and indeed to the elites who wield it. Expertise is increasingly identified with skill, the basis of trust and even of community. Two examples will make the point, and suggest what is at stake. A book by Randall Albury denies that much sense can be made out of the most common meanings of objectivity: knowledge unaffected by interests and idiosyncracies; "knowledge corresponding to reality"; or "value-free knowledge." He concludes that objectivity has no deeper meaning than "knowledge produced in conformity with the prevailing standards of scientific practice as determined by the current judgements of the scientific community." That is, there is no Archimedean point on which citizens can rest a useful critique of expert knowledge. They can merely insist that the interests of the scientists somehow reflect those of the public. Beyond that, they should stop lusting after impersonal standards, and learn to trust the specialists.

The other example is a critique of cost-benefit analysis by Mark Green. The search for quantitative rules is futile, he argues. Insistence on rigorous standards of knowledge has become a strategy of opposition, used by powerful industrialists to immobilize the regulatory agencies. To reject expert judgment, then, is to abandon all hope of constructive public action. The need for effective regulation, he concludes, requires "a presumption in favor of expert agencies appointed and confirmed by the president and Congress that, after hearing all evidence in due process hearings, arrive at a judgment."[67]

To be sure, Green does not favor bureaucratic elitism for its own sake. His opposition to methods claiming objectivity, such as cost-benefit analysis, derives also from a sense that they often measure the wrong thing. As an abstract proposition, rigorous standards promote public responsibility and may very well contribute to accountability, even to democracy. But if the real goals of public action must be set aside so that officials can be judged against standards that miss the point, something important has been lost. The drive to eliminate trust and judgment from the public domain will never completely succeed. Possibly it is worse than futile.

CHAPTER NINE

Is Science Made by Communities?

In defending the scientific community's just claims to knowledge I am also defending the moral superiority of that community relative to any other human association.
(Rom Harré, 1986)

OBJECTIVITY is one of the classic ideals of science. It refers to a cluster of attributes: first among them is truth to nature, but there is also impersonality, fairness, universality, and in general an immunity to all kinds of local distorting factors like nationality, language, personal interest, and prejudice. In some idioms, the ideal of rationality and objectivity has seemed to imply a thoroughgoing individualism in science. The classic figure here is Descartes, who wanted to build up a world using only materials demonstrated to be sound by the clear light of reason, a light in principle accessible to anyone, alone. David Hollinger argues that until some time after the Second World War the ethic of individualism was dominant in American writing about science.[1] He gives the example of Martin Arrowsmith, in Sinclair Lewis's novel. Social life of every kind, even the social life inside the laboratory, is for Arrowsmith at worst a temptation to fudge his results, at best a distraction from the serious business of science. The novel leaves the hero in the woods of New England, withdrawn from all company, pursuing his laboratory research with one male companion, in splendid isolation. In this way he was protected from the temptations of power and reputation, dedicating his life to the relentless pursuit of stony truth and ignoring the social graces represented by soft and deceitful women. Such a picture of the life of science now seems a fine if unwitting joke. Yet the idea of making individual rationality the foundation of objectivity in science dies hard—and not only among unreflecting scientists and starry-eyed novelists. Most philosophers, too, have not known quite how to embrace a social conception of rationality.

Meanwhile, the notion of scientific community has become a commonplace. Partly this is a matter of the emptying of content from the concept of community: we find in everyday journalistic usage now such locutions as the "business community" and the "black community." More inspired voices have spoken of the "intelligence community" (spies) and the vaguely oxymoronic "international community."[2] But talk of community has had a real purpose to fulfill in the rhetoric of science. Postwar American defenders of science, especially Vannevar Bush, posited a scientific community in order to make science self-regulating. He wanted to shore up its boundaries and to hold off the heavy hand of government science policy. In the event that scientific method failed to keep scientists from making errors, the community would step in to sift the good from the bad. Errors would be weeded out by reviewers or fail the test of replication and be expelled from the body of scientific knowledge. Also, the community was to judge what kind of work is worthwhile, and, with a soft touch if not an invisible hand, direct the available resources to those research areas where they would do the most good. Bush argued that it could do so much more effectively as a free community than would ever be possible under a centralized bureaucracy.

These ideals also inspired some postwar sociologists and philosophers to make science into a model of community life in our time. Science, it seemed, exemplifies formal democracy leading to a real meritocracy. It carries out its business through free discussion, and yet avoids logomachy as well as naked ideology, because scientists work so hard to put their ideas to the test. This seems at first a subtle alternative to the usual pieties uttered in praise of unprejudiced truth-seeking, but it has not always been stated with an excess of reserve. A well-known philosopher claims that the scientific community is superior to every other form of human association, enforcing "standards of honesty, trustworthiness and good work against which the moral quality of say Christian civilization stands condemned." If science were mere problem-solving, leading to a "historically conditioned consensus," then "the moral claims of the community to be the guardians of a kind of purity of practice against the blandishments of careerism and the temptations of wishful thinking" would be spurious. Science "is a disinterested institution for the gathering of trustworthy knowledge." Antirealism "is not only false but morally obnoxious as a denigration of that amazing moral phenomenon."[3]

The object of this scorn, paradoxically, was Thomas Kuhn, who wrote the most influential discussion of scientific community in recent times, and who treated it sympathetically. His disciplinary communities defined the standards, the tools, the concepts, and the problems that would be regarded as valid in any particular field. He argued that serious science is done only by those who have been well socialized into a body of specialists. Kuhn's book pointed unambiguously to the importance of social factors in science. These are now widely recognized. But there remains considerable question whether the groups that perform science are really best called "communities." Ferdinand Tönnies's well-worn sociological formulation distinguished between community and society. Societies are big, impersonal, and mechanistic; communities by contrast are small, intimate, and organic. The distinction points to some key issues in current discussions of science.

It shows up especially in the recent genre of microstudies of science, studies of particular laboratories or subdisciplines. Possibly the most influential term in what is called the new sociology of science is *negotiation*. It conveys the idea that general principles, so-called universal scientific laws, are never sufficiently definite or concrete to apply to the richly detailed circumstances of experience and experiment. Hence the meaning of experiments, even of theories, cannot be settled by general principles, but must be worked out by a narrow group of specialists. In this way, large problems and broad scientific questions are brought down to issues of detail, and, at the same time, abstract matters of truth are settled through close personal contact. This need not be literally true; negotiations can also take place by letter, telephone, broadsheet, lawsuit, or, in the extreme case, even publication. Still, they must involve communities in the strong sense of the term: the relevant deliberations are small-scale, close, and informal. The word "negotiation" demands this. This language of negotiation and community also suggests an affinity between science and other, less militantly objective pursuits. Stanley Fish argues that literary criticism cannot be a matter of demonstration, but always of "endlessly negotiated" meanings. And who does this negotiation? "Interpretive communities." Interpretation, suggests Fish, is in many ways informal, even implicit; it draws on tacit understandings, shared ideologies and expectations, and a common reserve of background knowledge.[4] According to current fashions, science is this way too. We might even say that it is made by interpretive communities.

To anyone who still feels a modest residual loyalty to pre-postmodernism, though, this seems a strange way of talking about science. Science is supposed to be about nature. It is supposed to yield knowledge that is impersonal, and in some way objective. And, not to persist too stubbornly with these ironic modalities, it succeeds. Knowledge in the sciences is widely shared, to the point that the same textbooks can be used all over the world. This is often taken as decisive evidence of the moral virtues of natural science, and it is real, even if it is often exaggerated. Scientists pride themselves on appealing to nature rather than opinion, and on using a neutral language of facts and laws, numbers and the logic of quantity. The universality of scientific knowledge is by no means complete, but the most skeptical sociology readily concedes that it is impressive. Is it not to the impersonal, objective methods of quantification and experimentation that we owe the universality of science?

I have argued that in a way it is. What makes science more impersonal helps it cross the boundaries of nation, language, experience, and discipline—or in a word, of community. But it is not at all clear that austere rigor and objectivity are intrinsic to the normal practice of science. Some of the best recent historical, sociological, and cultural work on science suggests that it is not.

NEGOTIATION AND AUTONOMY

Consider Martin Rudwick's *The Great Devonian Controversy*, subtitled "the shaping of scientific knowledge among gentlemanly specialists." This is, as advertised, the story of a controversy, and its gentlemen protagonists did not earn the honorific by a conspicuous display of even-tempered grace. But they were careful about decorum. One way of maintaining it was to allow vigorous, heated discussion inside the meetings of the Geological Society, and, at least as a matter of principle, to disguise all controversy in public. For Rudwick to get inside the arguments, he had to rely heavily on private documents: field notebooks, letters, and the minutes of meetings. For, as he puts it: "The role of formal published papers in relation to informal argument during the controversy could aptly be compared with the role of occasional—and generally unrevealing—press releases during the real hard work of diplomatic negotiations behind closed doors."[5] The metaphor applies equally well to people as to processes. The elite gentlemen of science behaved like diplomats, archetypal members of a closed and aristocratic culture. Gentlemen can often dispose of public business within their private world of clubs and informal contacts. For at least a century after 1870, the upper tier of the British civil service operated in much this fashion.[6] Communities of disciplinary specialists like Rudwick's geologists have done the same thing, settling among themselves the laws of nature. And this is not merely a matter of where scientific knowledge is worked out, but also of how. Arguments within a community of specialists can be made with a minimum of formality, only a modest concern for rigor, and with frequent recourse to shared, often tacit knowledge. This is of course all the more feasible when, as in Rudwick's case, nearly everybody who mattered inhabited a single city, London.

But this provincialism, or metropolitanism, suggests an important limitation to the argument. Rudwick's assessment of the relative places of public and private discussion in science might very well be accepted for the London geologists, and yet doubted in the more general case. It is by now a commonplace of social analysis that "modern" means the breakdown of community, and the intrusion of centralized institutions into almost every aspect of local and private life. Generally speaking, I accept this. My discussions of medical statistics and cost-benefit analysis are broadly consistent with Thomas Bender's formulation of the central problem faced by twentieth-century American intellectuals: "How does one achieve intellectual authority in a society of strangers? How does one locate an intellectual community with shared purposes, standards, and rules of discourse" in a heterogeneous city, nation, or world? He argues that university disciplines were a response to the breakdown in cities of a more unified social, economic, and intellectual elite. He points out that this has tended to open politics to the relatively weak. He adds, though, that the breakdown of community is not universal, and that (even) for American elites, power and community continue often to overlap.[7]

This identity of power and community does not apply well to scientific disciplines, if power is taken to refer to the domain of politics and statecraft. When scientists participate effectively in business or political decisions, this generally requires that they be able to join other kinds of communities. Even on their own terrain, their power is far from absolute; governmental intrusions have weakened the autonomy of scientific institutions. The effect on career patterns, for example, is striking, and not only in the United States. A former president of the Deutsche Forschungsgemeinschaft reminisced in 1989:

> When I came to Göttingen in 1931, everyone there knew who were the great scientists among the professors. Nearly everyone also knew who were the best young scientists, those with a great future. . . . The great men knew each other . . . and each had an influence in his faculty which went beyond his position in his own field. This influence reached into the committees on academic appointments, and thus it helped to contrive the appointment of professors of high quality. . . . I still find no fault with this system. But I know that today it can no longer work effectively except in exceptional circumstances. It is an informal system which requires unselfishness and self-criticism on the part of its main participants. This renders it defenseless against suspicion.

Ministry officials, who are responsible in Germany for university appointments, understood well the informal system, and supported it, but no longer. Our spokesman adds that formalized measures of quality in science may now be necessary as a defense against bureaucratic agencies seeking to wrest control over science from the scientists.[8]

Still, some disciplines that do not contribute much to explicitly political discussion have been able to maintain a large degree of intellectual autonomy. Where academic boundaries are sharp, the lack of clear geographical boundaries like those that defined the social world of the

London geologists may be no obstacle to the formation of an intimate scientific community. The private networks linking members of specialties can be very tight, even today. A compelling example is the community of high-energy physicists, as depicted by Sharon Traweek.[9] Particle physics is dominated by a small and elite group of highly mobile scientists. Experimentalists, in particular, can only carry out their research by making regular visits to one of a very few leading particle accelerators. Partly on that account everybody who's anybody knows everybody else. Mutual acquaintanceship is of course not the only basis for shared values and assumptions. Another is the long process of socialization. This includes formal study of physics in undergraduate and graduate courses, then a kind of research apprenticeship that extends through a long postdoctoral period in which most budding particle physicists are weeded out. Those who survive are remarkably homogeneous, not only in scientific commitments, but even in terms of personal habits, mannerisms, and dress. Their national origins and social backgrounds are more diverse.

This intense socialization, combined with a tight network of personal contacts, permits the high-energy physicists to operate with an astonishing degree of informality. Traweek's informants portrayed the written word as scarcely more consequential or revealing in modern, high-tech, high-energy experimental physics than in gentlemanly Victorian geology. Only graduate students pay much attention to published papers; mature scientists interact mainly by talking, not writing. Postdoctoral researchers are more likely to consult preprints than published reports, because these are at least current. Even so, the preprints are valuable mainly as a guide to the field, to find who is worth talking to. Those who are really well placed learn what they need to know informally. Publication is the responsibility of ghostwriters, mainly a matter of record, and contains only what is suitable for public consumption; and premature sharing of results or suspicions with outsiders is strongly discouraged.

Experimental high-energy physicists deploy the formalities of standard scientific prose in their writing, but much of it is not taken seriously. This does not mean that anything goes, or that physicists use mathematics only as a set of conventions. They aspire to get at a world that is more fundamental and less transitory than the one with which people such as historians and administrators content themselves. Partly on this account, they appreciate the impersonality of mathematics and like to make a joke of quantifying the ineffable and the personal. But in practice they do not believe that methodological rigidity is the best strategy for learning about their timeless world.

Traweek's informants told her that error bars are always informally multiplied by at least 3, as a rule of thumb, to decide whether somebody's results are likely to mean anything. More crucially, this mode of interpretation is highly nuanced and bound up with private knowledge. Researchers who are not known for conscientious, painstaking attention to detail will need even more decisive results to be taken seriously. And in fact informal judgments of character and reliability are crucial for interpreting their experiments. In other sciences, where equipment is relatively permanent or standardized, something like replication is always possible. But the particle physicists construct their own detectors, tinker with them constantly, adjust them and even rebuild them for new experiments. Hence it is exceptionally difficult to check the results of an experiment, and there is no alternative to judgment in determining how much faith to place in any particular report. Trevor Pinch makes the point explicitly in reference to a huge solar-neutrino detection experiment, which involved millions of gallons of chemicals in a deep underground cave. Nobody expected that this experiment would be performed twice. To interpret it, then, physicists had to assess the skill and trustworthiness of the experimenters. "In short, trust within science functions in a way similar to that in which it functions in any area of life involving human skills."[10]

Pinch implies that this is universal, that all of science depends on judgments of character and skill. No doubt he is right. But rarely is informal, personal knowledge so dominant as among the gentlemanly geologists and high-energy physicists. There are ways of making knowledge more rigid, standardized, and objective, and these have gone a long way to reduce the need for personal trust. Mathematics and quantification are of course not uniquely responsible for the increasing uniformity of knowledge, but their contribution has been impressive. Rigorous standards of mathematical proof, measurement systems, mathematical methods of statistics, and demographic, economic, and social numbers have been allies in a campaign to make knowledge more open and uniform.

Of course, not everybody has wanted open, uniform knowledge. The subjective form of measurement, discussed by Witold Kula, suited communities of peasants, and physicists, just fine. There were disagreements, but they could be negotiated face to face. Informal measurement was inseparable from the fabric of these relatively autonomous communities. It broke down with the intrusion of more centralized forms of power—both political and economic—into the relatively private domain of communal life. Relative autonomy, and frequent face-to-face interactions were also characteristic of the London geologists, and even of experimental particle physicists. The physicists are not spread over the globe, but share occupancy of a few islands. Their work has, at least until very recently, been so prestigious that they have had little responsibility except to each other. They have suffered a minimum of intervention by powerful outside interests. The physicists have wanted nothing from the government but money, and the government has, since the war, been content with the physicists' own marks of esteem, such as Nobel prizes. So they have been free to cultivate their own style, language, and traditions.

To be sure, these are not at all free of what is standardized and routinized. Objectivity is still required in some things, especially things of subordinate interest. It is clearly needed, for instance, in the initial screening of photographs and data. Until recently these were so abundant that relatively low-status labor had to be enlisted to do it; and low status labor was managed by imposing a relatively rigid standardization. Increasingly, computers have been made to filter out signals from noise, with almost complete reliability and an absolute lack of good sense or judgment. So when we say that most of what goes on in experimental particle physics is makeshift and negotiable, we are referring to the high-status work of the researchers themselves and of a few trusted technicians, resting securely atop a pyramid of objectified grunt work.[11]

We seem now to have reached the point where science is identified with a negotiated, local, private kind of knowledge, or really of skill, since the word "knowledge" generally presupposes a more rationalized form. This has the nice feature of being counterintuitive, but not everything that seems wrong is therefore true. We may recall that when the state obliterated all the local weights and measures, it did so by imposing the metric system, which was worked out by scientists. Science has been even more closely involved with other efforts to standardize instruments and measures, involving electricity, temperature, and forms of energy. Congresses of scientists, some of them of the very top rank, defined the basic units of electricity in the late nineteenth century. Almost every industrialized nation has a bureau of standards, staffed mainly by scientists and engineers. It was in the name of science that strict rules of calculation were defined to govern the decisions of actuaries and public engineers. There is more than a little reason to put science on the side of the state, the objectifiers and intruders, who imposed a more uniform, open language and drove out local customs and implicit conventions. There is something deeply right about Bruno Latour's phrase "center of calculation" to describe the point from which empires are administered, and about his emphasis on this in a book about technology and science.[12]

The role of science in standardizing and objectifying political and economic life is one of the most important reasons that humanists and social scientists should be concerned with it. But we may still be a little uncertain just why science has played this role, and for that matter why scientific knowledge itself is normally expressed in a highly objectified, rationalistic language. In this book I have emphasized two lines of response to this, though at bottom they are almost the same. One reflects the broader social and political relations of science, pressures from the outside. The other has to do with the social life of scientists in their own institutions, and the difficulties of forming a community of belief and practice.

STRONG AND WEAK COMMUNITIES

The remarkable openness and lack of rigid rules in high-energy physics is possible only under very special circumstances. It must be added that high-energy physics is scarcely a model of anti-objectivism in science. Every body of scientists, every disciplinary grouping, is subjected to strong pressures tending to the confinement of judgment in favor of mechanical and impersonal standards. Thomas Hobbes, the hero of Steven Shapin's and Simon Schaffer's book *Leviathan and the Air-Pump*, identified the problem clearly. Experiment, he held, provides no basis for acquiring public knowledge. Experiment is intrinsically private. Any particular one can only be properly witnessed by a few people. It is always possible to criticize an experiment by placing in the foreground details of construction and execution that, according to the logic of the experimental demonstration itself, must be relegated firmly to the background. Hobbes himself tried in this way to deconstruct, as we might say, Robert Boyle's air pump: he claimed it leaked, and pointed to the trials that didn't work as Boyle expected. Experimentation is futile, he suggested. The only firm basis for public knowledge, indeed for organizing a polity, is geometrical reasoning: solid demonstration, which brings its own evidence with it and depends on nothing more than writing on paper.[13]

Although his attack on experimentation failed spectacularly, the problems Hobbes identified were real. The products of special instruments and well-honed skills are not easily made a basis for public knowledge. Experimentalists developed a range of strategies for overcoming them. Some depended on forms of prestige that were widely accepted in the larger society. In the seventeenth century they drew on gentlemanly or courtly social codes, advertising dignity and independence or relationships to powerful patrons as a mark of disinterestedness. Men who had achieved a certain status, men who were above material concerns, presumably had no incentive to deceive. Other technologies of trust took a form that we can easily recognize as aspects of a community. Scientists drew professional boundaries, and excluded those outside them as amateurs, cranks, or charlatans. Degrees and professorships became signs of competence and integrity. For gentlemen and professional scientists alike, telling the truth was a matter of honor, and while it was always possible to doubt testimonies, this was not to be done lightly.[14]

To these formal evidences of integrity and competence were added personal ones. Insiders communicated regularly and intensely. The French and English capital cities, in particular, were attractive enough to permit many of the most important natural philosophers to deal with one another frequently and face to face. Where this was not possible, and especially across national boundaries, scholars and scientists formed a "republic of letters," maintained in the eighteenth and nineteenth centuries through extensive correspondences that often mixed the personal with the scientific.[15] Publication was ostensibly less personal, since it made knowledge available to almost everybody, but journals have often helped to define a more intimate kind of community. Many nineteenth-century journals were house organs, publishing mainly the research of a professor, his students, and other close associates. Even when this was not the case, to have a paper appear in a journal was often to be accepted into a club, so that the decision to publish it might involve explicit attention to the author's personal character. Still in the twentieth century, readers' reports suggest that the habits, methods, and backgrounds of the researchers themselves are judged along with their work.[16] It seems unlikely that recent attempts at anonymous reviewing have much diminished this reliance on the personal dimension. It now includes conferences, colloquia, and a kind of exogamy involving graduate students, junior faculty, sabbatical visits, and, especially, postdoctoral researchers. A gift culture based on the exchange of samples and techniques between laboratories greatly facilitates belief as well as replication, while relations within laboratories have sometimes involved the most intense mixing of private and professional, as a way of building trust.[17]

An equally—perhaps more—important basis for making shared knowledge was what I am calling "objectivity." Various strategies were cultivated to make knowledge less crafty and more open. A common device in early modern science was witnessing—turning judgment about the validity of an experiment into something like a court action. This was especially important in contexts where a real community had not yet formed itself. There were also outbreaks of witnessing in relatively well-established communities when consensus broke down. Lavoisier's experiments on oxygen, for example, were the occasion for a crisis in the German chemical community. It was especially troubling that various attempts at replication led to sharply different results. Experimental competence and integrity had been presumed as a standard of membership in the community of chemists, and it caused acute embarrassment that respected researchers reported contradictory outcomes of so crucial an experiment. There followed an epidemic of witnessings, which reached a peak when the pro-French Hermbstaedt listed thirteen of them, including chemists, counts, and doctors. This new fashion of parading witnesses, wrote an opponent of Hermbstaedt (and of Lavoisier), suggests that "no chemist trusts another anymore."[18] Which was true. Something like judicial objectivity became obligatory to heal the rifts in community, and occasionally does so even now. It should be clear that a community founded on objectivity is a weak or endangered community, one without sharp borders on the outside, and one without an effortless shared understanding—in short, a very modern kind of community.

The more customary strategies for communicating results to other expert researchers at a distance are reasonably familiar. Central among them is reporting experiments in sufficient detail that, in principle, other scientists in the same field, or community, ought to be able to replicate them. This in turn demands that one generally not announce results until a certain degree of experimental mastery has been achieved. That is, the demands of communication helped to define the subject matter of science, and to filter out most of what depended too much on faith in unreplicable observations. It also imposed constraints on what would count as acceptable methods and instruments. Late eighteenth-century British chemists regarded Lavoisier as uncollegial for relying on so elaborate and expensive an apparatus, since this had the effect of excluding the less well endowed from a natural philosophical conversation.[19]

Still, replication is not at all easy, even with the benefit of the modern standardization of measures and mass production of instruments. Notwithstanding the attractive solidity and impersonality of a rhetoric of "experimental fact," it may often be easier to form a community of theoretical than of experimental investigation. Theory, especially of a mathematical sort, has at least the virtues mentioned by Hobbes: the reasoning is explicit, and what appears on the printed page is largely self-contained. Theoretical agreement contributes greatly to the stability of experimental communities. There are also scientific communities committed mainly to mathematical theory: for example, eighteenth-century rational mechanics, or modern neoclassical economics. Both have been abstract and unworldly in order to be rigorous; and one advantage of mathematical rigor is that it helps to form and preserve scientific communities concerned with phenomena that are not yet well controlled in the laboratory or observatory, or with understandings that are contested.

Something like the rigor of mathematized theory is attainable in other ways. Uniform strategies of quantification and dictates on scientific method are among the most important. Normal science, wrote Kuhn, has little need of rules so long as paradigms are secure.[20] But new, weak, and exposed fields must often do without widely shared assumptions and meanings. Explicit standards address uncertainty and variability by attempting to control and standardize people. So also do stereotyped forms of presentation. In many fields, researchers are instructed to present in a prescribed order, almost according to a formula, their methods, results, and conclusions. American psychologists are the standard-bearers in this regard. Their handbook has grown to hundreds of pages on the points of style and rhetoric that define an acceptable research paper. Rigid insistence on statistical tests, now common in a wide range of scientific, social-scientific, and medical fields, is a related way of standardizing people, organizing a discourse, and imposing values that promote scientific unity, even if they may occasionally stand in the way of understanding the phenomena.[21]

Aspiring scientists are not alone in this. Nineteenth-century professionalizing historians, at least in America, redefined their field in terms of well-authenticated facts, preferably discovered in archives, to distinguish themselves from gentlemanly amateurs and to provide a basis for reaching consensus. Franklin Jameson feared the unrepresentativeness of documents required to do social history, and for that reason preferred to limit the discipline to political matters, where the documents were standard and the methods seemed clear. Implicitly, and sometimes explicitly, it was regarded as better to maintain objectivity at the cost of narrowness than to risk fracturing the discipline over large, intractable questions.[22]

I am suggesting here that the relative rigidity of rules for composing papers, analyzing data, even formulating theory, ought to be understood in part as a way of generating a shared discourse, of unifying a weak research community. Objective rules are like the witnesses brought forth by German chemists in a time of searing controversy; they serve as an alternative to trust. Results are to be assessed according to a protocol that is to be as nearly mechanical as possible. There should be little room for personal judgment, and hence also a minimum of opportunity for others to doubt the analysis.

INDISTINCT BOUNDARIES AND POWERFUL OUTSIDERS

These considerations go a long way toward explaining why the rules of right reasoning have generally been most explicitly defined and most rigorously enforced in weaker disciplines. Methodological strictness serves as an alternative to shared beliefs and as a check on the expression of idiosyncratic personal opinions. But an examination of disciplines from the inside can tell only part of the story. The insistence in scientific communication on objectivity and impersonality is partly also a response to pressures from outside. Or rather, mechanical objectivity is especially prominent where inside and outside are not sharply differentiated. Applied fields, at least those that bear on matters of policy, are almost always exposed to scrutiny and criticism by the interests they affect. As more and more science is supported by the state for practical purposes, the category of the "applied" field has expanded to include most research. Public responsibility, if it is even mildly enforced, breaks down the boundaries around the research community and makes it necessary to satisfy a larger audience.[23]

Such a situation encourages the greatest extremes of standardization and objectivity, a preoccupation with explicit, public forms of knowledge. This is naturally most evident where knowledge is to be shaped for policy purposes, as with the cost-benefit analysts discussed in chapter 7. But the boundary between public and private in science is increasingly threatened. As Gerald Holton remarks, with pardonable exaggeration, in America the laboratory notebook is being transformed into a repository of self-protective bookkeeping, since one never knows when the Secret Service might be called in to examine scientific results that some congressman doubts. The National Academy of Sciences has accepted the principle that scientists should declare their conflicts of interest and financial holdings before offering policy advice, or even information, to the government.[24] And while police inspections of notebooks remain exceptional, the personal and financial interests of scientists and engineers are often considered material, especially in legal and regulatory contexts.

Strategies of impersonality must be understood partly as defenses against such suspicions, and against their expansion into still more contexts. They generally take the form of objectivity claims. Objectivity means knowledge that does not depend too much on the particular individuals who author it. It provides no defense against critics who may wish to attack the foundations of an entire field. It does, however, tend to undermine claims that in this particular instance someone has bent knowledge to their own advantage or arbitrarily treated another unfairly. In a democratic political culture based above all on interests, such arguments often constitute the greater threat to the credibility of an applied discipline.

This is why the language of objectivity has been most compellingly attractive to people like intelligence testers, applied social researchers, and cost-benefit analysts. We find there a pervasive dread of "the prejudice of the investigator," often a willingness to leave untouched the most important issues in order to deal objectively with those that can be adequately quantified. Thus a member of Herbert Hoover's research committee on social trends wrote: "To safeguard the conclusions against bias, the researchers were restricted to the analysis of objective data. Since the available data do not cover all phases of many subjects studied, it was often impossible to answer questions of deep interest."[25] Army engineers congratulated themselves that important "intangibles" were omitted from water project justifications because they could not reliably be expressed in figures. Economists have often warned of the chaos that will ensue if the pure objectivity of quantitative rules is compromised by use of mere judgment. Expert judgment cannot easily be distinguished by outsiders from personal bias, and the two are often conflated. The solution is to ban subjectivity. Rules must rule, even if accepted truths must be supplemented by conventions. As Dickens's character Thomas Gradgrind put it: "Facts alone are wanted in life."

Naturally, not all researchers seek to absolve themselves of responsibility and divest themselves of judgment in favor of a parade of facts. One recent American commentator on social science in government argues forcefully that comprehensiveness should never be sacrificed in the interest of quantitative rigor. But he concedes that this is likely to weaken the impact of social science, and argues finally "that the scientific considerations outweigh these political arguments." Not science, but politics, demands narrow rigor.[26]

Is science made by communities? The answer is certainly yes. Who would dare deny it now? But that is only a very unsatisfactory part of the answer. In only a few disciplines is the dynamic of research activity so self-contained that interactions within the community are mainly responsible for the forms of approved knowledge. And in such fields, fields dominated by a relatively secure community, much of what we normally associate with the scientific mentality—such as an insistence on objectivity, on the written word, on rigorous quantification—is to a surprising degree missing. Scientific knowledge is most likely to display conspicuously the trappings of science in fields with insecure borders, communities with persistent boundary problems. That is, one has to look at a wider context for science to understand even the accepted forms of scientific production, the standards by which work is judged. So, science is indeed made by communities, but communities that are often troubled, insecure, and poorly insulated from outside criticism. Some of the most distinctive and typical features of scientific discourse reflect this weakness of community. The enormous premium on objectivity in science is at least partly a response to the resultant pressures.

Perhaps science does after all provide a model of democratic community, as the postwar sociologists hoped. But it is also a mirror of really existing political societies. This congruence goes a long way toward explaining the prestige of the scientific form of knowledge in modern public life. We have here not the stable, organic *Gemeinschaft*, but the impersonal and suspicious *Gesellschaft*, requiring a form of knowledge that, in important ways, is genuinely public in character.

Notes

The following abbreviations are used in the notes:

A.N.: Archives Nationales, Paris
BENPC: Bibliothèque de l'Ecole Nationale des Ponts et Chaussées, Paris
N.A.: National Archives (of the United States)
USGPO: U.S. Government Printing Office

Preface

1. Richards, *Mathematical Visions.*

Introduction
Cultures of Objectivity

Epigraph from Hofstadter, *Gödel, Escher, Bach*, 43–45.

1. On the meanings of objectivity, see Megill, "Introduction: Four Senses of Objectivity"; Daston and Galison, "Image of Objectivity"; Daston, "Objectivity and the Escape from Perspective"; Dear, "From Truth to Disinterestedness." For a philosophical consideration of objectivity and its alternatives, see Rorty, *Objectivity, Relativism, and Truth.*

2. Greenawalt, *Law and Objectivity.*

3. Porter, "Objectivity and Authority." Of course, even computer packages require choices, no single one of which may be unambiguously correct. And good social researchers generally have to edit their raw data.

4. Haskell, *Emergence of Professional Social Science.*

5. Church, "Economists as Experts"; Dumez, *L'economiste, la science et le pouvoir*, chaps. 3–4. See also Gigerenzer et al., *Empire of Chance.*

6. Polanyi, *Personal Knowledge.* Polanyi tended to idealize this "tacit dimension" because it made scientists seem more human and science a less mechanical activity. As Steve Fuller points out in "Social Epistemology," however, this has the decidedly undemocratic consequence of putting science wholly under the control of insiders.

7. One thinks naturally of the formerly communist countries of eastern Europe. But these also involve some crucial differences from political democracies, which I have not tried to address in this book.

8. Keyfitz, "Social and Political Context."

Chapter One
A World of Artifice

Epigraph from Abir-Am, "Politics of Macromolecules," 237.

1. Hacking, "Self-Vindication."

2. On its history see Burnham, "Editorial Peer Review," and Knoll, "Communities of Scientists."

3. Emblematic of the shift of interest in science studies from theories to practices is Pickering, *Science as Practice.*

4. Polanyi, *Personal Knowledge*, 207.

5. Collins, *Changing Order.*

6. Quoted in Heilbron and Seidel, *Lawrence and his Laboratory*, 318.

7. Polanyi, *Personal Knowledge*, 55.

8. Ibid., 31; Collins, *Artificial Experts.*

9. Funkenstein, *Theology and the Scientific Imagination*, 28, notes that "unequivocation" has been one of the enduring values of Western science and Western law since ancient times.

10. Hannaway, "Laboratory Design."

11. Lorraine Daston, "The Cold Facts of Light and the Facts of Cold Light," conference paper at UCLA workshop, February 1990, forthcoming in Daston and Park, *Wonders of Nature.*

12. Schaffer, "Glass Works."

13. Shapin and Schaffer, *Leviathan and the Air-Pump.*

14. Schaffer, "Late Victorian Metrology."

15. Shapin and Schaffer, *Leviathan and the Air-Pump.*

16. Hacking, *Representing and Intervening*; idem, "Self-Vindication."

17. Latour, *Science in Action*; idem, *Pasteurization of France*; quote from Zahar, "Role of Mathematics," p. 7.

18. Hudson, *Cult of the Fact*, 55–56; Gillispie, "Social Selection." On gender there is now a huge literature; the classic work is Keller, *Reflections on Gender and Science.*

19. Feldman, "Late Enlightenment Meteorology"; Heilbron, *Electricity in the 17th and 18th Centuries.*

20. Gillispie, *Edge of Objectivity*, chap. 5; Terrall, "Maupertuis and Eighteenth-Century Scientific Culture"; Jungnickel and McCormmach, *Intellectual Mastery of Nature*, vol. 1, 56.

21. Gillispie, *Science and Polity in France*, 65. On Condillac, see Rider, "Measures of Ideas," and, more generally, Foucault, *Order of Things.*

22. Roberts, "A Word and the World," and Heilbron, "Introductory Essay," in Frängsmyr, *Quantifying Spirit.*

23. Horkheimer and Adorno, *Dialektik der Aufklärung*, 11.

24. Cartwright, *Nature's Capacities and Their Measurement.*

25. Heilbron, "Fin-de-siècle Physics."

26. Biagioli, "Social Status of Italian Mathematicians."

27. Pauly, *Controlling Life.*

28. Porter, "Death of the Object"; Heidelberger, *Innere Seite der Natur.*

29. Pearson, *Grammar of Science*, 12, 77. On Pearson's use of method talk to expand the domain of science, see Yeo, "Scientific Method."

30. Pearson, *Grammar of Science*, 203–204, 353, 260.

31. For a fuller discussion, see Porter, "Death of the Object."

32. H. Johnson, *Order upon the Land.*

33. Duncan, *Social Measurement*, 36.

34. The seasonal cycle mentioned here reflects the experience of New En-

gland Indians: see Cronon, *Changes in the Land*, chap. 3; Merchant, *Ecological Revolutions.*

35. Landes, *Revolution in Time*, chap. 3.

36. Thompson, "Time, Work Discipline, and Industrial Capitalism"; O'Malley, *Keeping Watch.*

37. On the persistence of radically heterogeneous cultures of quantification, see Lave, "Values of Quantification."

38. Kula, *Measures and Men*, 39.

39. Ibid., 22.

40. Heilbron, "The Measure of Enlightenment"; Schneider, "Maß und Messen"; idem, "Forms of Professional Activity." On the difficulty of becoming numerate in an age of multifarious, nondecimal measures, see Cohen, *A Calculating People.*

41. Heilbron, *Dilemmas of an Upright Man*, 53–54; Mirowski, "Looking for Those Natural Numbers."

42. Kula, *Measures and Men*; Heilbron, "Measure of Enlightenment"; Alder, "Revolution to Measure."

43. Smith and Wise, *Industry and Empire*, chap. 20; Schaffer, "Late Victorian Metrology."

44. Cahan, *Institute for an Empire.*

45. Lundgreen, "Measures for Objectivity," 94, 45.

46. Friedman, *Appropriating the Weather*, 62–66; Latour, *We Have Never Been Modern*, 113.

47. Hunter, "National System of Scientific Measurement," 869.

48. Liebenau, *Medical Science and Medical Industry*, 6–8, 21, 41.

49. Hatcher and Brody, "Biological Standardization of Drugs," 361, 369, 370.

50. Burn, "Errors of Biological Assay," p. 146; on digitalis, see Stechl, "Biological Standardization of Drugs," 132–149. Methods were also subdivided by duration: 1-hour, 4-hour, and 12-hour.

51. Stechl, "Biological Standardization," chap. 9.

52. Miles, "Biological Standards."

53. League of Nations, Health Organisation, *The Biological Standardisation of Insulin* (Geneva: League of Nations, April 1926); quotes from Henry H. Dale, "Introduction," 5–8.

54. Miles, "Biological Standards," 289, 287.

55. Burn et al., *Biological Standardization*, 5.

Chapter Two
How Social Numbers Are Made Valid

Epigraph from Dupuit, "De la mesure," 375. This chapter draws from my "Objectivity as Standardization."

1. On the census, see Alonso and Starr, *Politics of Numbers*; Anderson, *American Census.*

2. Converse, *Survey Research in the United States*, 138, 194, 267–304; Flem-

ing, "Attitude: The History of a Concept," Porter, "Objectivity as Standardization," from which a few of the sentences here have been copied.

3. First quote from the 1903 edition of the international nomenclature of disease; second from a paper by Constance Perry and Alice Dolman; both in Fagot-Largeault, *Causes de la mort*, 204, 229.

4. Bru, "Estimations laplaciennes."

5. Bourguet, *Déchiffrer la France*, 216; also Brian, *Mesure de l'état*, part 2.

6. Balzac, *Le curé de village*, chap. 12, at 131.

7. Porter, *Rise of Statistical Thinking*, chap. 2; Hacking, "Statistical Language."

8. Himmelfarb, *Poverty and Compassion*, 41.

9. Hacking, *Taming of Chance*, 47; Tompkins, "Laws of Sickness and Mortality"; "Editorial Note," *Assurance Magazine*, 3 (1852–53), 15–17, at 15.

10. *Report from the Select Committee on the Friendly Society Bill*, Parliamentary Papers, 1849, XIV, 1, testimony by William Sanders, 43–56, at 46. Accident statistics also varied depending on forms of intervention: see Bartrip and Fenn, "Measurement of Safety."

11. Trebilcock, *Phoenix Assurance*, 605.

12. Dickens, *Martin Chuzzlewit*, chap. 27, at 509–510. He also let his readers know that there was no paid-up capital, that the company put false numbers in its advertisements, and that its premiums were too low. The shadiness of the enterprise was most glaringly evident from the character of its officers.

13. Babbage, *Institutions for the Assurance of Lives*, 125.

14. Trebilcock, *Phoenix Assurance*, 211–212, 419, 552.

15. Ibid., 607–608.

16. Supple, *Royal Exchange Assurance*, 176–177, 99.

17. *Report of the Select Committee on Joint Stock Companies, together with the Minutes of Evidence*, British Parliamentary Papers, House of Commons, 1844, VII, 147–148.

18. Ward, "Medical Estimate of Life," 252, 338, 336; reprinted from *American Life Assurance Magazine*. In this section I have drawn from Porter, "Precision and Trust."

19. Desrosières and Thévenot, *Catégories socioprofessionelles*, 39.

20. Peterson, "Politics and the Measurement of Ethnicity."

21. Desrosières, "How to Make Things Which Hold Together"; Desrosières, "Specificités de la statistique"; also Desrosières and Thévenot, *Catégories socioprofessionelles*; Boltanski, *Les cadres*.

22. Thévenot, "La politique des statistiques."

23. Anderson, *Imagined Communities*; Revel, "Knowledge of the Territory"; Patriarca, *Numbers and the Nation*.

24. Zinoviev, *Homo Sovieticus*, 96.

25. Adorno, "European Scholar in America," 347, 366.

26. Quote from Miller and O'Leary, "Accounting and the Governable Person," 253; Chandler, *Strategy and Structure*; Johnson, "Management Accounting."

27. Hopwood, *Accounting System*, 2–3.

28. Richard Brown of the National Wildlife Federation, cited in Caufield,

"The Pacific Forest," 68; also Hays, "Politics of Environmental Administration," 48.

29. Miller and O'Leary, "Accounting and the Governable Person"; Miller and Rose, "Governing Economic Life"; Rose, *Governing the Soul.* Two classic works concerned with accounting in the multicentered firm are Brown, *Centralized Control*; Sloan, *My Years with General Motors.*

30. Sewell, *Work and Revolution.*

31. For example, the Bank of England, which created "moving averages" in the early nineteenth century as a strategy of deception when Parliament required it to reveal its holdings of specie; see Klein, *Time and the Science of Means.*

32. On eighteenth-century information, see Brown, *Knowledge Is Power.* It would be rash to generalize too widely from Brown's observations on colonial America, yet much of this rings true for Europe as well.

33. Palmer, *Age of the Democratic Revolution*; Habermas, *Structural Transformation of the Public Sphere.*

34. Cronon, *Nature's Metropolis*, chap. 3. See also my "Information, Power, and the View from Nowhere."

Chapter Three
Economic Measurement and the Values of Science

Epigraph from Popper, *Open Society*, 22. This chapter is based on my "Rigor and Practicality."

1. Feldman, "Applied Mathematics."

2. Terrall, "Representing the Earth's Shape"; Greenberg, "Mathematical Physics."

3. Latour, *Science in Action*; Desrosières, *Politique des grands nombres.*

4. See Legoyt, remarks, 284. On insurance, see Duhamel, "De la necessité d'une statistique."

5. Hollander, "Whewell and Mill on Methodology."

6. Whewell to Jones, July 23, 1831, in Todhunter, *Whewell*, vol. 2, 353, 94. Whewell's negative intentions are also clear from two letters of 1829 to Jones, quoted in Henderson, "Induction," 16.

7. Whewell review of Jones, *Essay*, 61.

8. Whewell, "Mathematical Exposition," 2, 32.

9. Wise, "Exchange Value."

10. Jenkin, "Trade-Unions," 9, 15.

11. Jenkin, "Graphic Representation," 93, 87.

12. Jenkin, "Incidence of Taxes."

13. Babbage, *On the Economy of Machinery and Manufactures.*

14. Lexis, "Zur mathematisch-ökonomischen Literatur," 427.

15. Wise, "Work and Waste."

16. Ibid., 417.

17. Quote (1855) from Wise and Smith, "The Practical Imperative," 245.

18. Quoted in Wise, "Work and Waste," 224.

19. See Perrot's introduction and text in Lavoisier, *De la richesse territoriale de France.*

20. Fox, "Introduction" to Carnot, *Reflections.*

21. Grattan-Guinness, "Work for the Workers"; Grattan-Guinness, *Convolutions in French Mathematics*, chap. 16. Comparative measurements of human and machine labor power go back to the beginning of the eighteenth century, especially in France; see Lindqvist, "Labs in the Woods."

22. F. Caquot, quoted in Divisia, *Exposés d'économique*, x.

23. Picon, *L'invention de l'ingénieur moderne*, 396, 452–453.

24. Quoted in Fourcy, *Histoire de l'Ecole Polytechnique*, 350.

25. On the Laplacian reforms, see chapter 6. Crépel, *Arago*, reconstructs the course from student notes. See also Grison, "François Arago."

26. Picon, *L'invention de l'ingénieur moderne*, e.g., 346; C. Smith, "The Longest Run."

27. See Navier, "Comparaison des avantages."

28. See Léon's review of Chevalier, *Travaux publics de la France.*

29. For example, Coriolis, "Durée comparative de différentes natures de grés"; Reynaud, "Tracé des routes," who, however, concluded that the formulas connecting grades with costs of operation were too imperfect to be relied upon, and that informal techniques of quantification were best.

30. Dupuit, *Titres scientifiques*, 3–10 (copy in Bibliothèque Nationale, Paris); also Tarbé de Saint-Hardouin, *Quelques mots sur M. Dupuit*, and Dupuit Dossier 10, Correspondance, II (uncatalogued), both in Bibliothèque de l'Ecole Nationale des Ponts et Chaussées (BENPC), Paris.

31. Dupuit, "Sur les frais d'entretien des routes," 74. He was criticized by Garnier, "Sur les frais d'entretien des routes." Dupuit's arguments for systematic maintenance were not original, though his quantitative treatment was. See Etner, *Calcul économique en France*, chap. 2.

32. Jullien, "Du prix des transports"; Ribeill, *Révolution ferroviaire*, 87–101.

33. Belpaire, *Traité des dépenses d'exploitation aux chemins de fer*, 26.

34. Ibid., 577–578.

35. Navier, "De l'exécution des travaux publics." See Etner, *Calcul économique en France*, and Ekelund and Hébert, "French Engineers."

36. Kranakis, "Social Determinants of Engineering Practice," 32. Much later, in 1887, Veron Duverger characterized state engineers as "those theoreticians . . . who exhibit a marked tendency toward oppressive regulation and a mathematical tyranny so absolutely contrary to the spirit of commercial and industrial enterprise." Quoted in Elwitt, *Making of the Third Republic*, 150.

37. Kranakis, "Affair of the Invalides Bridge"; Picon, *L'Invention de l'ingénieur moderne*, 371–384. It was Marc Seguin, victim of Navier's high-handedness in 1822, who put up the replacement bridges. See Gillispie, *Montgolfier*, 154–177, on the Seguin family and bridge building.

38. Dupuit, "Influence des péages," 213.

39. Dupuit, *Titres scientifiques*, 31; Dupuit, *La liberté commerciale*, 230; Dupuit, "Mesure de l'utilité des travaux publics."

40. Bordas, "Mesure de l'utilité des travaux publics," 257, 279.

41. Dupuit, "Influence des péages," 375; Dupuit, "Mesure de l'utilité des travaux publics," 342, 372; Dupuit, "De l'utilité et de sa mesure," in *De l'utilité*, 191.

42. Elwitt, *Third Republic Defended*, 51; Cheysson quote at 67.

43. There was a continuous though relatively inconspicuous tradition of energeticist economics dating from about the 1870s. For the most part it was deliberately subversive of the classical mainstream. See Martinez-Alier, *Ecological Economics.*

44. Cheysson, "Cadre, objet et méthode de l'économie politique," 48.

45. Cheysson, "Statistique géométrique," On graphical methods in French engineering, see Lalanne, "Tables graphiques."

46. Divisia, *Exposés d'économique*, 101.

47. Mirowski, *More Heat than Light.* For a more favorable view, see Schabas, *A World Ruled by Number.*

48. Etner, *Calcul économique en France*, 199, 238–239.

49. Ménard, *Cournot*, quote at 12; Dumez, *Walras*; Ingrao and Israel, *The Invisible Hand.*

50. L. P. Williams, "Science, Education, and Napoleon I," 378.

51. Shinn, *Savoir scientifique et pouvoir social.* On Laplace as scientific empire-builder, see Fox, "Rise and Fall of Laplacian Physics."

52. Dhombres, "L'Ecole Polytechnique," 30–39.

53. Zwerling, "The Ecole Normale Supérieure," proposes 1840 as the date when the Ecole Normale surpassed the Ecole Polytechnique as the place to get an education in research science.

54. Ménard, *Cournot*, 63–64.

55. Ibid., 44, 93–110, 139, 200.

56. Ibid., 5, 15.

57. Cournot, *Théorie des richesses*, 22–25.

58. Cournot also wrote on measuring the work of machines: see Ménard, "La machine et le coeur," 142.

59. Walras to Cournot, March 20, 1874, letter 253 in Jaffé, *Correspondence of Léon Walras.*

60. Walras to Ferry, March 11, 1878, letter 403 in ibid. See also letter 444 to Ferry.

61. Letters from Hippolyte Charlon, September 22, 1873, and to Charlon, October 15, 1873, numbers 234 and 236 in ibid.

62. Charlon to Walras, January 30, 1876, letter 347 in ibid.

63. Ibid., vol. 3. The correspondence with Laurent begins in 1898 with letter 1374.

64. Laurent, *Petit traité d'économie politique.*

65. Jaffé, *Correspondence*, letter 1380.

66. Laurent, *Statistique mathématique*, iv, 1.

67. Laurent to Walras, November 29, 1898, and reply December 3, 1898, letters 1374 and 1377; also Walras to Georges Renard, July 1899, letter 1409, all in Jaffé, *Walras.* On the Circle (later Institute) of Actuaries, see Zylberberg, *L'économie mathématique en France.*

68. Alcouffe, "Institutionalization of Political Economy."

69. Renouvier to Walras, May 18, 1874, letter 274 in Jaffé, *Walras.*

70. See Walras to Jevons, May 25, 1877, letter 357 in ibid.

71. Newcomb, *Principles of Political Economy*; Moyer, *Simon Newcomb.*

72. Quoted in Wise and Smith, "Practical Imperative," 327–328.

73. Foxwell, "Economic Movement in England," 88, 90.

74. McCloskey, "Economics Science."

75. Mehrtens, *Moderne—Sprache—Mathematik.* This idea of mathematics as release came from Carl Friedrich Gauss in 1802.

Chapter Four
The Political Philosophy of Quantification

Epigraph from *Dialektik der Aufklärung*, 13.

1. Westbrook, *John Dewey*, 141–144, 170; Popper, *The Open Society and Its Enemies*, vol. 1, at 1; vol. 2, at 218.

2. Defoe, *The Complete English Tradesman*, 23; Ziman, *Reliable Knowledge*, 12.

3. Daston and Galison, "Image of Objectivity"; Dennis, "Graphic Understanding." On the objectivity of statistical images, see Brautigam, *Inventing Biometry*, chap. 6.

4. Pearson, "Ethic of Freethought," 19–20.

5. Pearson, *Grammar of Science*, 6, 8.

6. Gillispie, *Edge of Objectivity*, especially preface to second (1990) edition; Worster, *Nature's Economy.* On the experimental control of the self, see E. Keller, "Paradox of Scientific Subjectivity."

7. Ringer, *Decline of the German Mandarins*; Goldstein, "Psychological Modernism in France"; Porter, "Death of the Object."

8. From a letter to Charles Merriam, quoted in Ross, *Origins of American Social Science*, 403–404.

9. Traweek, *Beamtimes and Lifetimes*, 162. A still more important issue than openness, which Traweek's study raises movingly, is the sacrifice that women, and men, must make to succeed in this culture. Among the experimental physicists, it must be added, commitment to quantification is only a small part of this story.

10. Bulmer et al., eds., *Social Survey in Historical Perspective*, esp. 35–38 of editors' introduction; also Sklar, "*Hull-House Maps and Papers*"; Lewis, "'Webb and Bosanquet."

11. Gigerenzer, *Empire of Chance*, chap. 7. This preference for the quantitative study of the powerless remains, though by now the quantifier's net has spread out to include everyone.

12. Hacking, *Taming of Chance*; Rose, *Governing the Soul.*

13. Hilts, "*Aliis exterendum*"; Porter, *Rise of Statistical Thinking*, chaps. 2, 4; Dear, "*Totius in verba.*"

14. Wiener, *Reconstructing the Criminal.*

15. Himmelfarb, *Poverty and Compassion*, 116; on the Statistical Society, see Cullen, *The Statistical Movement.*

16. Funkenstein, *Theology and the Scientific Imagination*, 358.

17. Legoyt, "Congrès de statistique," 271.

18. Lécuyer, "L'hygiène en France"; Lécuyer, "Statistician's Role"; Brian, "Prix Montyon."

19. Brian, "Moyennes," 122; Kang, "Lieu de savoir social," 253; Coleman, *Death Is a Social Disease.*

20. Excerpt from Statistical Society of Paris, statutes, 7.

21. Chevalier, opening address, 2.

22. Balzac, *Les Employés*, 1112.

23. Foville, "Rôle de la statistique," 214.

24. Chevalier, opening address, 2–3. In 1894 E. Levasseur noted the characteristic American fondness for statistics and gave a similar explanation: "Département du travail." But American public statistics were notably haphazard until after the Civil War; see M. Anderson, *American Census.*

25. Porter, *Statistical Thinking*, 172–173.

26. Loua, "A nos lecteurs."

27. Liesse, *Statistique*, 57.

28. Faure, "Organisation de l'enseignement"; Cheysson, report of prize commission, 1883; Laurent, *Statistique mathématique*; Liesse, *Statistique*, 47 and *passim.*

29. Starr, "Sociology of Official Statistics."

30. See Daston, *Classical Probability*; Baker, *Condorcet.*

31. Bertillon, "Durée de la vie humaine," 45, 47.

32. Bertillon, "Mortalité d'une collectivité," 29.

33. Loua, comment.

34. Foville, "La statistique et ses ennemis," 448. Gondinet, *Le Panache*, 112; Labiche and Martin, *Capitaine Tic*, 18, 21.

35. Le Play, "Vues générales sur la statistique," 10.

36. Jules Simon, éloge for Hippolyte Carnot, quoted in Charle, *Elites de la république*, 27.

37. Oakeshott, "Rationalism in Politics," 31.

38. Horkheimer and Adorno, *Dialektik der Aufklärung*, 11; Marcuse, *Reason and Revolution*. The argument remains appealing: Merchant, *Ecological Revolutions*, 266–267.

39. Daston and Galison, "Image of Objectivity," 82; Dear, "From Truth to Disinterestedness."

40. Peter Galison, "In the Trading Zone," paper given at UCLA, December 1989.

41. Ezrahi, *Descent of Icarus.* On the political, philosophical, and aesthetic resonances of a commitment to transparency, see Galison, "Aufbau/Bauhaus."

Chapter Five
Experts against Objectivity

Epigraph from testimony in *Report from the Select Committee on Assurance Associations* (hereafter *SCAA*), British Parliamentary Papers, 1853, vol. 21, 246. This chapter draws from my "Quantification and the Accounting Ideal," and "Precision and Trust."

1. See Pearson's remarkable book, published under the pseudonym Loki, *The New Werther.*

2. Goody, *Domestication of the Savage Mind*, 15; Goody, *Literacy in Traditional Societies*; Graff, *Legacies of Literacy*, 54–55.

3. R. H. Parker, *Accountancy Profession in Britain*, 4, 26–29; Jones, *Accountancy and the British Economy*; Gourvish, "Rise of the Professions."

4. Lavoie, "Accounting of Interpretations," criticizes accounting positivism, as do Ansari and McDonough, "Intersubjectivity."

5. Wagner, "Defining Objectivity in Accounting," 600, 605.

6. Johnson and Kaplan, *Relevance Lost*, esp. chap. 6; quote at 125.

7. Chandler, *Visible Hand*, 267–269, 273–281; H. T. Johnson, "Nineteenth-Century Cost Accounting"; Johnson and Kaplan, *Relevance Lost.*

8. Temin, *Inside the Business Enterprise*, esp. Lamoreaux, "Information Problems and Banks"; also my review essay, "Information Cultures."

9. Zeff, "Evolution of Accounting Principles."

10. May, "Introduction" (1938) to session of American Institute of Accountants, on "A Statement of Accounting Principles" in Zeff, *Accounting Principles*, 1–2; Wilcox, "What is Lost" (1941), ibid., 96, 101; Werntz, "Progress in Accounting" (1941), ibid., 315–323.

11. Flamholtz, "Measurement in Managerial Accounting." The rule doubtless seemed less objectionable to companies during the Depression, a time of deflation.

12. Chambers, "Measurement and Objectivity," 268.

13. Quoted in Arnett, "Objectivity to Accountants," 63.

14. Burke, "Objectivity and Accounting," 842.

15. Bierman, "Measurement and Accounting," 505–506.

16. Ijiri and Jaedicke, "Reliability and Objectivity," quotes at 474, 476.

17. Ashton, "Objectivity of Accounting Measures," 567; also Parker, "Testing Comparability and Objectivity."

18. Wojdak, "Levels of Objectivity."

19. Ashton, "Objectivity of Accounting Measures."

20. On following rules, see Bloor, "Left and Right Wittgensteinians."

21. Loft, "Cost Accounting in the U.K"; see also Burchell et al., "Value Added in the United Kingdom."

22. Power, "After Calculation."

23. Barnes, "Authority and Power."

24. American Psychological Association, *Publication Manual*, 19.

25. Gowan, "Origins of Administrative Elite"; Perkin, *Rise of Professional Society*; Reader, *Professional Men.*

26. MacLeod, *Government and Expertise*; Smith and Wise, *Energy and Empire*, chap. 19; Hunt, *The Maxwellians*, chap. 6; Harris, "Economic Knowledge."

27. Heclo and Wildavsky, *Private Government of Public Money*, 2, 15, 61–62.

28. Harris, "Economic Knowledge," 394.

29. Self, *Econocrats and the Policy Process.*

30. Colvin, *Economic Ideal in British Government*; Ashmore et al., *Health and Efficiency.*

31. Wynne, *Rationality and Ritual*, 65–66; Williams, "Cost-Benefit Analysis," 200.

32. Gowan, "Origins of Administrative Elite"; Greenleaf, *A Much Governed Nation*; Brundage, *England's Prussian Minister*; Hamlin, *Science of Impurity.*

33. Farren, "Life Contingency Calculation," 185–187, 121.

34. Bailey and Day, "Rate of Mortality," 318; Lance, "Marine Insurance," 364.

35. Alborn, "A Calculating Profession."

36. Two anonymous letters: "Solution of Problem" and "The Same Subject," *Assurance Magazine*, 12 (1864–66), 301–2.

37. Campbell-Kelly, "Data Processing in the Prudential."

38. *Report from the Select Committee on Friendly Societies*, British Parliamentary Papers, 1849, XIV, testimony of Francis G. P. Neison, 8.

39. Jellicoe, "Rates of Mortality"; Day, "Assuring against Issue."

40. Curtin, *Death by Migration*. An example of such inquiry is Jellicoe, "Military Officers in Bengal." Trebilcock, *Phoenix Assurance*, 552–565, discusses the surcharges imposed by the Pelican.

41. Brown, "Fires in London"; Lance, "Marine Insurance," 362. Still, it was well known among the professionals that there will be fluctuations from mean values. Probability arguments were used to argue that a friendly society should have at least 150 to 300 members to be reasonably safe. Charles Babbage explained this to a select committee (see *Report from the Select Committee on the Laws respecting Friendly Societies*, British Parliamentary Papers, House of Commons, 1826–27, III, 28–33). The committee undercut him by asking about the effects of epidemic diseases, to which the standard probability calculus could not apply.

42. McCandlish, "Fire Insurance," 163.

43. Trebilcock, *Phoenix Assurance*, 355, 419, 446 and *passim*.

44. Testimony of Francis G. P. Neison in *SCAA*, 204.

45. Gray, "Survivorship Assurance Tables," 125–126.

46. H. Porter, "Education of an Actuary," 108–111.

47. Ibid., 108, 112, 116, 117.

48. Farren, "Reliability of Data," 204.

49. Alborn, "The Other Economists," 236.

50. Testimony of William S. D. Pateman in *SCAA*, 282.

51. Testimony of John Finlaison in *SCAA*, 49–64.

52. Ryley in *SCAA*, 246. See also testimonies of George Taylor at 30, Charles Ansell at 70, and James John Downes at 105.

53. *Report of the Select Committee on Joint Stock Companies, together with the Minutes of Evidence*, British Parliamentary Papers, House of Commons, 1844, VII, testimony of Charles Ansell (1841), 49.

54. *SCAA*, testimony of Charles Ansell, at 69, 74, 82.

55. Samuel Ingall in *SCAA*, 158–159, 165.

56. Ansell at 81; Downes at 105, 107, 108; Neison at 197; Farr at 303, all in *SCAA*.

57. *Report of the Select Committee on the Laws Respecting Friendly Societies*, British Parliamentary Papers, House of Commons, 1824, IV, 18, referring to 59th Geo. 3. c. 128. The committee reported that this provision was deleted before passage, but several witnesses spoke as if the measure were in effect.

58. Testimony in ibid. by William Morgan, 52, and Thomas John Becher, 30. Becher determined finally that an actuary was a mathematician, someone

who could solve a "question in fluxions" and "make . . . calculations algebraically as well as arithmetically."

59. Testimony in Select Committee on Joint Stock Companies, 1844, 81.

60. Downes, 108, and Jellicoe, 188, 184, in *SCAA*.

61. Thomson, 85–104, and Edmonds, 138, both in *SCAA*.

62. Francis G. P. Neison in *SCAA*, 196.

63. John Adams Higham in *SCAA*, 213, 220.

64. Thomson in *SCAA*, 97; also Alborn, "The Other Economists," 239.

Chapter Six
French State Engineers

Epigraph from Divisia, *Exposés d'économique*, 47

1. Etner, *Calcul économique en France*, 22, 115.

2. Gillispie, "Enseignement hégemonique"; Kranakis, "Social Determinants of Engineering Practice."

3. Picon, "Ingénieurs et mathématisation"; Picon, *L'invention de l'ingénieur moderne*, 371–388, 424–442.

4. Gispert, "Enseignement scientifique."

5. Picon, *L'invention de l'ingénieur moderne*, 393.

6. Ibid., 442, 511–512; Couderc, *Essai sur l'Administration*, 54.

7. This general council was made up of the inspectors general plus some other officers in Paris. The hierarchy was not perfectly simple, and the numbers varied from time to time, but during most of the nineteenth century there were about 5 inspectors general. Below them were about 15 divisional inspectors, perhaps 105 chief engineers (most in charge of one département each), 300 or more ordinary engineers, plus a cohort of aspiring engineers and students. See Picon, *L'invention de l'ingénieur moderne*, 314–317; Gustave-Pierre Brosselin, *Note sur l'origine, les transformations, et l'organisation du Conseil général des Ponts et Chaussées*, Bibliothèque de l'Ecole National des Ponts et Chaussées (hereafter BENPC), c1180 x27084. A detailed listing of the top officers in the Corps can be found for any given year in the (variously titled) *Almanach National/Almanach Royal/Almanach Impérial*, under Ministère des Travaux Publics.

8. For example, H. Sorel, président, "Embranchements de Livarot à Lisieux et de Dozulé à Caen: Observations de la Chambre de commerce de Honfleur," séance du 16 juin 1874, BENPC c672 x12022. Honfleur, like every town, tried to get good rail service and block competing lines.

9. "Enquêtes relatives aux travaux publics," reprinted in Picard, *Chemins de fer français*, vol. 4, "Documents Annexes," 1–3; Henry, *Formes des Enquêtes administratives*; Thévenez, *Legislation des chemins de fer*, 78–85; and especially Tarbé des Vauxclairs, *Dictionnaire des travaux publics*, s.v. "Enquête," 237–239, and "De commodo et incommodo," 195.

10. Tarbé, *Dictionnaire*, s.v. "Concours de l'Etat aux travaux particuliers, concours des particuliers aux travaux publics," 153–154.

11. The Corps des Ponts paid homage to British leadership in railroad projecting by printing translations of British commercial evaluations of railroads: e.g., Booth, "Chemin de fer de Liverpool à Manchester." Such work reflected

not just Britain's technological lead in railroads, but also the expectations of sophisticated British investors. The Parliament also had to be convinced, since every canal, railroad, or other construction involving the violation of property rights required a private bill. This normally meant a genuinely adversarial procedure, since opposing the line was a good way to win a larger indemnity and involved testimonies as to the value of the line followed by cross-examination. The ability to work this system was an essential qualification of a high-level engineer. See Christopher Hamlin, "Engineering Expertise and Private Bill Procedure in Nineteenth-Century Britain," paper at Joint Anglo-American History of Science meetings, Toronto, July 25–28, 1992.

12. John H. Weiss, "Careers and Comrades," unpublished manuscript, chap. 4.

13. Geiger, "Planning the French Canals." Railway estimates were no better; in the 1830s the Corps underestimated the cost of the line from Paris to Le Havre by 50 percent. See Dunham, "How the First French Railways Were Planned," 19.

14. Minard, "Tableau comparatif."

15. Chardon, *Travaux publics*, 24 and *passim.*

16. Quoted in Etner, *Calcul économique en France*, 129.

17. Picon, *L'invention de l'ingénieur moderne*, 321; Fichet-Poitrey, *Le Corps des Ponts et Chaussées.* E. Weber, *Peasants into Frenchmen*, ranks railroads beside schools as key agents in the creation of a French national identity in the countryside.

18. Quoted without attribution (and then rebutted) by Louvois ("président du comité central du chemin de fer de Paris à Lyon par la Bourgogne"), "Au rédacteur," 1. At issue was the construction of a direct line from Paris to Strasbourg. Louvois favored the diversion of lines to pass through cities and towns. On French railway planning in the 1840s, see Pinkney, *Decisive Years in France.*

19. Courtois, *Questions d'économie politique*, quote at 1; formulas at 4–6.

20. Courtois, *Choix de la direction*, 59, 9, 15. See also Etner, *Calcul économique en France*, 127–128.

21. Courtois, *Choix*, 52; Jouffroy, *Ligne de Paris à la Frontière*, 76, 190–191.

22. Minard, *Second Mémoire.* On railroad planning, see C. Smith, "The Longest Run."

23. Daru, *Chemins de fer*, 121, 136.

24. "Rapport de la commission chargée de l'examen des project du chemin de fer de Paris à Dijon. Resumé," unpublished manuscript, BENPC x6329. The report was signed by Fèvre, Kermaingant, Hanvilliers, Mallet, and Le Masson.

25. Teisserenc, "Principes généraux," 6–8.

26. Albrand, *Rapport de la Commission*; Lepord, "Rapport de l'Ingénieur en chef du Finistère," October 11, 1854, BENPC, manuscripts, 2833.

27. Jean-Auguste Philipert Lacordaire, "Chemin de fer de Dijon à Mulhouse," 3 parts dated March 20 and March 24, 1845, BENPC c394 x6248–6250; Lacordaire, "Chemin de fer de Dijon à Mulhouse, Ligne mixte dite de Conciliation, par Gray et Vallée de l'Ognon; Avantages et Désavantages de cette Ligne," dated April 2, 1845, BENPC c394 x6247.

28. France, Chambre des deputés, Annexe au procès verbal de la séance du 4 juin 1878, no. 794, "Projet de Loi relatif au classement du réseau complémentaire des chemins de fer d'intérêt general, presenté . . . par M. C. de Freycinet," 3.

29. France, Assemblée Nationale, Annexe au procès verbal de la séance du 7 juillet 1875, no. 3156, Krantz, rapporteur, "Rapport . . . ayant pour objet la déclaration d'utilité publique et la concession d'un chemin de fer sous-marin entre la France et l'Angleterre."

30. Chardon, *Travaux publics*, 171–180. Bureaucratic delays and an antiquated commitment to canals at the Corps des Ponts have often been blamed for the relatively slow pace of railroad construction in France. Its activities are placed in a more favorable light by Ratcliffe, "Bureaucracy."

31. For example, Assemblée Nationale, Annexe au procès-verbal de la séance du 13 juillet 1875, Aclogue, rapporteur, "Rapport . . . pour objet la déclaration d'utilité publique et la concession de certaines lignes de chemin de fer à la Compagnie du Midi."

32. "Déliberations du Conseil Général des Ponts et Chaussées. Minutes. 4me Trimestre, 1869," meeting of December 23, AN F14 15368.

33. France, Assemblée Nationale, Annexe au procès verbal de la séance du 3 février 1872, no. 1588, Ernest Cézanne, rapporteur, "Rapport au nom de la Commission d'Enquête sur les chemins de fer et autres voies de transports sur diverses pétitions relatives à la concession d'une ligne directe de Calais à Marseille."

34. France, Assemblée Nationale, Annexe au procès verbal de la séance du 23 février 1875, no. 2905, Ernest Cézanne, rapporteur, "Rapport . . . relatif à la déclaration d'utilité publique de plusieurs chemins de fer, et à la concession de ces chemins à la Compagnie de Paris à Lyon et à la Mediterranée."

35. This concerned a line from Amiens to Dijon, proposed in 1869; Picard, *Chemins de fer français*, vol. 3, 326.

36. Ibid., vol. 1, 273–276, 294–295 and *passim*. Picard worked his way up through the Corps des Ponts to inspector general, head of railroads in the ministry of public works, and membership in the Conseil d'Etat, of which he eventually became vice president.

37. E.g., MM. Mellet et Henry (identified as "adjudicataires du chemin de fer de Paris à Rouen et à la mer"), *L'Arbitraire administratif.*

38. Baum, "Longueurs virtuelles."

39. Baum, "Etude sur les chemins de fer."

40. Michel, "Trafic probable." An analysis of Michel's methods appeared as Anon., "Moyens de déterminer l'importance du trafic d'un chemin de fer d'intérêt local," *Journal de la Société de Statistique de Paris,* 8 (1867), 132–33. See also Etner, *Calcul économique en France*, 185–190.

41. Baum, *Chemins de fer d'intérêt local du Département du Morbihan: Rapport de l'Ingénieur en chef* (Vannes: Imprimerie Galles, 1885); I used the copy in BENPC c1006 x18978.

42. Fournier de Flaix, "Canal de Panama." These estimates of costs and revenues for the Panama Canal fluctuated greatly. But they were validated by a committee on statistics in the Congrès international des études du Canal in-

terocéanique, headed by a distinguished statistician, E. Levasseur. See Simon, *The Panama Affair*, 30–31. But the glory of France sold as many bonds as did predictions of costs and revenues, especially after signs of trouble appeared.

43. From a summary by inspector general Schérer of a report by chief engineer Laterrade on a proposed line from Villeneuve to Falgueyrat, see item no. 113 in Section du Conseil Général des Ponts et Chaussées, Chemins de fer, Registre des delibérations du 4 jan. au 25 mars 1879 inclusivement, AN F14 15564. This section of the council sent the report back for further studies, calling for a less exclusive attention to purely local benefits.

44. See "Rapport de M. l'Inspecteur général Deslandes sur la concession et la demande en déclaration d'utilité publique" for lines from Le Mans to Grand-Lucé and Ballon to Antoigné, no. 126 in ibid.

45. Elwitt, *Making of the Third Republic*, chaps. 3–4.

46. Etner, *Calcul économique en France*, 148, 193ff.

47. C. de Freycinet, "Discours prononcé à la Chambre des Deputés le 14 mars 1878," extrait du *Journal Officiel* du 15 mars 1878, 25–26, cited by Etner in ibid.

48. Georges-Médéric Léchalas, on a line in the Seine-Inférieure, cited in (n.a.), "La mesure de l'utilité des chemins de fer," *Journal des économistes*, November 1879, offprint. Two other engineers who weighed in on the measurement of utility in the wake of Freycinet were Hoslin, *Limites de l'intérêt public*, who argued that these measures of public utility were of decisive importance; and La Gournerie, *Etudes Economiques*, appendix D, 65–68, who was skeptical of any measures that greatly exceeded revenue.

49. Christophle, *Discours sur les travaux publics*, préface; Lavollée, "Chemins de fer et le budget."

50. Labry, "A Quelles conditions"; Labry, "Profit des travaux"; Doussot, "Observations sur une note de Labry"; Labry, "Outillage national."

51. Considère, "Utilité: Nature et valeur," 217–348.

52. Colson, "Formule d'exploitation de M. Considère"; Considère, "Utilité: Examen des observations"; Colson, "Note sur le nouveau mémoire," 153.

53. Colson, *Cours*, vol. 6, *Travaux publics*, chap. 3.

54. On state support and regulation of railroads, see Doukas, *French Railroads and the State.*

55. E.g., France, Conseil d'Etat, *Enquête sur l'application des tarifs des chemins de fer* (Paris: Imprimerie National, 1850), BENPC c336 x5779; Poirrier, *Tarifs des chemins de fer*; Noël, *Question des tarifs.*

56. Tézenas du Montcel and Gérentet (of the Saint-Etienne chamber of commerce), *Rapport de la commission* is typical. On rate-setting, see Ribeill, *Révolution ferroviaire*, 282–292.

57. Dupuit, "Influence des péages," 225–229.

58. Picard, *Chemins de fer*, chap. 3 ("Mesure de l'utilité des chemins de fer"), 280.

59. Proudhon, *Des réformes dans l'exploitation des chemins de fer*, cited in Tavernier, "Note sur tarification," 575.

60. Baum, "Des prix de revient"; Baum, "Note sure les prix de revient"; Baum, "Le prix de revient"; also La Gournerie, "Essai sur le principe des tarifs

dans l'exploitation des chemins de fer" (1879), in his *Etudes économiques*; Ricour, "Répartition du trafic"; Ricour, "Prix de revient."

61. Tavernier, "Exploitation des grandes compagnies"; Tavernier, "Notes sur les principes"; also some earlier, less systematic critiques: Menche de Loisne, "Influence des rampes"; Nordling, "Prix de revient."

62. Baum, "Note sur les prix de revient" (1889); Tavernier, "Note sur les principes," 570. On arguments about mean values, see Feldman et al., *Moyenne, milieu, centre.*

63. Armatte, "L'économie à l'Ecole polytechnique"; Picon, *L'Invention de l'ingénieur moderne*, 452–453.

64. Colson, *Cours*, vol. 1, *Théorie générale des phénomènes économiques*, 1–2 and 38–39, quote at 39.

65. Before Colson took it over, the course in political economy was given by liberal nonengineers: Joseph Garnier, from 1846 to 1881, and Henri Baudrillart, from 1881 to 1892 (see *Cours d'économie politique. Notes prises par les éleves. Ecole Nationale des Ponts et Chaussées*, 1882, BENPC 16034).

66. Colson, *Cours*, vol. 6, *Les travaux publics*, 183.

67. Colson, *Transports et tarifs*, chap. 2; idem, *Cours*, vol. 6, *Les travaux publics*, chap. 3.

68. Not without reason, Walras blamed the engineering economists for the French railway monopolies. This was one reason for his antagonism; see Etner, *Calcul économique en France*, 106–107.

69. Considère, "Utilité: Nature et valeur," 349–354; Colson, *Transports et tarifs*, 44. Considère, however, argued on the basis of indirect benefits that it is often wise for the state to lower fares below this point, whereas Colson treated it as the minimum.

70. Colson, *Cours*, vol. 6, *Les travaux publics*, 209–211.

71. Caron, *Histoire de l'exploitation*, 370–372.

72. Colson, *Cours*, vol. 6, *Les travaux publics*, 210–211, 198–199.

73. Weisz, *Emergence of Modern Universities.*

74. Tudesq, *Grands notables*, vol. 2, 636.

75. Brunot and Coquand, *Corps des Ponts et Chaussées*, 407; trans. in Kranakis, "Social Determinants," 33–34.

76. Balzac, *Le curé de village*, chap. 23; Gaston Darboux, "Eloge historique de Joseph Bertrand," in Bertrand, *Eloges académiques*, x–xi.

77. Arago, *Histoire de ma jeunesse*, 46.

78. Picon, *L'invention de l'ingénieur moderne*, 92–93; Shinn, *L'Ecole Polytechnique*, 24–35. On the Saint-Simonian connection, see Picon, *L'invention*, 455, 595–597; also Hayek, *Counterrevolution of Science*, who exaggerates the narrow scientific orientation of polytechnicians.

79. Quoted in Fourcy, *Histoire de l'Ecole Polytechnique*, 351.

80. Tudesq, *Grands notables*, vol. 1, 352.

81. Weiss, "Bridges and Barriers," 19–20. The fullest discussion of these debates is in Shinn, *L'Ecole Polytechnique.* Weiss notes that the Ecole Centrale, designed to provide a more practical technological education than the Ecole Polytechnique, and to train engineers for private industry rather than state service, also became strongly elitist and, in the later nineteenth century, even took on some features of a state *corps.* See Weiss, *Making of Technological Man.*

82. Conseil d'Instruction, Ecole Polytechnique, minutes of meetings, t. 5, meeting of September 27, 1812, in Bibliothèque de l'Ecole Polytechnique, archives, Lozère, France.

83. From a tract by the engineer A. Léon published in 1849, and written to justify the gulf between mere *conducteurs* and engineers; see Kranakis, "Social Determinants," 28–29; Weiss, "Careers and Comrades," chap. 6.

84. "Rapport fait par le citoyen Stourm, au nom du comité des travaux publics, sur le project de loi relatif à des changes dans l'organisation du corps des conducteurs des ponts et chaussées et dans le mode de recrutement des ingénieurs," *Le Moniteur universel*, December 19, 1848, 3606–3610; Jules Dupuit, "Comment doit-on recruter le corps des Ponts et Chaussées," undated manuscript response to the above report, Dupuit papers, dossier 7, BENPC, uncatalogued.

85. See Shinn, *L'Ecole Polytechnique*, 119.

86. Ibid., *passim*; Charle, *Elites de la République*; Picon, "Années d'enlisement."

87. Suleiman, *Elites in French Society*, 163, 165. On the British Civil Service see, chapter 5 above.

88. Suleiman, *Elites in French Society*, 168.

89. Suleiman, *Politics, Power, and Bureaucracy*, 262; C. Day, *Education for the Industrial World*, 10.

90. Suleiman, *Politics, Power, and Bureaucracy*, 246. Here he is criticizing Thoenig, *L'ère des technocrates.*

91. See correspondence of Dupuit in dossier 9 of Dupuit file, BENPC, uncatalogued. For example, in December 1827 Comoy sent condolences because Dupuit was bored to death with Le Mans. In October 1827 Jullien wrote that Le Mans could scarcely be duller than Nevers, but that he was so engaged by his canal that he didn't miss the pleasures of Paris: prefects, functionaries, high society, women. He added that it must be different for Dupuit, who cut such a figure in the salons of (inspectors general) Becquey and Navier.

92. Berlanstein, *Big Business*, chap. 3, quote at 113.

93. On the careers of polytechnicians, see Kindleberger, "Technical Education"; Zeldin, *France, 1848–1945*, vol. 1; Berlanstein, *Big Business.*

94. Elwitt, *Making of Third Republic*, 155.

95. Quoted in Sharp, *French Civil Service*, 33.

96. Chardon, *Administration de la France*, 56, 58; also Chardon, *Pouvoir administratif*, 34. Chardon objected to political interference, but especially to the fracturing of authority implied by the existence of eighty-six prefects.

97. Fayol, *General and Industrial Management*, 33.

98. On the *concours*, see Zeldin, *France, 1848–1945*, vol. 1, 118ff.; Gilpin, *France*, 103; Sharp, *French Civil Service*. The *concours* was also favored during the brief life of the Second Republic; see Thuillier, *Bureaucratie et bureaucrates*, 334–339; Hippolyte Carnot, quoted in Charle, *Hauts fonctionnaires.*

99. Courcelle-Seneuil, "Etude sur le mandarinat français" (1872), in Thuillier, *Bureaucratie*, 104–113. Oddly, he wanted to solve the problem with still greater reliance on the *concours.*

100. Quoted in Thuillier, *Bureaucratie*, 346. Joan Richards remarks in "Rigor and Clarity," 303, that in the early decades of Polytechnique, mathemat-

ics was valued there as an objective test of intellectual strength, contributing to "a fair, nonaristocratic, meritocratic society."

101. Though this might be solved, thought Joseph Bertrand, if Polytechnique gave its own examination in *études littéraires*; see minutes of Conseil de Perfectionnement, Ecole Polytechnique, t. 8 (1856–1874), meeting of April 28, 1874, 342–343, Bibliothèque de l'Ecole Polytechnique, archives, Lozère, France. Also Fayol, *General and Industrial Management*, 86.

102. Ibid. (Conseil de Perfectionnement), 359.

103. Fougère, "Introduction générale," to Fougère, *L'administration française*, 3–9.

104. Hoffmann, "Paradoxes," 17; Suleiman, "From Right to Left."

105. Luethy, *France against Herself*, 38; Legendre, *Histoire de l'administration*, 536–537; Hoffmann, "Paradoxes," 9; Grégoire, *Fonction publique*, 70; Osborne, *A Grande Ecole*, 82, 86.

106. Grégoire, *Fonction publique*, 101–104.

107. Suleiman, *Politics, Power, and Bureaucracy*, 280–281.

108. Sharp, *French Civil Service*, vii.

109. Suleiman, *Elites in French Society*, 171.

110. Ibid., 173.

111. Fayol, *General and Industrial Management*, 82, 86.

112. Brun, *Technocrates et Technocratie*, 49, 74.

113. Kuisel, *Capitalism and the State.*

114. Zeldin, *France, 1848–1945*, vol. 2, 1128, notes that even in 1963 most "technocrats" thought general culture rather than technical knowledge was the basis of their success. The opposite view, that technocrats are pure theorists locked in their study, is often expressed, as in Bauchard, *Technocrates et pouvoir*, 9–11, but this is unconvincing.

115. Quoted in Brun, *Technocrates et technocratie*, 82.

116. Jouvenel, *Art of Conjecture.* See also Meynaud, "Spéculations sur l'avenir"; Gilpin, *France*, 231ff.

117. Fourquet, *Comptes de la puissance.* The work of Hackett and Hackett, *Economic Planning in France*, suggests that econometric modeling and other advanced quantitative techniques were much less used up to 1960.

118. Servos, "Mathematics in America."

Chapter Seven
U.S. Army Engineers and the Rise of Cost-Benefit Analysis

Epigraph from *Congressional Record*, 80 (1936), 7685. Need I stipulate here that my use of this bit of grandiloquence implies no fondness for the notorious racist who uttered it?

1. Shallat, "Engineering Policy"; Calhoun, *American Civil Engineer*, 141–181.

2. Harold L. Ickes, "Foreword," to Maass, *Muddy Waters*, ix.

3. Lundgreen, "Engineering Education"; Porter, "Chemical Revolution of Mineralogy."

4. Lewis, *Charles Ellet*, 11; Shallat, "Engineering Policy," 12–14.

5. Chandler, *Visible Hand*; Hoskin and Macve, "Accounting and the Examination."

6. Lewis, *Charles Ellet*, 17–20, 54; Calhoun, *Intelligence of a People*, 301–304.

7. E.g., Fink, *Argument* (1882).

8. Pingle, "Early Development." On American developments, beginning with the Gallatin report of 1808 on the value of western lands, see Hines, "Precursors to Benefit-Cost Analysis."

9. Hays, "Preface, 1969," in *Conservation*; Wiebe, *Search for Order*; Haskell, *Emergence of Professional Social Science*.

10. House of Representatives, Committee on Flood Control, *Flood Control Plans and New Projects: Hearings* . . . , April 20 to May 14, 1941, 495. To be sure, McSpadden was playing to the crowd, and in fact used economic quantification rather effectively in his testimony. It didn't hurt that he was supported by Oklahoma oil interests. For once, a dam was moved—far enough to save the house.

11. Wright, "The Value and Influences of Labor Statistics" (1904), quoted in Brock, *Investigation and Responsibility*, 154.

12. W. Nelson, *Roots of American Bureaucracy*, chap. 4; Schiesl, *Politics of Efficiency*.

13. Glaeser, *Public Utility Economics*, chap. 6. Glaeser agreed with the ICC that an objective valuation of utility capital was impossible, and called instead for reliance on the judgment of expert professionals (438, 500, 638–639, 696).

14. Keller, *Regulating a New Economy*, 50, 63; also Brock, *Investigation and Responsibility*, 192–200.

15. M. Keller, *Affairs of State*, 428; Skowronek, *Building a New American State*, 144–151.

16. M. Keller, *Affairs of State*, 381–382; Hays, *Conservation*, 93, 213; Reuss and Walker, *Financing Water Resources Development*, 14.

17. 61st Cong., 2d sess., 1910, H.D. 678 [5732], Channel from Aransas Pass Harbor through Turtle Cove to Corpus Christi, Texas. For another example, see R. Gray, *National Waterway*, 222–223.

18. N.A. 77/496/3, Board of Engineers for Rivers and Harbors, Administrative Files, 91, 125.

19. 69th Cong., 1st sess. (1925), H.D. 125, Skagit River, Washington, 21. Or the calculation might be reversed: the capitalized value of expected flood damages defined the limit of permissible expenditure. This method was criticized by G. White, "Limit of Economic Justification."

20. 69th Cong., 1st sess. (1925), H.D. 123.

21. 73d Cong., 1st sess. (1933), H.D. 31 [9758], Kanawha River, West Virginia.

22. 73d Cong., 1st sess. (1933), H.D. 45, Bayou Lafourche, Louisiana.

23. Hammond, "Convention and Limitation."

24. Barber, *New Era to New Deal*, 21.

25. The mean calculated benefit-cost ratio for flood-control projects varied enormously by region. In most places (according to an optimistic accounting) it was between 1.6 and 3.0 or 4.0, but in the lower Mississippi it was 4.8, and in

the upper Mississippi 13.7. The need to spread its largesse more evenly explains why the Corps refused to assign priority to projects according to their benefit-cost ratios. See H.R., Committee on Public Works, *Costs and Benefits of the Flood Control Program*, 85th Cong., 1st sess., House Committee Print no. 1, April 17, 1957.

26. Arnold, *1936 Flood Control Act.*

27. Lowi, "State in Political Science," 5.

28. *Congressional Record*, 90 (1944), 8241, 4221.

29. John Overton in *Congressional Record*, 83 (1938), 8603.

30. H.R., Committee on Flood Control, *Comprehensive Flood Control Plans: Hearings*, 76th Cong., 3d sess., 1940, 13; idem, *Flood Control Plans and New Projects: 1943 and 1944 Hearings*, 78th Cong., 1st and 2d sess., 20.

31. The 1.03 ratio refers to a project on the Lehigh River, Pennsylvania: H.R., Committee on Flood Control, *Flood Control Bill of 1946*, 79th Cong., 2d sess., April–May 1946, 23–36. On the Neches-Angelina: H.R., Committee on Public Works, Subcommittee on Rivers and Harbors, *Rivers and Harbors Bill of 1948*, 80th Cong., 2d sess., February–April 1948, 189. On Prescott Bush: Senate, Committee on Public Works, Subcommittee on Flood Control—Rivers and Harbors, *Hearings: Rivers and Harbors—Flood Control, 1954*, 83d Cong., 2d sess., July 1954, 20.

32. All from *Congressional Record*, 80 (1936), 8641, 7758, 7576.

33. Cited in Ferejohn, *Pork Barrel Politics*, 21.

34. H.R., Committee on Public Works, Subcommittee to Study Civil Works, *Study of Civil Works: Hearings*, 82d Cong., 2d sess., March–May 1952, 3 vols., part 1, at 31, 11.

35. H.R., Committee on Flood Control, *Hearings: Comprehensive Flood Control Plans*, 75th Cong., 1st sess., March–April 1938, 306–307; H.R., Committee on Public Works, Subcommittee on Flood Control, *Hearings: Deauthorize Project for Dillon Dam, Licking River, Ohio*, 80th Cong., 1st sess., June 1947, 81. Of those who opposed the dam, Corry hinted darkly: "Frankly, we question their motives. We do not know what their motives are. We do not believe that they have been brought out at this hearing." The opponents were, in fact, the unfortunates upstream, who were about to find their houses and farms under a hundred feet of water.

36. On the current politics of Mississippi River control, see McPhee, *Control of Nature*, part 1, "Atchafalaya."

37. H.R., Committee on Flood Control, *Hearings*, 1938, at 914, 927–928.

38. Ibid., 927, 912.

39. H.R., Committee on Flood Control, *Flood Control Plans and New Projects: Hearings* . . ., April–May 1941, at 728–729, 732.

40. Ibid., 824, 825. Whittington, who was less inclined to grandstanding, finally became impatient with this exchange: "I don't believe the Army engineers are any freer from political pressure than we are here; but I don't think there is any unrighteous or crooked pressure."

41. H.R., Committee on Public Works, Subcommittee on Rivers and Harbors, *Rivers and Harbors Bill, 1948: Hearings*. . ., 80th Cong., 2d sess., February–April 1948, at 198–199, 201.

42. N.A. 77/111/1552/7249, "Outline of Review. Project Application," dated August 31, 1935; H.R., Committee on Flood Control, *1946 Hearings*, 119–122.

43. H.R., Committee on Flood Control, *1946 Hearings*, 392, 675; Senate, Committee on Public Works, Subcommittee on Flood Control and River and Harbor Improvements, *Hearings: Rivers and Harbors—Flood Control Emergency Act*, 80th Cong., 2d sess., May–June 1948, at 77–82. Such language may also be found in project reports: for example, 76th Cong., 2d sess., H.D. 655 [10504], *Fall River and Beaver Creek, S. Dak.* Once, at least, the Senate even recommended a new dam without a Corps report. This came at the end of years of disagreement between Massachusetts and Connecticut, which wanted flood control on the Connecticut River, and Vermont, which would get most of the submerged land and very little of the benefit. Overton took great care to mark this compromise as exceptional, so it would not be a precedent. See *Congressional Record*, 90 (1944), 8557.

44. H.R., Committee on Flood Control, *1943 and 1944 Hearings*, vol. 1, 1943, 190–233, quotes at 196, 225.

45. H.R., Committee on Flood Control, *Hearings, 1941*, 512–521, excerpted from some exceptional hearings of the Board of Engineers on a flood control project for the Little Missouri River. Meeting in Washington, the Board had turned it down. The new round of hearings in Arkansas, explained board chairman Thomas M. Robins, was "due to the very strenuous efforts of your very able senator from Arkansas, Senator Miller." Those efforts had evidently been exerted throughout the process, for the official Corps report of 1940 had already adopted "irregular" methods in order to quantify recreational benefits and to predict generous increases in farm income, and hence to get the benefit-cost ratio even as high as 0.92. See 76th Cong., 2d sess., H.D. 837 [10505], *Little Missouri River, Ark.*, 50.

46. See H.R., Committee on Flood Control, *Hearings, 1938*, 270–275, in which a representative of the Chamber of Commerce of Chicopee, Massachusetts, complained that a levee on the Connecticut River in the town really was economically justified, and that the Corps' negative report resulted from an inadequate appreciation of the indirect damages due to the closing of factories and consequent unemployment.

47. The Corps is routinely charged with fierce resistance to multiple-use water management until sometime in the 1950s. That interpretation seems to have originated in attacks on the Corps by supporters of the Bureau of Reclamation, particularly Maass, *Muddy Waters*. He made the Bureau stand for rational, systematic management within the executive branch, the Corps for narrowminded pork barrel politics under the patronage of Congress. I find that the Corps was remarkably bold during the 1940s and even 1930s in pursuing new goals of river control (if only to improve its benefit-cost ratios), especially given that its mandate limited it to navigation and flood control. In the late 1950s, the Corps enlisted Maass as a consultant. The charge that he was bought off is unfair, but professional involvement with the Corps certainly improved his opinion of the agency. He claimed that it at last embraced multiple-use river planning in those years. See Reuss, *Interview with Arthur Maass*, 6.

48. Testimony of E. W. Opie, H.R., Committee on Flood Control, *1946 Hearings*, 86–90.

49. See ibid. Quotes are from Senate, Committee on Commerce, *Hearings: Flood Control*, 79th Cong., 2d sess., June 1946, at 157, 228. Upstream interests were slightly more effective. A request from the governor of Virginia at least convinced the Senate to lower the dam by 20 feet, at some cost to the calculated benefit-cost ratio. See *Congressional Record*, 92 (1946), 7087. In 1934, a Corps report called flooding on the Rappahannock "inconsequential." N.A. 77/111/1418/7249.

50. The 1946 hearings are H.R., Committee on Rivers and Harbors, *Hearings . . . on . . . the Improvement of the Arkansas River and Tributaries . . .*, 79th Cong., 2d sess., May 8–9, 1946, quotes at 3, 113; see also Moore and Moore, *The Army Corps*, 31–33.

51. Senate, Committee on Commerce, *Hearings: Rivers and Harbors*, 79th Cong., 2d sess., June 1946, at 2, 39–45.

52. Ibid., 61, 75, 86, 142–143.

53. Ibid., 121–122, 125–126, 131. The Corps rebuttal report is printed at 143–153.

54. Senate, Committee on Commerce, Subcommittee on Rivers and Harbors, *Hearings: Construction of Certain Public Works on Rivers and Harbors*, 66th Cong., 1st sess., June 1939, at 6, 10.

55. H.R., Committee on Rivers and Harbors, *Hearings . . . on the Improvement of Waterway Connecting the Tombigbee and Tennessee Rivers, Ala. and Miss.*, 79th Cong., 2d sess., May 1–2, 1946. Pp. 3–117 contain the 1939 report; 119–178 the 1946 revision; 179ff. the hearings; quote at 185.

56. H.R., Committee on Appropriations, *Investigation of Corps of Engineers Civil Works Programs: Hearings before the Subcommittee on Deficiencies and Army Civil Functions*, 82d Cong., 1st sess., 1951; 2 vols., vol. 2, quote at 154–155. Later there was a struggle over how to put environmental values into the analysis; see Stine, "Environmental Politics."

57. See, among many possible examples, H.R., Committee on Public Works, Subcommittee on Flood Control, *Deauthorize Dillon Dam* (1947), 8–11; Senate, Committee on Public Works, Subcommittee on Flood Control, *1948 Hearings*, 100–112; E. Peterson, *Big Dam Foolishness*; Leuchtenberg, *Flood Control Politics*, 49.

58. H.R., Committee on Public Works, Subcommittee to Study Civil Works, *Study of Civil Works: Hearings* (1952), part 2; idem, *The Flood Control Program of the Department of Agriculture; Report*, 82d Cong., 2d sess., December 5, 1952.

59. See H.R., Committee on Flood Control, *1946 Hearings*, 114.

60. Leopold and Maddock, *Flood Control Controversy*.

61. H.R., Committee on Flood Control, *1943 and 1944 Hearings*, vol. 2 (1944), 621.

62. "Memorandum from Harlan H. Barrows, Director CVPS to Commissioner [Harry Bashore], March 15, 1944," in N.A. 115/7/639/131.5.

63. Memorandum from John Page to Secretary of Interior Harold Ickes,

dated March 28, 1939, N.A. 115/7/639/131.5. His instances of cooperation included planning for both the Pine Flat (Kings River) and Friant dams.

64. Memorandum, Ickes to Roosevelt, July 19, 1939, N.A. 115/7/639/131.5.

65. Worster, *Rivers of Empire*, chap. 5; Reisner, *Cadillac Desert.*

66. The files of the Bureau of Reclamation begin with a letter from W. P. Boone of the Kings River Water Association, dated January 2, 1936 (N.A. 115/7/1643/301). The possibilities of government dams on the Kings had already been noticed by the Federal Power Commission: see Ralph R. Randell, *Report to the Federal Power Commission on the Storage Resources of the South and Middle Forks of Kings River, California* (Washington, D.C.: Federal Power Commission, June 5, 1930), a copy of which is in N.A. 115/7/1643/ B. W. Gearhart's bill was H.R. 1972, dated February 7, 1939.

67. "Kings Park and Pine Flat Tie Up Fails," *San Francisco Chronicle*, March 30, 1939, 12; letter, L. B. Chambers to Harry L. Haehl, August 18, 1938, N.A., San Bruno, Calif., R.G. 77, uncatalogued general administrative files (1913–1942) of main office, South Pacific Division, Corps of Engineers, Box 17, FC 501. For further correspondence between water companies and the Corps, see N.A. (Suitland) 77/111/678/7402/1.

68. See memorandum from McCasland to unnamed "Hydraulic Engineer," dated July 22, 1939; S. P. McCasland, *Kings River, California. Project Report No. 29*, dated June 1939; both in N.A. 115/7/642/301.

69. See penciled memorandum of telephone call "by R. A. Sterzik" dated February 25, 1939, which the unnamed recipient of the call recorded as a chart giving annual benefits, annual cost, and "degree of protection" (measured inversely as frequency of floods exceeding capacity) for three reservoir volumes. This pertained to the Kern River, which soon afterward was swept up in the same controversy that surrounded the Kings. N.A., San Bruno, R.G. 77, accession no. 9NS-77-91-033, Box 3, folder labeled "Kern River Survey."

70. Memorandum, May 6, 1939, by B. W. Steele (principal engineer) to the Board of Engineers, recommending a reservoir with capacity of 780,000 acre feet; letter, May 16, 1939, M. C. Tyler, assistant chief of engineers, to Warren T. Hannum, division engineer; "Comment" (on this letter), May 29, 1939, by L. B. Chambers, district engineer, to chief of engineers via division engineer; letter, June 1939?, Hannum to Board of Engineers; all in N.A. 77/111/678/7402/1; memorandum, May 18, 1939, Chambers to Hannum, in N.A. (San Bruno), general administrative files, main office of South Pacific Division of Corps, Box 17, FC 501. The location of the (undated) graphs in the files suggests they were prepared by, or under, Steele. Since they supported the smaller dam, it is significant and perhaps surprising that the files of the Bureau of Reclamation also contain a copy, along with the district engineer's original report.

71. See letter, December 11, 1939, R. A. Wheeler (chief of engineers) to John Page (commissioner of reclamation). Page recommended the change on the same day in a memorandum to the chief engineer in Denver. See N.A. 115/7/642/301.

72. See Page memorandum in previous note; also a letter, October 28, 1939, Denver chief engineer to Page, which I found in the files of the Corps, N.A. 77/111/678/7402.

73. In California they wanted to credit only half of this quarter to irrigation; see memoranda, June 15, 1939, district engineer Chambers to division engineer Hannum; and June 16, 1939, Hannum to Board of Engineers, in N.A. (San Bruno), R.G. 77, general administrative files of main office, South Pacific Division, Box 17, FC 501.

74. This equal division of benefits was proposed in McCasland's *Kings River Project* (note 68); see especially the report summary in the files of the Corps of Engineers. It is attached there to some critical comments by division engineer Hannum, and an equally critical letter (dated January 16, 1940) to the assistant chief of engineers, Thomas M. Robins. Hannum argued that the benefits of flood control far exceeded those of irrigation for the project. But the Corps engineers in Washington were by then feeling pressure from the president, and were eager to advise him that an agreement had been reached; see letter, January 16, 1940, Robins to Page, all in N.A. 77/111/678/7402/1.

75. 76th Cong., 3d sess., H.D. 630 [10503], *Kings River and Tulare Lake, California . . .: Preliminary Examination and Survey* [by Corps of Engineers], February 2, 1940; idem, H.D. 631 [10501], *Kings River Project in California. . .: Report of the Bureau of Reclamation*, February 12, 1940. On the premature release of the Corps report, see memoranda by Ickes and Frederic Delano (of the National Resources Planning Board) to Roosevelt, and the explanation by Harry Woodring, secretary of war, in N.A. (San Bruno), general administrative files of main office, South Pacific Division of Corps, Box 17, FC 501.

76. Roosevelt's decision is printed in the Bureau's report (H.D. 631). On his understanding of the mission of the Corps, and for the Ickes remark, see memoranda by Roosevelt to Woodring, June 6, 1940, and Ickes to Roosevelt, received in the White House the same day; copies of both in N.A. (San Bruno), general administrative files, main office, South Pacific Division of Corps, Box 17, FC 501.

77. Maass and Anderson, *Desert Shall Rejoice*, 264–265; Hundley, *Great Thirst*, 261.

78. H.R., Committee on Flood Control, *Hearings, 1941*, 97ff.; idem, *1943 and 1944 Hearings*, vol. 1, 249ff.; vol. 2, 588ff.; *Congressional Record*, 90 (1944), 4123–4124.

79. Maass and Anderson, *Desert Shall Rejoice*, 260. The Corps changed its allocation of benefits in the Kings River from time to time, evidently for reasons of political convenience: see Maass, *Muddy Waters*, chap. 5. But these changes also reflected genuine uncertainty, as evidenced by the appearance of abstract discussions of allocation methods in internal documents, such as "Summary of Cost Allocation Studies on Authorized Pine Flat Reservoir and Related Facilities, Kings River, California," a report by the Sacramento District, Corps of Engineers, dated October 28, 1946. I thank Allen Louie of the Sacramento District Planning Division for providing me with a copy of this document.

80. Senate, Committee on Irrigation and Reclamation, Subcommittee on Senate Resolution 295, *Hearings: Central Valley Project, California*, 78th

Cong., 2d sess., July 1944; *Fresno Bee*, issues of April 25, 26, 29, May 29, September 27, 30, October 5, 23, all 1941; also several papers in June 1943. A particularly early expression of opposition to the big water interests is a leaflet, the *Pine Flat News*, dated April 15, 1940, in N.A. 115/7/639/023.

81. H.R., Committee on Public Works, Subcommittee to Study Civil Works, *Economic Evaluation of Federal Water Resource Development Projects: Report . . . by Mr. [Robert] Jones of Alabama*, 82d Cong., 2d sess., House Committee Print no. 24, December 5, 1952, at 14–18. Sometimes the Bureau didn't even convert from gross to net agricultural revenue: see A. B. Roberts, *Task Force Report on Water Resources Projects: Certain Aspects of Power, Irrigation and Flood Control Projects*, prepared for the Commission on Organization of the Executive Branch of the Government, Appendix K (Washington, D.C.: USGPO, January 1949), 21.

82. Leslie A. Miller et al., *Task Force Report on Natural Resources: Organization and Policy in the Field of Natural Resources*, prepared for the Commission on Organization of the Executive Branch of Government, Appendix K (Washington, D.C.: USGPO, January 1949), 23.

83. H.R., Committee on Public Works, Subcommittee to Study Civil Works, *Economic Evaluation*, 7; idem., *Hearings*, 489–490; H.R., Committee on Flood Control, *1943 and 1944 Hearings*, vol. 2 (1944), 640, 633. The Bureau even applied this form of accounting to whole river basins, so that better projects could cover for the worst ones: Reisner, *Cadillac Desert*, 140–141. Elizabeth Drew, "Dam Outrage," 56, cited a remark that cost-benefit "measurements are pliant enough to prove the feasibility of growing bananas on Pike's Peak." This remark could only have been inspired by the Bureau of Reclamation, which indeed had to be particularly inventive in the mountains and high plains of Colorado.

84. John M. Clark, Eugene L. Grant, Maurice M. Kelso, *Report of Panel of Consultants on Secondary or Indirect Benefits of Water-Use Projects*, dated June 26, 1952, 3, 12. The occasion for the manual was the Bureau's refusal to endorse the F.I.A.R.B.C. *Proposed Practices*, discussed below. There is a copy of this report in N.A. 315/6/4.

85. H.R., Committee on Public Works, Subcommittee to Study Civil Works, *The Civil Functions Program of the Corps of Engineers, United States Army. Report . . . by Mr. Jones of Alabama*, 82d Cong., 2d sess., December 5, 1952, at 6, reported that since 1930, the Board of Engineers had decided unfavorably on 55.2 percent of surveys and preliminary reports. Many rejected projects were later approved, as benefits came to be defined more expansively. Even so, the Corps used its cost-benefit yardstick to delay the more doubtful projects.

86. In 1938, Sacramento district engineer L. B. Chambers decided against a project on the Humboldt River in Nevada. The interests protested to Warren T. Hannum, the division engineer in San Francisco, complaining that their water had been valued at only $1 per acre-foot, while the city folk in southern California were being credited with values twenty or more times higher. Hannum, seemingly convinced, asked Chambers to justify the analysis, which he then did in some detail. N.A. (San Bruno), general administrative files, main office, South Pacific Division of Corps, Box 17, FC 501. The grasshopper calculation is men-

tioned in an unpublished autobiography by William Whipple, Jr., dated 1987, held by the archives of the Office of History, Army Corps of Engineers. For the number of engineers, see U.S. Commission on Organization of the Executive Branch of Government, *The Hoover Commission Report* (New York: McGraw-Hill, 1949; reprinted, Westport, Conn.: Greenwood Press, 1970), 279.

87. From River and Harbor Circular Letter no. 39, June 9, 1936, in N.A. 77/142/11. Other early circulars on economic analysis include R&H 43 (June 22, 1936); R&H 46 (August 12, 1938); R&H 49 (August 23, 1938); R&H 42 (August 11, 1939); R&H 43 (August 14, 1939); R&H 62 (December 27, 1939); R&H 29 (June 1, 1940); R&H 43 (August 30, 1940). These may be found in N.A. 77/142/11–16. Many of the circulars of 1939 and 1940 are concerned with interagency harmonization of economic procedures. Some manuals from the late 1950s and early 1960s may be found in Office of History, Army Corps of Engineers, XIII–2, 1956–62 Manuals.

88. Quoted in a mimeographed pamphlet by J. R. Brennan, written for the War Department, Corps of Engineers, Los Angeles Engineer District, *Benefits from Flood Control. Procedure to be followed in the Los Angeles Engineer District in appraising benefits from flood control improvements*, December 1, 1943 (earlier editions, October 1, 1939, April 15, 1940), N.A., Pacific Southwest Region (Laguna Niguel, California), 77/800.5. The Chief of Engineers approved this pamphlet for circulation to other districts without sanctioning it as generally binding.

89. H.R., Committee on Flood Control, *Hearings on Levees and Flood Walls, Ohio River Basin*, 75th Cong., 1st sess., June 1937, at 140–141.

90. An example of historical damages ($13,888) averaging much less than "potential damages" ($43,000) is in H.R., 76th Cong., 3d sess. (1940), H.D. 719 [10505], *Walla Walla River and Tributaries, Oregon and Washington*, 17. For a formal discussion of these methods, see Corps of Engineers, Los Angeles Engineer District, *Benefits from Flood Control*, chaps. 1–2. These general methods were often cited in project reports, and occasionally even in congressional hearings, e.g., H.R., Committee on Flood Control, *Hearings, 1938*, 207.

91. 76th Cong., 2d sess., 1940, H.D. 479 [10503], *Chattanooga, Tenn. and Rossville, Ga.*, 29–30, 33.

92. "Memorandum of the States of Colorado, Kansas, and Nebraska with Reference to a Flood Control Plan for the Republican River Basin," July 13, 1942; Memorandum from Kansas City district engineer A. M. Neilson to division engineer, April 11, 1941, both in N.A. 77/111/1448/7402.

93. Letter, C. L. Sturdevant, division engineer, to Thomas M. Robins, office of chief of engineers, December 11, 1939, N.A., 77/111/1448/7402; also H.R., 76th Cong., 3d sess. (1940), H.D. 842 [10505], *Republican River, Nebr. and Kans.* (Preliminary Examination and Survey).

94. The agreement consisted mainly of heaping together most projects that either agency had ever considered. There is extensive discussion in *Congressional Record*, 90 (1944), e.g., at 4132, on the Republican River. For project surveys see 81st Cong., 2d sess. (1949–50), H.D. 642 [11429a], *Kansas River and Tributaries, Colorado, Nebraska, and Kansas*; also Wolman et al., *Report.*

95. H.R., Committee on Public Works, Subcommittee to Study Civil Works, *Study of Civil Works*, 25; idem, *Civil Functions of Corps*, 34 (both 1952).

96. J. L. Peterson of the Ohio River Division of the Corps, 1954, cited in Moore and Moore, *Army Corps*, 37–39.

97. For Wheeler, see H.R., Committee on Rivers and Harbors, *Hearings on Tombigbee and Tennessee*, 185. On Isabella reservoir: *Definite Project Report. Isabella Project. Kern River, California. Part VII—Recreational Facilities* (August 27, 1948), Appendix A. "Preliminary Report of Recreational Facilities by National Park Service," in N.A. (San Bruno), R.G. 77, accession no. 9NS-77-91-033, Box 2. On the survey of experts: U.S. Department of the Interior, National Park Service, *The Economics of Public Recreation: An Economic Study of the Monetary Value of Recreation in the National Parks* (Washington, D.C.: Land and Recreational Planning Division, National Park Service, 1949). The Congress had authorized the Corps to support waterway traffic by yachts, houseboats, etc., in legislation of 1932; see Turhollow, *Los Angeles District*.

98. Senate, Committee on Public Works, Subcommittee on Flood Control—Rivers and Harbors, *Hearings: Evaluation of Recreational Benefits from Reservoirs*, 85th Cong., 1st sess., March 1957, at 33.

99. The need for "objectivity" was invoked by Edmund Muskie of Maine after Elmer Staats of the Bureau of the Budget spoke of the element of judgment: Senate, Committee on Public Works, Subcommittee on Flood Control—Rivers and Harbors, *Hearings: Land Acquisition Policies and Evaluation of Recreation Benefits*, 86th Cong., 2d sess., May 1960, 151; see also U.S. Water Resources Council, *Evaluation Standards for Primary Outdoor Recreation Benefits* (Washington, D.C.: USGPO, June 4, 1964).

100. N.A. 315/2/1, first file, called "Interdepartmental Group," 1943–1945. Price's paper concerned a proposed dam on the Alabama-Coosa river system.

101. N.A. 315/2/1, 1st meeting, January 26, 1944.

102. N.A. 315/2/1, meetings 12 (January 25, 1945), 23 (December 27, 1945), 24 (January 31, 1946), 27 (April 25, 1946).

103. N.A. 315/6/1, 1st meeting, April 24, 1946. The members were G. L. Beard, chief of flood control division, Corps of Engineers; J. W. Dixon, director of project planning, Bureau of Reclamation; F. L. Weaver, chief of river basin division, Federal Power Commission; E. H. Wiecking, office of the secretary of agriculture. Two members of the staff were identified as economists: N. A. Back of Agriculture and G. E. McLaughlin of the Bureau of Reclamation. To them should be added M. M. Regan of Agriculture, who appeared at the second meeting. R. C. Price was also on the staff.

104. The first progress report on "Qualitative Aspects of Benefit-Cost Practices" used by the four agencies is attached to the minutes of the 29th meeting of the subcommittee; the second progress report on "Measurement Aspects of Benefit-Cost Practices" was distributed for the 50th meeting, in N.A. 315/6/1 and 315/6/3. See Federal Inter-Agency River Basin Committee, Subcommittee on Benefits and Costs, *Proposed Practices for Economic Analysis of River Basin Projects* (Washington, D.C.: USGPO, 1950), 58–70, 71–85.

105. A copy is in N.A. 315/6/3, 55th meeting. The principal authors, identified in an almost illegible carbon copy of assignments of tasks in 315/6/5, were evidently M. M. Regan and E. H. Wiecking, with assistance from E. C. Weitsell and N. A. Back.

106. The second edition of *Proposed Practices* (1958) took an even stronger line and denied their legitimacy altogether.

107. FIARBC, *Proposed Practices*, 7, 27. The book retreated some in the 1958 edition, deleting the quoted sentence and listing the value of life with scenic values among "intangibles." But it added a footnote declaring that "it may be desirable in some cases to provide uniform allowances of justifiable expenditure values for certain intangibles" (p. 7) On this topic, see Porter "Objectivity as Standardization."

108. See "First Progress Report of the Work Group on Benefits and Costs: Arkansas-White-Red Report" (by a subcommittee of an interagency committee on the Arkansas-White-Red rivers), in N.A. 315/6/5; also Wallace R. Vawter, "Case Study of the Arkansas-White-Red Basin Inter-Agency Committee," in U.S. Commission on Organization of the Executive Branch of Government [second Hoover Commission], Task Force on Water Resources and Power, *Report on Water Resources and Power* (n.p. June 1955), 3 vols., vol. 3, 1395–1472. Asked for its advice, the F.I.A.R.B.C. subcommittee first recommended an index of 150 for prices received and 175 for prices paid by farmers, then decided to set both at 215, so that the ratios remained constant and the projection was without effect.

The agencies were driven to cooperate in these river basin committees by fear that independent bureaucracies comparable to the Tennessee Valley Authority might be created; see Goodwin, "Valley Authority Idea."

109. H.R., Committee on Public Works, Subcommittee to Study Civil Works, *Study of Civil Works* (1952), 7. Pick dismissed Arthur Maass's severe criticism of the Corps in *Muddy Waters* as an attempt to build up "his philosophy of government, which is a greater centralized authority in the executive branch." Among Maass's crimes was to use the National Archives: "Criticism of the Corps of Engineers is the vehicle selected through which to peddle the philosophy of Government of a small and effective group who have been able to gain access to the archives of this great Government of ours to select and use to their advantage any information which can be found in the writings and sayings of the leaders of those various sections of Government, that is not generally available to all of the people of the United States." In response to a critical article by the governor of Wyoming, he wrote: "Apparently Mr. Miller would have one believe that the corps can influence the vote of the United States Senate. This is, of course, an absurd position." Ibid., 84, 107.

110. See Bureau of Budget files held by Office of History, Corps of Engineers, file labeled "Bureau Projects with Issues. 1947–1960. Corps Projects with Issues. 1948–1960," e.g., a report critical of recreation benefits dated May 31, 1960, and another opposing a project with a calculated benefit-cost ratio of 0.93: this number should be decisive except in the case of "unusual and major intangible benefits" such as loss of life. On the fruitless efforts of the Budget Bureau to rein in the Corps, see Ferejohn, *Pork Barrel Politics*, 79–86. The Budget Bureau's successor agency, the Office of Management and Budget, has become the most outspoken advocate of cost-benefit analysis in the federal government. Its power, on paper at least, reached a peak under Ronald Reagan, who required all new regulations to be supported by cost-benefit analyses. This dis-

couraged new regulations, as intended, but was too unwieldy for O.M.B. to enforce in detail. See Smith, *Environmental Policy.*

111. From a draft manuscript, dated December 15, 1949, in Office of History, Army Corps of Engineers, files on Civil Works Reorganization, 1943–49, First Hoover Commission, III 3–13, "corresp: fragments, MG Pick. 1949." U.S. Commission On Organization of the Executive Branch of Government, *The, [first] Hoover Commission Report on Organization of the Executive Branch of Government* (New York: McGraw-Hill, 1949), chap. 12. For Corps rebuttal by C. H. Chorpening, see H.R., Committee on Public Works, Subcommittee to Study Civil Works, *Study of Civil Works* (1952), 61.

112. U.S. Commission [2d Hoover Commission], *Report on Water Resources* (1955), vol. 1, 24, 104–110; vol. 2, 630, 652–653. The call for an objective panel to evaluate projects was echoed by Engineers Joint Council, *Principles of a Sound Water Policy* (1951 and) *1957 Restatement*, Report No. 105, May 1957, and later by Carter, "Water Projects." On the Hoover commissions and the effort to streamline American bureaucracy, see Crenson and Rourke, "American Bureaucracy."

113. Moore and Moore, *Army Corps*; Reuss, "Coping with Uncertainty."

114. E.g., Clark, *Economics of Public Works*, a work prepared during the Depression under the National Planning Board and the National Resources Board of the Public Works Administration. George Stigler, who wrote an influential paper on the "new welfare economics" in 1943, cut this set of teeth while apportioning benefits for the National Resources Planning Board; see his *Unregulated Economist*, 52.

115. Hammond, *Benefit-Cost Analysis*; idem, "Convention and Limitation."

116. The highway officials developed a "Red Book" to match the water analysts' green one: American Association of State Highway Officials (AASHO), Committee on Planning and Design Policies, *Road User Benefit Analysis for Highway Improvements* (Washington, D.C.: AASHO, 1952); see also Kuhn, *Public Enterprise Economics.*

117. Fortun and Schweber, "Scientists and the Legacy."

118. Leonard, "War as Economic Problem"; Orlans, "Academic Social Scientists."

119. Some economists have wanted to deny that their specialty could have been born impure, and have proposed instead that it was a natural outgrowth of welfare economics, specifically the Kaldor-Hicks reading of Pareto optimality. But histories of cost-benefit analysis by practitioners often recognize its bureaucratic origins; this applies not only to Hammond's critical history, but also Prest and Turvey, "Cost-Benefit Analysis"; Dorfman, "Forty Years." Although the former is a British paper, both identify the origins of cost-benefit analysis specifically with the Army Corps of Engineers.

120. U.S. Bureau of Agricultural Economics, "Value and Price of Irrigation Water," typescript, marked for administrative use only, by the California Regional Office (Berkeley), dated October 1943, no authors named, University of California, Berkeley, Water Resources Library Archives, G4316 G3–1. My impression that most BAE planning was not based on cost-benefit considerations is informed by a cursory inspection of its archives, which suggest that even with respect to water projects it did not habitually undertake to quantify benefits be-

fore the late 1940s. See, for example, U.S. Department of Agriculture, *Water Facilities Area Planning Handbook*, January 1, 1941, in N.A. 83/179/5. After 1950, agricultural economists began publishing regularly on the costs and benefits of water projects, later expanding to the analysis of other programs; for example: Regan and Greenshields, "Benefit-Cost Analysis"; Gertel, "Cost Allocation"; Ciriacy-Wantrup, "Cost Allocation"; Griliches, "Research Costs." On the BAE's history, see Hawley, "Economic Inquiry," 293–299.

121. Regan and Greenshields, "Benefit-Cost," relies on sources like Clark, *Public Works*, and Grant, *Engineering Economy.*

122. Margolis, "Secondary Benefits"; Eckstein, *Water-Resource Development*; Krutilla and Eckstein, *Multiple-Purpose River Development*; McKean, *Efficiency in Government*; Margolis, "Economic Evaluation"; U.S. Bureau of the Budget, Panel of Consultants [Maynard M. Hufschmidt, chairman, Krutilla, Margolis, Stephen Marglin], *Standards and Criteria for Formulating and Evaluating Federal Water Resource Developments* (Washington, D.C.: Bureau of the Budget, June 30, 1961); Haveman, *Water Resource Investment.*

123. See Sagoff, *Economy of the Earth*, 76.

124. Weisbrod, *Economics of Public Health*; Weisbrod, "Costs and Benefits of Medical Research"; Hansen, "Investment in Schooling"; Dodge and Stager, "Economic Returns to Graduate Study."

125. Topics covered in Dorfman, *Measuring Benefits.* Quote from Fritz Machlup, "Comment," on Burton Weisbrod, "Preventing High School Dropouts," at 155. His intention was to disparage Weisbrod's neglect of "noneconomic" values (Machlup's scare quotes).

126. A detailed and well-argued example, the attempt to control pollution in the Delaware River basin, is presented by Ackerman, *Uncertain Search.* His criticism is by no means limited to economic quantification.

127. Fischhoff, *Acceptable Risk*, xii, 55–57, quote at 57. While resisting the use of risk analysis by interested parties, they regard more favorably the implicit codes of professional judgment. "Narrow solutions are to be expected when professionals have a limited perspective on their own and little influence on higher-level policymaking" (p. 64).

128. Written by Partha Dasgupta, Amartya Sen, and Stephen Marglin. They called this ambition unrealizable, but aimed to pursue it as far as possible: United Nations Industrial Development Organization, *Guidelines for Project Evaluation* (Project Formulation and Evaluation Series, no. 2; New York: United Nations, 1972), 172.

Chapter Eight
Objectivity and the Politics of Disciplines

Epigraph from Finney, *Statistical Method*, 170.

1. Weber, *Economy and Society*, vol. 1, 225–226; also vol. 2, 983–985. Habermas, *Structural Transformation*, takes Weber for granted and interprets the push for calculability and impersonality in state administration as a consequence of the requirements of bourgeois capitalism.

2. Heclo, *Government of Strangers*, 158, 171.

3. Wilson, *Bureaucracy*, x, 31.

4. Ibid., 342; also Price, *Scientific Estate*, 57–75. On the bureaucratic-legal mode, see White, "Rhetoric and Law."

5. Hammond, "Convention and Limitation," 222. The use of quantitative analysis for legitimation has been widely emphasized, though often with the implication that the analysis is powerless, even fraudulent, and that decisions are really made on other grounds. There is a useful discussion in Benveniste, *Politics of Expertise*, 56ff.

6. Jasanoff, *Fifth Branch*, 9; also Balogh, *Chain Reaction*, 34.

7. Hofstadter, *Anti-Intellectualism*, 428. On Jacksonian bureaucracy, see Wood, *Radicalism of American Revolution*, 303–305.

8. The reluctance of American courts to face the complexities of statistics has been criticized by statisticians as well as by legal scholars: see DeGroot et al., *Statistics and the Law*, esp. Maier, Sacks and Zabell, "Hazelwood," and Finkelstein and Levenbach, "Price-Fixing Cases"; also Tribe, "Trial by Mathematics."

9. Wilson, *Bureaucracy*, 280, 286, citing Nathan Glazer. Martin Bulmer, "Governments and Social Science," contrasts the U.S. with Britain on theoretical vs. practical expertise.

10. Roger Smith and Brian Wynne, "Introduction," and Wynne, "Rules of Laws," in Smith and Wynne, *Expert Evidence*, quote at 51. On expert testimony in the United States, see Freidson, *Professional Powers*, 100–102.

11. McPhee, *Control of Nature*, 147–148.

12. MacKenzie, "Negotiating Arithmetic."

13. Jasanoff, "Misrule of Law"; idem, *Fifth Branch*, 58; idem, "Problem of Rationality." Cairns and Pratt, "Bioassays," 6, remark that American regulators long preferred chemical and physical tests to biological ones for evaluating hazards to the environment, because the former were readily quantified and relatively well standardized, while the latter were merely germane. When at last outside researchers worked out routines for single-species and then multispecies biological assays, the regulators slowly accepted them.

14. Brickman et al., *Controlling Chemicals*, 304.

15. From a report of the National Academy of Sciences, 1983, cited in Jasanoff, *Risk Management*, 26; National Research Council, *Regulating Chemicals*, 33.

16. Jasanoff, "Science, Politics."

17. Bledstein, *Culture of Professionalism*, 90; Starr, *Social Transformation of American Medicine.*

18. Ackerman and Hassler, *Clean Coal, Dirty Air*, 4; Vogel, "New Social Regulation"; Shabman, "Water Resources Management." For a wider perspective, see Lowi, *End of Liberalism.*

19. Ralph Frammolino, "Getting Grades for Diversity," *Los Angeles Times*, February 23, 1994, A15; last quote from *Notice* of University of California Academic Senate, 18 (4), February 1994, 2.

20. See Bulmer et al., "Social Survey," and Gorges, "Social Survey in Germany." Bulmer shows how objectivity was pursued by Chicago sociologists to escape politics in "Decline of Social Survey."

21. Stigler, *History of Statistics*, 28.

22. "The only trades which it seems possible for a joint stock company to carry on successfully without an exclusive privilege are those of which all the operations are capable of being reduced to what is called a Routine, or to such a uniformity of method as admits of little or no variation." In every other case, he argued, the "superior vigilance and attention of private adventurers" would surely defeat the organization men. Smith, *Wealth of Nations*, vol. 2, 242.

23. Swijtink, "Objectification of Observation," 278; Schaffer, "Astronomers Mark Time"; Daston, "Escape from Perspective."

24. From an exchange of letters, 1860–61, quoted in Rothermel, "Images of the Sun," 157–158.

25. Written in 1866, quoted in Blanckaert, "Méthodes des moyennes," 225.

26. Olesko, *Physics as a Calling*; Gooday, "Precision Measurement." The spread of error theory in experimental physics was sporadic, however, in part because of persistent doubts about the homogeneity of data. An example: C. V. Goys, seeking a better measurement of the gravitational constant in 1895, found that weekend traffic shook his apparatus and gave a different mean value. So he ignored weekend measures. Mendoza, "Theory of Errors."

27. Makeham, "Law of Mortality," 301–302.

28. "Proceedings of the Institute of Actuaries of Great Britain and Ireland," *Assurance Magazine*, 1 (1850–51), no. 1, 103–112; Porter, "Education of an Actuary," 125.

29. Lawrence, "Incommunicable Knowledge," 507.

30. Geison, "Divided We Stand."

31. Frank, "Telltale Heart," 212; see also Evans, "Losing Touch." Warner, *Therapeutic Perspective*, shows that quantification was already promoting the objectification of medical practice in America by the late nineteenth century.

32. Desrosières, "Masses, individus, moyennes"; also Armatte, "Moyenne."

33. Review dated 1840, cited in Matthews, *Mathematics*, 75.

34. Weisz, "Academic Debate."

35. See, for example, Hill, "Observation and Experiment," and, "Smoking and Carcinoma." The debate was a serious one; R. A. Fisher himself was one of the skeptics. Berkson, "Smoking and Lung Cancer," argued that since the control group in a leading study of smoking and cancer was healthier than the smokers in almost every respect, it had to be an unusual population and not a fair control.

36. Matthews, *Mathematics*, chap. 5.

37. Hill, "Clinical Trial-II" (from a talk at Harvard Medical School, 1952), 29–31.

38. Marc Daniels, quoted in Hill, "Clinical Trial-I," 27.

39. Hill, "Clinical Trial-II," 34, 38; idem, "Philosophy of Clinical Trial" (1953), 12, 13; Sutherland, "Statistical Requirements," 50; see also Marks, "Notes from Underground," 318.

40. Hill, *Principles of Medical Statistics*; Marks, *Ideas as Reforms*, 15–16.

41. Hill, "Aims and Ethics," 5; idem, "Clinical Trial-II," 38.

42. Marks, "Notes from Underground."

43. Davis, "Life Insurance"; Sellers, "Office of Industrial Hygiene."

44. Trevan, "Determination of Toxicity," which introduced the idea and notation of LD 50; Finney, *Statistical Method.*

45. Mainland, *Clinical and Laboratory Data*, 145–147, discusses the testing of digitalis. The pharmacologist, he explained, reports to the manufacturer a mean and standard error for each sample.

46. Marks, "Ideas as Reforms," chap. 2.

47. Bodewitz et al., "Regulatory Science."

48. Quirk, "Food and Drug Administration," 222; also Temin, *Taking Your Medicine.* This approval was subsequently revoked. Such determinations are not made frivolously by the FDA; to impose a higher standard than statistical significance invites legal challenge by the drug companies.

49. A recent example is an effort to oblige physicians to make diagnoses according to programmable criteria, which was defended (unsuccessfully) by researchers in the name of science and of openness: Anderson, "Reasoning of the Strongest."

50. Marks, *Ideas as Reforms*, chaps. 3–4. On statistical tact, see p. 175: Major Greenwood used the phrase, and A. B. Hill cited it. Examples of the effort to standardize clinical practice, especially in research, include Cochrane et al., "Observers' Errors"; Hoffmann, *Clinical Laboratory Standardization.*

51. The drive for medical objectivity is of course not limited to statistics. Tests of organ compatibility, for example, have been valued as an objective basis for deciding who gets scarce organs, despite their failure to help much in predicting transplant success: Löwy, "Tissue Groups."

52. Justice John W. Goff of the Superior Court of New York, quoted in Kevles, "Testing the Army," 566.

53. Zenderland, "Debate over Diagnosis"; Carson, "Army Alpha."

54. Von Mayrhauser, "Manager, Medic, and Mediator"; Samelson, "Mental Testing"; Sutherland, *Ability, Merit and Measurement*, chap. 10.

55. Resnick, "Educational Testing"; Chapman, *Schools as Sorters.*

56. Danziger, *Constructing the Subject*, quotes at 79, 109.

57. Ibid., 81–83; Gigerenzer and Murray, *Cognition as Intuitive Statistics*, 27.

58. Ash, "Historicizing Mind Science"; Mauskopf and McVaugh, *Elusive Science*; Hacking, "Telepathy."

59. Danziger, *Constructing the Subject*, 148–149, 153–155; Gigerenzer, "Probabilistic Thinking"; Coon, "Standardizing the Subject."

60. Gigerenzer, "Probabilistic Thinking"; Gigerenzer et al., *Empire of Chance*, chaps. 3, 6; Hornstein, "Quantifying Psychological Phenomena." Critics of this mechanical ideal of inference have often pointed to the fear of subjectivity as a reason for its perpetuation: for example, see Parkhurst, "Statistical Hypothesis Tests."

61. Gigerenzer, "Superego, Ego, and Id."

62. Gigerenzer and Murray, *Cognition*; Gigerenzer et al., *Empire of Chance*, chap. 6.

63. Burgess, "Success or Failure on Parole," 245.

64. Kleinmuntz, "Clinical Judgment," 553. The *locus classicus* for this de-

bate is Meehl, *Clinical versus Statistical Prediction.* The most influential defender of clinical judgment has been Robert R. Holt; see his *Prediction and Research.* A history of early debates is Gough, "Clinical versus Statistical Prediction."

65. Collins, *Artificial Experts*; Ashmore et al., *Health and Efficiency*; Mirowski and Sklivas, "Why Econometricians Don't Replicate."

66. Balogh, *Chain Reaction.*

67. Albury, *Politics of Objectivity*, 36; Green, "Cost-Benefit Analysis as Mirage"; also Shapin, *Social History of Truth.* On skill and community, see the decidedly nostalgic book by Harper, *Working Knowledge.*

Chapter Nine
Is Science Made by Communities?

Epigraph from Harré, *Varieties of Realism*, 1.

1. Hollinger, "Free Enterprise." On individualism and community in Vannevar Bush's rhetoric, see also Owens, "Patents."

2. "A Word up Your Nose," *The Economist*, August 7, 1993, 20.

3. Harré, *Varieties of Realism*, 1–2, 6–7. On community talk in recent science studies, see Jacobs, "Scientific Community."

4. Fish, *Is There a Text in This Class*, 14–17.

5. Rudwick, *Great Devonian Controversy*, 448.

6. Heclo and Wildavsky, *Private Government of Public Money.*

7. Bender, "Erosion of Public Culture," 89; Bender, *Community and Social Change*, 149.

8. Leibnitz, "Measurement of Quality," 483–485.

9. Traweek, *Beamtimes and Lifetimes.*

10. Pinch, *Confronting Nature*, 207.

11. And not just in physics. Haraway, *Primate Visions*, 170–171, describes how Jane Goodall standardized her notetaking onto forms before entrusting it to a Tanzanian field assistant and then to an assortment of students.

12. Latour, *Science in Action.*

13. Shapin and Schaffer, *Leviathan and the Air-Pump.*

14. Biagioli, *Galileo Courtier*; Shapin, *Social History of Truth.*

15. Daston, "Republic of Letters."

16. Nyhart, "Writing Zoologically"; Harry M. Marks, "Local Knowledge: Experimental Communities and Experimental Practices, 1918–1950," paper given at University of California, San Francisco, May 1988. Readers will, I hope, forgive me for citing an unpublished paper on this point; I thank Harry Marks for making it available to me.

17. Holton, "Fermi's Group"; Holton, "On Doing One's Damndest: The Evolution of Trust in Scientific Findings," forthcoming. On research schools, see Geison, "Research Schools," and Geison and Holmes, *Research Schools.*

18. W. B. Trommsdorff, quoted in Hufbauer, *German Chemical Community*, 139.

19. Golinski, *Science as Public Culture*, 138.

20. Kuhn, *Structure of Scientific Revolutions*, 47.

21. McCloskey, *Rhetoric of Economics*, chaps. 9–10; Bazerman, *Shaping Written Knowledge*, chap. 9.

22. Novick, *That Noble Dream*, 4, 52–53, 89–90.

23. On the importance of external pressures in giving shape to the methods of science, see Knorr-Cetina, *Manufacture of Knowledge*, chap. 4.

24. Holton, "On Doing One's Damndest"; Hammond and Adelman, "Science, Values, and Human Judgment," 390–391. They lament: "Thus [the National Academy of Sciences] has already fallen victim to the ethic of the lawyer (and the journalist). Trust no one, is the rule, unless they can offer this negative proof."

25. Bulmer, "Social Indicator Research," 112, 119; first quote by William F. Ogburn, second by Leonard White. Foundations like the Rockefeller, wanting to support research that would not generate controversy, encouraged this attitude; see Craver, "Patronage."

26. Nathan, *Social Science in Government*, 94.

Bibliography

Note on Sources

The bibliography is intended to be complete, with two important exceptions. It omits most government publications that do not appear under the name of a particular author, except for a very few important books that are cited repeatedly in the notes. It also omits manuscript materials, including ephemeral publications that are unlikely to be available outside the archives in which I found them. In every case, archival locations are provided in the notes. At the Archives Nationales in Paris I have relied mainly on materials pertaining to public works in the F14 series. At the National Archives of the United States I have used four record groups: 77 (Army Corps of Engineers), 83 (Bureau of Agricultural Economics), 115 (Bureau of Reclamation), and 315 (Federal Inter-Agency River Basin Committee). Materials that have been properly catalogued are identified only by numbers and slashed lines: e.g., N.A. 77/111/642/301. Here 77 refers to the record group, 111 to the "entry," 642 to the box number, and all subsequent numbers to internal filing schemes. All archives I have consulted are listed in the acknowledgments.

Abir-Am, Pnina, "The Politics of Macromolecules: Molecular Biologists, Biochemists, and Rhetoric," *Osiris*, 7 (1992), 210–237.

Ackerman, Bruce, and William T. Hassler, *Clean Coal, Dirty Air.* (New Haven, Conn.: Yale University Press, 1981).

Ackerman, Bruce, et al., *The Uncertain Search for Environmental Quality* (New York: Free Press, 1974).

Adorno, Theodor W., "Scientific Experiences of a European Scholar in America," Donald Fleming, trans., in Donald Fleming and Bernard Bailyn, eds., *The Intellectual Migration: Europe and America, 1930–1960* (Cambridge, Mass.: Harvard University Press, 1969), 338–370.

Alborn, Timothy, "The Other Economists: Science and Commercial Culture in Victorian England" (Ph.D. dissertation, Harvard University, 1991).

———, "A Calculating Profession: Victorian Actuaries among the Statisticians," *Science in Context*, forthcoming, 1994.

Albrand, rapporteur, *Rapport de la Commission Spéciale des Docks au Conseil Municipal de la Ville de Marseille* (Marseille: Typographie des Hoirs Feissat Ainé et Demonchy, 1836).

Albury, Randall, *The Politics of Objectivity* (Victoria, N.S.W.: Deakin University Press, 1983).

Alcouffe, Alain, "The Institutionalization of Political Economy in French Universities, 1819–1896," *History of Political Economy*, 21 (1989), 313–344.

Alder, Ken, "A Revolution to Measure: The Political Economy of the Metric System in France," in Wise, *Values of Precision*, 39–71.

Alonso, William, and Paul Starr, eds., *The Politics of Numbers* (New York: Russell Sage Foundation, 1987).

American Psychological Association, *Publication Manual*, 2d ed. (Washington, D.C.: American Psychological Association, 1974).

Anderson, Benedict, *Imagined Communities: Reflections on the Origin and Spread of Nationalism*, 2d ed. (New York: Verso, 1991).

Anderson, Margo, *The American Census: A Social History* (New Haven, Conn.: Yale University Press, 1989).

Anderson, Warwick, "The Reasoning of the Strongest: The Polemics of Skill and Science in Medical Diagnosis," *Social Studies of Science*, 22 (1992), 653–684.

Ansari, Shahid L., and John J. McDonough, "Intersubjectivity—The Challenge and Opportunity for Accounting," *Accounting, Organizations, and Society*, 5 (1980), 129–142.

Arago, François, *Histoire de ma jeunesse* (Paris: Christian Bourgeois, 1985).

Armatte, Michel, "La moyenne à travers les traités de statistique au XIXè siècle," in Feldman et al., *Moyenne*, 85–106.

———, "L'économie à l'Ecole Polytechnique," in Belhoste et al., *Formation*, 375–396.

Arnett, H. E., "What Does 'Objectivity' Mean to Accountants," *Journal of Accountancy*, May 1961, 63–68.

Arnold, Joseph L., *The Evolution of the 1936 Flood Control Act* (Fort Belvoir, Va.: Office of History, U.S. Army Corps of Engineers, 1988).

Ash, Mitchell, "Historicizing Mind Science: Discourse, Practice, Subjectivity," *Science in Context*, 5 (1992), 193–207.

Ashmore, Malcolm, Michael Mulkay, and Trevor Pinch, *Health and Efficiency: A Sociology of Health Economics* (Milton Keynes, U.K.: Open University Press, 1989).

Ashton, Robert H., "Objectivity of Accounting Measures: A Multirule-Multimeasure Approach," *Accounting Review*, 52 (1977), 567–575.

Babbage, Charles, *On the Economy of Machinery and Manufactures*, 3d ed. (London: Charles Knight, 1833).

———, *A Comparative View of the Various Institutions for the Assurance of Lives* (1826; reprinted New York: Augustus M. Kelley, 1967).

Bailey, Arthur, and Archibald Day, "On the Rate of Mortality amongst the Families of the Peerage," *Assurance Magazine*, 9 (1860–61), 305–326.

Baker, Keith, *Condorcet: From Natural Philosophy to Social Mathematics* (Chicago: University of Chicago Press, 1975).

Balogh, Brian, *Chain Reaction: Expert Debate and Public Participation in American Commercial Nuclear Power, 1945–1975* (New York: Cambridge University Press, 1991).

Balzac, Honoré de, *Le curé de village* (1st ed., 1841; Paris: Société d'Editions Littéraires et Artistiques, 1901).

———, *Les Employés* in *La Comédie humaine*, vol. 7 (Paris: Gallimard, 1977).

Barber, William J., *From New Era to New Deal: Herbert Hoover, the Economists, and American Economic Policy, 1921–1933* (Cambridge, U.K.: Cambridge University Press, 1985).

Barnes, Barry, "On Authority and Its Relation to Power," in Law, *Power*, 180–195.

Bartrip, P.W.J., and P. T. Fenn, "The Measurement of Safety: Factory Accident Statistics in Victorian and Edwardian Britain," *Historical Research*, 63 (1990), 58–72.

Bauchard, Philippe, *Les technocrates et le pouvoir* (Paris: Arthaud, 1966).

Baum, Charles, "Des prix de revient des transport par chemin de fer," *Annales des Ponts et Chaussées* [5] 10 (1875), 422–481.

———, "Etude sur les chemins de fer d'intérêt local," *Annales des Ponts et Chaussées* [5], 16 (1878), 489–546.

———, "Des longueurs virtuelles d'un tracé de chemin de fer," *Annales des Ponts et Chaussées* [5], 19 (1880), 455–578.

———, "Note sur les prix de revient des transports par chemin de fer, en France," *Annales des Ponts et Chaussées* [6] 6 (1883), 543–594.

———, *Chemins de fer d'intérêt local du Département du Morbihan: Rapport de l'Ingénieur en chef* (Vannes: Imprimerie Galles, 1885).

———, "Le prix de revient des transports par chemin de fer," *Journal de la Société de Statistique de Paris*, 26 (1885): 199–217.

———, "Note sur les prix de revient des transports," *Annales des Ponts et Chaussées* [6] 15 (1888), 637–683.

Bazerman, Charles, *Shaping Written Knowledge* (Madison: University of Wisconsin Press, 1988).

Belhoste, Bruno, Amy Dahan Dalmedico, and Antoine Picon, eds., *La formation polytechnicienne* (Paris: Dunod, 1994).

Belpaire, Alphonse, *Traité des dépenses d'exploitation aux chemins de fer* (Brussels: J. F. Buschmann, 1847).

Bender, Thomas, *Community and Social Change in America* (New Brunswick, N.J.: Rutgers University Press, 1978).

———, "The Erosion of Public Culture: Cities, Discourses, and Professional Disciplines," in Thomas Haskell, ed., *The Authority of Experts* (Bloomington: Indiana University Press, 1984).

Benveniste, Guy, *The Politics of Expertise*, 2d ed. (San Francisco: Boyd & Fraser, 1977).

Berkson, Joseph, "The Statistical Study of Association between Smoking and Lung Cancer," *Proceedings of the Staff Meetings of the Mayo Clinic*, 30 (15), July 27, 1955, 319–348.

Berlanstein, Lenard, *Big Business and Industrial Conflict in Nineteenth-Century France* (Berkeley: University of California Press, 1991).

Bertillon, Louis-Adolphe, "Des diverses manières de mesurer la durée de la vie humaine," *Journal de la Société de Statistique de Paris*, 7 (1866), 45–64.

———, "Méthode pour calculer la mortalité d'une collectivité pendant son passage dans un milieu déterminé . . . ," *Journal de la Société de Statistique de Paris*, 10 (1869), 29–40, 57–65.

Bertrand, Joseph, *Eloges académiques. Nouvelle série* (Paris: Hachette, 1902).

Biagioli, Mario, "The Social Status of Italian Mathematicians, 1450–1600," *History of Science*, 27 (1989), 41–95.

———, *Galileo Courtier: Science, Patronage, and Political Absolutism* (Chicago: University of Chicago Press, 1993).

Bierman, Harold, "Measurement and Accounting," *Accounting Review*, 38 (1963), 501–507.

Blanckaert, Claude, "Méthodes des moyennes et notion de série suffisante en anthropologie physique (1830–1880)," in Feldman et al., *Moyenne*, 213–243.

Bledstein, Burton, *The Culture of Professionalism* (New York: Norton, 1976).

Bloor, David, "Left and Right Wittgensteinians," in Pickering, *Science*, 266–282.

Bodewitz, J.H.W., Henk Buurma, and Gerard H. deVries, "Regulatory Science and the Social Management of Trust in Medicine," in Wiebe E. Bijker, Thomas P. Hughes, and Trevor Pinch, eds., *The Social Construction of Technological Systems* (Cambridge, Mass.: MIT Press, 1987), 243–259.

Boltanski, Luc, *Les cadres: La formation d'un groupe social* (Paris: Editions de Minuit, 1982).

Booth, Henry, "Chemin de fer de Liverpool à Manchester: Notice historique," *Annales des Ponts et Chaussées*, 1 (1831), 1–92.

Bordas, Louis, "De la mesure de l'utilité des travaux publics," *Annales des Ponts et Chaussées* [2], 13 (1847), 249–284.

Bourguet, Marie-Noëlle, *Déchiffrer la France: La statistique départementale à l'époque napoléonienne* (Pairs: Edition des Archives Contemporaines, 1988).

Brautigam, Jeffrey, "Inventing Biometry, Inventing 'Man': Biometrika and the Transformation of the Human Sciences," (Ph.D. dissertation, University of Florida, 1993).

Brian, Eric, "Les moyennes à la Société de Statistique de Paris (1874–1885)," in Feldman et al., *Moyenne*, 107–134.

———, "Le Prix Montyon de statistique à l'Académie des Sciences pendant la Restauration," *Revue de synthèse*, 112 (1991), 207–236.

———, *La mesure de l'état: Administrateurs et géomètres au XVIIIe siècle* (Paris: Albin Michel, 1994).

Brickman, Ronald, Sheila Jasanoff, and Thomas Ilgen, *Controlling Chemicals: The Politics of Regulation in Europe and the United States* (Ithaca, N.Y.: Cornell University Press, 1985).

Brock, William, *Investigation and Responsibility: Public Responsibility in the United States, 1865–1900* (Cambridge, U.K.: Cambridge University Press, 1984).

Brown, Donaldson, *Centralized Control with Decentralized Responsibilities* (New York: American Management Association, 1927).

Brown, Richard D., *Knowledge Is Power: The Diffusion of Information in Early America, 1700–1865* (New York: Oxford University Press, 1989).

Brown, Samuel, "On the Fires in London During the 17 Years from 1833 to 1849," *Assurance Magazine*, 1, no. 2 (1851), 31–62.

Bru, Bernard, "Estimations laplaciennes," in Jacques Mairesse, ed., *Estimations et sondages* (Paris: Economica, 1988), 7–46.

Brun, Gérard, *Technocrates et technocratie en France, 1918–1945* (Paris: Editions Albatross, 1985).

Brundage, Anthony, *England's Prussian Minister: Edwin Chadwick and the Politics of Government Growth* (University Park: Pennsylvania State University Press, 1988).

Brunot, A., and R. Coquand, *Le Corps des Ponts et Chaussées* (Paris: Editions du Centre National de la Recherche Scientifique, 1982).

Bud-Frierman, Lisa, ed., *Information Acumen: The Understanding and Use of Knowledge in Modern Business* (London: Routledge, 1994).

Bulmer, Martin, "The Methodology of Early Social Indicator Research: William Fielding Ogburn and 'Recent Social Trends,' 1933," *Social Indicators Research*, 13 (1983), 109–130.

———, "Governments and Social Science: Patterns of Mutual Influence," in Bulmer, *Social Science Research*, 1–23.

———, ed., *Social Science Research and the Government: Comparative Essays on Britain and the United States* (Cambridge, U.K.: Cambridge University Press, 1987).

———, "The Decline of the Social Survey Movement and the Rise of American Empirical Sociology," in Bulmer et al., *Social Survey*, 291–315.

Bulmer, Martin, Kevin Bales, and Kathryn Kish Sklar, "The Social Survey in Historical Perspective," in Bulmer et al., *Social Survey*, 13–48.

———, eds., *The Social Survey in Historical Perspective* (Cambridge, U.K.: Cambridge University Press, 1991).

Burchell, Stuart, Colin Clubb, and Anthony Hopwood, "Accounting in Its Social Context: Towards a History of Value Added in the United Kingdom," *Accounting, Organizations, and Society*, 17 (1992), 477–499.

Burgess, Ernest W., "Factors Determining Success or Failure on Parole," in Andrew A. Bruce et al., *The Workings of the Indeterminate Sentence Law and the Parole System in Illinois* (1928; reprinted Montclair, N.J.: Patterson Smith, 1968).

Burke, Edward J., "Objectivity and Accounting," *Accounting Review*, 39 (1964), 837–849.

Burn, J. H., "The Errors of Biological Assay," *Physiological Review*, 10 (1930), 146–169.

Burn, J. H., D. J. Finney, and L. G. Goodwin, *Biological Standardization*, (1937; Oxford: Oxford University Press, 2d ed., 1950).

Burnham, John C., "The Evolution of Editorial Peer Review," *Journal of the American Medical Association*, 263, no. 10, March 9, 1990, 1323–1329.

Cahan, David, *An Institute for an Empire: The Physikalische-Technische Reichsanstalt, 1871–1918* (Cambridge, U.K.: Cambridge University Press, 1989).

Cairns, John Jr., and James R. Pratt, "The Scientific Basis of Bioassays," *Hydrobiologia*, 188/189 (1989), 5–20.

Calhoun, Daniel, *The American Civil Engineer: Origins and Conflict* (Cambridge, Mass.: MIT Press, 1960).

———, *The Intelligence of a People* (Princeton, N.J.: Princeton University Press, 1973).

Campbell-Kelly, Martin, "Large-Scale Data Processing in the Prudential, 1850–1930," *Accounting, Business, and Financial History*, 2 (1992), 117–139.

Carnot, Sadi, *Reflections on the Motive Power of Fire* (Manchester, U.K.: Manchester University Press, 1986).

Caron, François, *Histoire de l'exploitation d'un grand réseau: La Compagnie de Fer du Nord, 1846–1937* (Paris: Mouton, 1973).

Carson, John, "Army Alpha, Army Brass, and the Search for Army Intelligence," *Isis*, 84 (1993), 278–309.

Carter, Luther J., "Water Projects: How to Erase the 'Pork Barrel' Image," *Science*, 182, October 19, 1973, 267–269, 316.

Cartwright, Nancy, *Nature's Capacities and Their Measurement* (Oxford: Clarendon Press, 1989).

Caufield, Catherine, "The Pacific Forest," *New Yorker*, May 14, 1990, 46–84.

Chambers, R. J., "Measurement and Objectivity in Accounting," *Accounting Review*, 39 (1964), 264–274.

Chandler, Alfred, Jr., *Strategy and Structure: Chapters in the History of Industrial Enterprise* (Cambridge, Mass.: MIT Press, 1962).

———, *The Visible Hand: The Managerial Revolution in American Business* (Cambridge, Mass.: Harvard University Press, 1977).

Chapman, Paul Davis, *Schools as Sorters: Lewis M. Terman, Applied Psychology, and the Intelligence Testing Movement* (New York: New York University Press, 1988).

Chardon, Henri, *Les travaux publics: Essai sur le fonctionnement de nos administrations* (Paris: Perrin et Cie., 1904).

———, *L'administration de la France: Les fonctionnaires* (Paris: Perrin, 1908).

———, *Le pouvoir administratif* (Paris: Perrin, nouvelle ed., 1912).

Charle, Christophe, *Les hauts fonctionnaires en France au XIXe siècle* (Paris: Gallimard, 1980).

———, *Les élites de la République* (Paris: Fayard, 1987).

Chevalier, Michel, opening address, *Journal de la Société de Statistique de Paris*, 1 (1860), 1–6.

Cheysson, Emile, "Le cadre, l'objet et la méthode de l'économie politique" (1882), in Cheysson, *Oeuvres*, vol. 2, 37–66.

———, report of prize commission, 1883, *Journal de la Société de Statistique de Paris*, 25 (1884), 50–57.

———, "La statistique géométrique" (1887), in Cheysson, *Oeuvres*, vol. 1, 185–218.

———, *Oeuvres choisies*, 2 vols., Paris: A. Rousseau, 1911).

Christophle, Albert, *Discours sur les travaux publics prononcés . . . dans les sessions législatives de 1876 et 1877* (Paris: Guillaumin et Cie., n.d. [ca. 1888].

Church, Robert, "Economists as Experts: The Rise of an Academic Profession in the United States, 1870–1920," in Lawrence Stone, ed., *The University in Society* (Princeton, N.J.: Princeton University Press, 1974).

Ciriacy-Wantrup, S. V., "Cost Allocation in Relation to Western Water Policies," *Journal of Farm Economics*, 36 (1954), 108–129.

Clark, John M., *Economics of Planning Public Works* (1935; reprinted New York: Augustus M. Kelley, 1965).

Cochrane, A. L., P. J. Chapman, and P. D. Oldham, "Observers' Errors in Taking Medical Histories," *The Lancet*, May 5, 1951, 1007–1009.

Cohen, Patricia Cline, *A Calculating People: The Spread of Numeracy in Early America* (Chicago: University of Chicago Press, 1982).

Coleman, William, *Death Is a Social Disease: Public Health and Political Econ-*

omy in Early Industrial France (Madison: University of Wisconsin Press, 1982).

Collins, Harry, *Changing Order* (Los Angeles: Russell Sage Foundation, 1985).

———, *Artificial Experts: Social Knowledge and Intelligent Machines* (Cambridge, Mass.: MIT Press, 1990).

Colson, Clément-Léon, "La formule d'exploitation de M. Considère," *Annales des Ponts et Chaussées* [7], 4 (1892), 561–616.

———, "Note sur le nouveau mémoire de M. Considère," *Annales des Ponts et Chaussées* [7], 7 (1894), 152–164.

———, *Transports et tarifs*, 2d ed. (Paris: J. Rothschild, 1898).

———, *Théorie générale des phénomènes économiques*, vol. 1 of his *Cours.*

———, *Les travaux publics et les transports*, vol. 6 of his *Cours.*

———, *Cours d'économie politique professé à l'Ecole Nationale des Ponts et Chaussées*, 2d ed., 6 vols. (Paris: Gauthier-Villars et Félix Alcan, 1907–10).

Colvin, Phyllis, *The Economic Ideal in British Government: Calculating Costs and Benefits in the 1970s* (Manchester, U.K.: Manchester University Press, 1985).

Considère, Armand, "Utilité des chemins de fer d'intérêt local: Nature et valeur des divers types de convention," *Annales des Ponts et Chaussées* [7], 3 (1892), 217–485.

———, "Utilité des chemins de fer d'intérêt local: Examen des observations formulées par M. Colson," *Annales des Ponts et Chaussées* [7], 7 (1894), 16–151.

Converse, Jean M., *Survey Research in the United States: Roots and Emergence, 1890–1960* (Berkeley: University of California Press, 1987).

Coon, Deborah J., "Standardizing the Subject: Experimental Psychologists, Introspection, and the Quest for a Technoscientific Ideal," *Technology and Culture*, 34 (1993), 261–283.

Coriolis, G., "Premiers résultats de quelques expériences relatives à la durée comparative de différentes natures de grés," *Annales des Ponts et Chaussées*, 7 (1834), 235–240.

Couderc, J., *Essai sur l'Administration et le Corps Royal des Ponts et Chaussées* (Paris: Carillan-Goeury, 1829).

Courcelle-Seneuil, J.-G., "Etude sur le mandarinat français," in Thuillier, *Bureaucratie.*

Cournot, A. A., *Recherches sur les principes mathématiques de la théorie des richesses* (Paris: Hachette, 1838).

———, *Exposition de la théorie des chances et des probabilités* (Paris: Hachette, 1843).

Courtois, Charlemagne, *Mémoire sur différentes questions d'économie politique relatives à l'établissement des voies de communication* (Paris: Carillan-Goeury, 1833).

———, *Mémoire sur les questions que fait naître le choix de la direction d'une nouvelle voie de communication* (Paris: Imprimerie Schneider et Langrand, 1843).

Craver, Earlene, "Patronage and the Directions of Research in Economics: The

Rockefeller Foundation in Europe, 1924–1938," *Minerva*, 24 (1986), 205–222.

Crenson, Matthew A., and Francis E. Rourke, "By Way of Conclusion: American Bureaucracy since World War II," in Galambos, *New American State*, 137–177.

Crépel, Pierre, ed., *Arago*, vol. 4 of *Sabix, Bulletin de la Société des Amis de la Bibliothèque de l'Ecole Polytechnique*, May 1989.

Cronon, William, *Changes in the Land* (New York: Hill and Wang, 1983).

———, *Nature's Metropolis: Chicago and the Great West* (New York: Norton, 1991).

Cullen, Michael, *The Statistical Movement in Early Victorian Britain* (Hassocks, U.K.: Harvester, 1975).

Curtin, Philip, *Death by Migration: Europe's Encounter with the Tropical World in the Nineteenth Century* (Cambridge, U.K.: Cambridge University Press, 1989).

Danziger, Kurt, *Constructing the Subject: Historical Origins of Psychological Research* (Cambridge, U.K.: Cambridge University Press, 1990).

Daru, M. le comte, *Des chemins de fer et de l'application de la loi du 11 juin 1842* (Paris: Librairie Scientifique-Industrielle de L. Mathias, 1843).

Daston, Lorraine, *Classical Probability in the Enlightenment* (Princeton, N.J.: Princeton University Press, 1988).

———, "The Ideal and Reality of the Republic of Letters in the Enlightenment," *Science in Context*, 4 (1991), 367–386.

———, "Objectivity and the Escape from Perspective," *Social Studies of Science*, 22 (1992), 597–618.

Daston, Lorraine, and Peter Galison, "The Image of Objectivity," *Representations* (1992).

Daston, Lorraine, and Katherine Park, *Wonders of Nature: The Culture of the Marvelous, 1500–1740* (Cambridge, Mass.: Harvard University Press, forthcoming).

Davis, Audrey B., "Life Insurance and the Physical Examination: A Chapter in the Rise of American Medical Technology," *Bulletin of the History of Medicine*, 55 (1981), 392–406.

Day, Archibald, "On the Determination of the Rates of Premiums for Assuring against Issue," *Assurance Magazine*, 8 (1858–60), 127–138.

Day, Charles R., *Education for the Industrial World: The Ecoles d'Arts et Métiers and the Rise of French Industrial Engineering* (Cambridge, Mass.: MIT Press, 1987).

Dear, Peter, "*Totius in verba*: The Rhetorical Construction of Authority in the Early Royal Society," *Isis*, 76 (1985), 145–161.

———, "From Truth to Disinterestedness in the Seventeenth Century," *Social Studies of Science*, 22 (1992), 619–631.

Defoe, Daniel, *The Complete English Tradesman* (1726; Gloucester: Alan Sutton, 1987).

DeGroot, Morris H., Stephen E. Fienberg, and Joseph B. Kadane, eds., *Statistics and the Law* (New York: John Wiley & Sons, 1986).

Dennis, Michael Aaron, "Graphic Understanding: Instruments and Interpretation in Robert Hooke's *Micrographia*," *Science in Context*, 3 (1989), 309–364.

Desrosières, Alain, "Les specificités de la statistique publique en France: Une mise en perspective historique," *Courier des Statistiques*, no. 49, January 1989, 37–54.

———, "How to Make Things Which Hold Together: Social Science, Statistics, and the State," in P. Wagner, B. Wittrock, and R. Whitley, eds., *Discourses on Society, Sociology of the Sciences Yearbook*, 15 (1990), 195–218.

———, "Masses, individus, moyennes: La statistique sociale au XIXè siècle," in Feldman et al., *Moyenne*, 245–273.

———, *La politique des grands nombres: Histoire de la raison statistique* (Paris: Editions La Découverte, 1993).

Desrosières, Alain, and Laurent Thévenot, *Les catégories socioprofessionelles* (Paris: Editions La Découverte, 1988).

Dhombres, Jean, "L'Ecole Polytechnique et ses historiens," in Fourcy, *Histoire*, 30–39.

Dickens, Charles, *Martin Chuzzlewit* (New York: Penguin, 1968).

Divisia, François, *Exposés d'économique: L'apport des ingénieurs français aux sciences économiques* (Paris: Dunod, 1951).

Dodge, David A., and David A. A. Stager, "Economic Returns to Graduate Study in Science, Engineering, and Business," *Benefit-Cost Analysis: An Aldine Annual, 1972* (Chicago: Aldine, 1973).

Dorfman, Robert, "Forty Years of Cost-Benefit Analysis," in Richard Stone and William Peterson, eds., *Econometric Contributions to Public Policy* (London: Macmillan, 1978), 268–288.

———, ed., *Measuring Benefits of Government Investments* (Washington, D.C.: Brookings Institution, 1965).

Doukas, Kimon A., *The French Railroads and the State* (1945; reprinted New York: Farrar, Straus, & Giroux, 1976).

Doussot, Antoine, "Observations sur une note de M. l'ingénieur en chef Labry relative à l'utilité des travaux publics," *Annales des Ponts et Chaussées* [5], 20 (1880), 125–130.

Drew, Elizabeth, "Dam Outrage: The Story of the Army Engineers," *Atlantic*, 225, April 1970, 51–62.

Duhamel, Henry, "De la necessité d'une statistique des accidents," *Journal de la Sociéte de Statistique de Paris*, 29 (1888), 127–168.

Dumez, Hervé, *L'économiste, la science et le pouvoir: Le cas Walras* (Paris: Presses Universitaires de France, 1985).

Duncan, Otis Dudley, *Notes on Social Measurement: Historical and Critical* (New York: Russell Sage Foundation, 1984).

Dunham, Arthur L., "How the First French Railways Were Planned," *Journal of Economic History*, 1 (1941), 12–25.

Dupuit, Jules, "Sur les frais d'entretien des routes," *Annales des Ponts et Chaussées* [2], 3 (1842), 1–90.

———, "De la mesure de l'utilité des travaux publics," *Annales des Ponts et Chaussées* [2], 8 (1844), 332–375 (English translation in *International Economic Papers*, 2 (1952), 83–110).

Dupuit, Jules, "De l'influence des péages sur l'utilité des voies de communication," *Annales des Ponts et Chaussées* [2], 17 (1849), 170–249 (English translation in part in *International Economic Papers*, 11 (1962), 7–31).

———, *Titres scientifiques de M. J. Dupuit* (Paris: Mallet-Bachelier, 1857).

———, *La liberté commerciale: Son principe et ses conséquences* (Paris: Guillaumin, 1861).

———, *De l'utilité et de sa mesure: Ecrits choisis et republiés*, Mario de Bernardi, ed. (Turin: La Riforma Sociale, 1933).

Eckstein, Otto, *Water-Resource Development: The Economics of Project Evaluation* (Cambridge, Mass.: Harvard University Press, 1958).

Ekelund, Robert B., and Robert F. Hébert, "French Engineers, Welfare Economics, and Public Finance in the Nineteenth Century," *History of Political Economy*, 10 (1978), 636–668.

Elwitt, Sanford, *The Making of the Third Republic: Class and Politics in France, 1868–1884* (Baton Rouge: Louisiana State University Press, 1975).

———, *The Third Republic Defended: Bourgeois Reform in France, 1880–1914* (Baton Rouge: Louisiana State University Press, 1986).

Etner, François, *Histoire du calcul économique en France* (Paris: Economica, 1987).

Evans, Hughes, "Losing Touch: The Controversy over the Introduction of Blood Pressure Instruments into Medicine," *Technology and Culture*, 34 (1993), 784–807.

Ezrahi, Yaron, *The Descent of Icarus: Science and the Transformation of Contemporary Democracy* (Cambridge, Mass.: Harvard University Press, 1990).

Fagot-Largeault, Anne, *Les causes de la mort: Histoire naturelle et facteurs de risque* (Paris: Vrin, 1989).

Farren, Edwin James, "On the Improvement of Life Contingency Calculation," *Assurance Magazine*, 5 (1854–55), 185–196; 8 (1858–60), 121–127.

———, "On the Reliability of Data, when tested by the conclusions to which they lead," *Assurance Magazine*, 3 (1852–53), 204–209.

Faure, Fernand, "Observations sur l'organisation de l'enseignement de la statistique," *Journal de la Société de Statistique de Paris*, 34 (1893), 25–29.

Fayol, Henri, *General and Industrial Management* [1916] (London: Sir Isaac Pitman and Sons, 1949).

Federal Inter-Agency River Basin Committee, Subcommittee on Benefits and Costs, *Proposed Practices for Economic Analysis of River Basin Projects* (Washington, D.C.: USGPO, 1950; revised ed., 1958).

Feldman, Jacqueline, Gérard Lagneau, and Benjamin Matalon, eds., *Moyenne, milieu, centre: histoires et usages* (Paris: Editions de l'Ecole des Hautes Etudes en Sciences Sociales, 1991).

Feldman, Theodore S., "Applied Mathematics and the Quantification of Experimental Physics: The Example of Barometric Hypsometry," *Historical Studies in the Physical Sciences*, 15 (1985), 127–197.

———, "Late Enlightenment Meteorology," in Frängsmyr et al., *Quantifying Spirit*, 143–177.

Ferejohn, John A., *Pork Barrel Politics: Rivers and Harbors Legislation, 1947–1968* (Stanford, Calif.: Stanford University Press, 1974).

Fichet-Poitrey, F., *Le Corps des Ponts et Chaussées: Du génie civil à l'aménagement du territoire* (Paris, 1982).

Fink, Albert, *Argument . . . before the Committee on Commerce of the United States House of Representatives*, March 17–18, 1882 (Washington, D.C.: USGPO, 1882).

Finkelstein, Michael, and Hans Levenbach, "Regression Estimates of Damages in Price-Fixing Cases," in DeGroot et al., *Statistics*, 79–106.

Finney, Donald J., *Statistical Method in Biological Assay*, (London: Charles Griffin and Co., 1952).

Fischhoff, Baruch, et al., *Acceptable Risk* (Cambridge, U.K.: Cambridge University Press, 1981).

Fish, Stanley, *Is There a Text in This Class? The Authority of Interpretive Communities* (Cambridge, Mass.: Harvard University Press, 1980).

Flamholtz, Eric, "The Process of Measurement in Managerial Accounting: A Psycho-Technical Systems Perspective," *Accounting, Organizations, and Society*, 5 (1980), 31–42.

Fleming, Donald, "Attitude: The History of a Concept," *Perspectives in American History*, 1 (1967), 287–365.

Fortun, M., and S. S. Schweber, "Scientists and the Legacy of World War II: The Case of Operations Research," *Social Studies of Science*, 23 (1993), 595–642.

Foucault, Michel, *The Order of Things* (New York: Vintage, 1973).

Fougère, Louis, "Introduction générale," *Histoire de l'Administration française dépuis 1800* (Geneva: Droz, 1975).

Fourcy, Ambroise, *Histoire de l'Ecole Polytechnique* (1837; Paris: Belin, 1987).

Fournier de Flaix, E., "Le canal de Panama," *Journal de la Société de Statistique de Paris, 22 (1881), 64–70.*

Fourquet, François, *Les comptes de la puissance: Histoire de la comptabilité nationale et du plan* (Paris: Encres Recherches, 1980).

Foville, Alfred de, "La statistique et ses ennemis," *Journal de la Société de Statistique de Paris*, 26 (1885), 448–454.

———, "Le rôle de la statistique dans le présent et dans l'avenir," *Journal de la Société de Statistique de Paris*, 33 (1892), 211–214.

Fox, Robert, "The Rise and Fall of Laplacian Physics," *Historical Studies in the Physical Sciences*, 4 (1974), 89–136.

Fox, Robert, and George Weisz, eds., *The Organization of Science and Technology in France, 1808–1914* (Cambridge, U.K.: Cambridge University Press, 1980).

Foxwell, H. S., "The Economic Movement in England," *Quarterly Journal of Economics*, 1 (1886–87), 84–103.

Frängsmyr, Tore, John Heilbron, and Robin Rider, eds., *The Quantifying Spirit in the Eighteenth Century* (Berkeley: University of California Press, 1990).

Frank, Robert, "The Telltale Heart: Physiological Instruments, Graphic Methods, and Clinical Hopes, 1865–1914," in William Coleman and Frederic L. Holmes, eds., *The Investigative Enterprise: Experimental Physiology in Nineteenth-Century Medicine* (Berkeley: University of California Press, 1988), 211–290.

Freidson, Eliot, *Professional Powers: A Study of the Institutionalization of Formal Knowledge* (Chicago: University of Chicago Press, 1986).

Friedman, Robert Marc, *Appropriating the Weather: Vilhelm Bjerknes and the Construction of a Modern Meteorology* (Ithaca, N.Y.: Cornell University Press, 1989).

Fuller, Steve, "Social Epistemology as Research Agenda of Science Studies," in Pickering, *Science*, 390–428.

Funkenstein, Amos, *Theology and the Scientific Imagination from the Middle Ages to the Seventeenth Century* (Princeton, N.J.: Princeton University Press, 1986).

Furner, Mary O., and Barry Supple, eds., *The State and Economic Knowledge: The American and British Experiences* (Cambridge, U.K.: Cambridge University Press, 1990).

Galambos, Louis, ed., *The New American State: Bureaucracies and Policies since World War II* (Baltimore: Johns Hopkins University Press, 1987).

Galison, Peter, "Aufbau/Bauhaus: Logical Positivism and Architectural Modernism," *Critical Inquiry*, 16 (1990), 709–752.

Garnier, Joseph, "Sur les frais d'entretien des routes en empierrement," *Annales des Ponts et Chaussées* [2], 10 (1845), 146–196.

Geiger, Reed, "Planning the French Canals: The 'Becquey Plan' of 1820–1822," *Journal of Economic History*, 44 (1984), 329–339.

Geison, Gerald L., "'Divided We Stand': Physiologists and Clinicians in the American Context," in Morris J. Vogel and Charles Rosenberg, eds., *The Therapeutic Revolution: Essays in the Social History of American Medicine* (Philadelphia: University of Pennsylvania, Press, 1979), 67–90.

———, "Scientific Change, Emerging Specialties, and Research Schools," *History of Science*, 19 (1981), 20–40.

———, ed., *Professions and the French State, 1700–1900* (Philadelphia: University of Pennsylvania Press, 1984).

Geison, Gerald L., and Frederic L. Holmes, eds., *Research Schools: Historical Reappraisals*, *Osiris*, 8 (1993).

Gertel, Karl, "Recent Suggestions for Cost Allocation of Multiple Purpose Projects in the Light of Public Interest," *Journal of Farm Economics*, 33 (1951), 130–134.

Gigerenzer, Gerd, "Probabilistic Thinking and the Fight against Subjectivity," in Krüger et al., *Probabilistic Revolution*, vol. 2, 11–33.

———, "The Superego, the Ego, and the Id in Statistical Reasoning," in Gideon Keren and Charles Lewis, eds., *A Handbook for Data Analysis in the Behavioral Sciences: Methodological Issues* (Hillsdale, N.J.: Erlbaum, 1993).

Gigerenzer, Gerd, and David J. Murray, *Cognition as Intuitive Statistics* (Hillsdale, N.J.: Erlbaum, 1987).

Gigerenzer, Gerd, et al., *The Empire of Chance: How Probability Changed Science and Everyday Life* (Cambridge, U.K.: Cambridge University Press, 1989).

Gillispie, Charles, *The Edge of Objectivity* (Princeton, N.J.: Princeton University Press, 1960).

———, "Social Selection as a Factor in the Progressiveness of Science," *American Scientist*, 56 (1968), 438–450.

———, *Science and Polity in France at the End of the Old Regime* (Princeton, N.J.: Princeton University Press, 1980).

———, *The Montgolfier Brothers and the Invention of Aviation* (Princeton, N.J.: Princeton University Press, 1983).

———, "Un enseignement hégémonique: Les mathématiques," in Belhoste et al., *Formation*, 31–43.

Gilpin, Robert, *France in the Age of the Scientific State* (Princeton, N.J.: Princeton University Press, 1968).

Gispert, Hélène, "L'enseignement scientifique supérieure et les enseignants, 1860–1900: Les mathématiques," *Histoire de l'Education*, no. 41, January, 1989, 47–78.

Glaeser, Martin G., *Outlines of Public Utility Economics* (New York: Macmillan, 1927).

Goldstein, Jan, "The Advent of Psychological Modernism in France: An Alternative Narrative," in Ross, *Modernist Impulses*, 190–209.

Golinski, Jan, *Science as Public Culture: Chemistry and Enlightenment in Britain* (New York: Cambridge University Press, 1992).

Gondinet, Edmond, *Le Panache. Comédie en trois actes* (Paris: Michel Levy Fréres, 1876).

Gooday, Graeme, "Precision Measurement and the Genesis of Teaching Laboratories in Victorian Britain," *British Journal for the History of Science*, 23 (1990), 25–51.

Gooding, David, Trevor Pinch, and Simon Schaffer, eds., *The Uses of Experiment* (Cambridge, U.K.: Cambridge University Press, 1989).

Goodwin, Craufurd D., "The Valley Authority Idea—The Failing of a National Vision," in Erwin C. Hargrove and Paul K. Conkin, eds., *TVA: Fifty Years of Grass-Roots Bureaucracy* (Urbana: University of Illinois Press, 1983), 263–296.

Goody, Jack, *The Domestication of the Savage Mind* (Cambridge, U.K.: Cambridge University Press, 1977).

———, ed., *Literacy in Traditional Societies* (Cambridge, U.K.: Cambridge University Press, 1968).

Gorges, Irmela, "The Social Survey in Germany before 1933," in Bulmer et al., *Social Survey*, 316–339.

Gough, Harrison G., "Clinical versus Statistical Prediction in Psychology," in Leo Postman, ed., *Psychology in the Making: Histories of Selected Research Problems* (New York: Alfred A. Knopf, 1964).

Gourvish, T. R., "The Rise of the Professions," in T. R. Gourvish and Alan O'Day, eds., *Later Victorian Britain, 1867–1900* (New York: St. Martin's Press, 1988), 13–35.

Gowan, Peter, "The Origins of the Administrative Elite," *New Left Review*, 61 (March–April 1987), 4–34.

Graff, Harvey J., *The Legacies of Literacy* (Bloomington: Indiana University Press, 1987).

Grant, Eugene L., *Principles of Engineering Economy* (New York: Ronald Press, 1930).

Grattan-Guinness, Ivor, "Work for the Workers: Advances in Engineering Mechanics and Instruction in France, 1800–1930," *Annals of Science*, 41 (1984), 1–33.

———, *Convolutions in French Mathematics*, 3 vols. (Basel, Switzerland: Birkhäuser, 1990).

Gray, Peter, "On the Construction of Survivorship Assurance Tables," *Assurance Magazine*, 5 (1854–55), 107–126.

Gray, Ralph D., *The National Waterway: A History of the Chesapeake and Delaware Canal, 1769–1985*, 2d ed. (Urbana: University of Illinois Press, 1989).

Green, Mark J., "Cost-Benefit Analysis as a Mirage," in Timothy B. Clark, Marvin H. Kosters, and James C. Miller III, eds., *Reforming Regulation* (Washington, D.C.: American Enterprise Institute, 1980).

Greenawalt, Kent, *Law and Objectivity* (New York: Oxford University Press, 1992).

Greenberg, John, "Mathematical Physics in Eighteenth-Century France," *Isis*, 77 (1986), 59–78.

Greenleaf, W. H., *The British Political Tradition*, vol. 3: *A Much Governed Nation* (London: Methuen, 1987).

Grégoire, Roger, *La fonction publique* (Paris: Armand Colin 1954).

Griliches, Zvi, "Research Costs and Social Returns: Hybrid Corn and Related Innovations," *Journal of Political Economy*, 66 (1958), 419–431.

Grison, Emmanuel, "François Arago et l'Ecole Polytechnique," in Crépel, *Arago*, 1–28.

Habermas, Jürgen, *The Structural Transformation of the Public Sphere*, Thomas Burger, trans. (1962; Cambridge, Mass.: MIT Press, 1989).

Hackett, John, and Anne-Marie Hackett, *Economic Planning in France* (Cambridge, Mass.: Harvard University Press, 1963).

Hacking, Ian, *Representing and Intervening* (Cambridge, U.K.: Cambridge University Press, 1983).

———, "Telepathy: Origins of Randomization in Experimental Design," *Isis*, 79 (1988), 427–451.

———, *The Taming of Chance* (Cambridge, U.K.: Cambridge University Press, 1990).

———, "Statistical Language, Statistical Truth, and Statistical Reason: The Self-Authentification of a Style of Scientific Reasoning," in Ernan McMullin, ed., *The Social Dimensions of Science* (Notre Dame: University of Notre Dame Press, 1992), 130–157.

———, "The Self-Vindication of the Laboratory Sciences," in Pickering, *Science as Practice, 29–64.*

Hamlin, Christopher, *A Science of Impurity: Water Analysis in Nineteenth-Century Britain* (Berkeley: University of California Press, 1990).

Hammond, Kenneth R., and Leonard Adelman, "Science, Values, and Human Judgement," *Science*, 194, 22 October 1976, 389–396.

Hammond, Richard J., *Benefit-Cost Analysis and Water-Pollution Control* (Stanford, Calif.: Food Research Institute of Stanford University, 1960).

———, "Convention and Limitation in Benefit-Cost Analysis," *Natural Resources Journal*, 6 (1966), 195–222.

Hannaway, Owen, "Laboratory Design and the Aim of Science: Andreas Libavius versus Tycho Brahe," in *Isis*, 77 (1986), 585–610.

Hansen, W. Lee, "Total and Private Rates of Return to Investment in Schooling," *Journal of Political Economy*, 71 (1963), 128–140.

Haraway, Donna, *Primate Visions: Gender, Race, and Nature in the World of Modern Science* (New York: Routledge, 1989).

Harper, Douglas, *Working Knowledge: Skill and Community in a Small Shop* (Chicago: University of Chicago Press, 1987).

Harré, Rom, *Varieties of Realism* (New York: Basil Blackwell, 1986).

Harris, Jose, "Economic Knowledge and British Social Policy," in Furner and Supple, *State and Economic Knowledge*, 379–400.

Haskell, Thomas, *The Emergence of Professional Social Science* (Urbana: University of Illinois Press, 1977).

Hatcher, Robert A., and J. G. Brody, "The Biological Standardization of Drugs," *American Journal of Pharmacy*, 82 (1910), 360–372.

Haveman, Robert, *Water Resource Investment and the Public Interest* (Nashville: Vanderbilt University Press, 1965).

Hawley, Ellis R., "Economic Inquiry and the State in New Era America: Antistatist Corporatism and Positive Statism in Uneasy Coexistence," in Furner and Supple, *State and Economic Knowledge*, 287–324.

Hayek, Friedrich, *The Counterrevolution of Science* (Indianapolis: Liberty Press, 1979 reprint).

Hays, Samuel P., *Conservation and the Gospel of Efficiency: The Progressive Conservation Movement, 1890–1920*, 2d ed. (Cambridge, Mass.: Harvard University Press, 1969).

———, "The Politics of Environmental Administration," in Galambos, *New American State*, 21–53.

Heclo, Hugh, *A Government of Strangers: Executive Politics in Washington* (Washington, D.C.: The Brookings Institution, 1977).

Heclo, Hugh, and Aaron Wildavsky, *The Private Government of Public Money: Community and Policy inside British Politics* (Berkeley: University of California Press, 1974).

Heidelberger, Michael, *Die innere Seite der Natur: Gustav Theodor Fechners wissenschaftlich-philosophische Weltauffassung* (Frankfurt am Main: Vittorio Klostermann, 1993).

Heilbron, John L., *Electricity in the 17th and 18th Centuries* (Berkeley: University of California Press, 1979).

———, "Fin-de-siècle Physics," in Carl-Gustav Bernhard, Elisabeth Crawford, and Per Sèrböm, eds., *Science, Technology, and Society in the Time of Alfred Nobel* (Oxford: Pergamon Press, 1982), 51–71.

———, *The Dilemmas of an Upright Man: Max Planck as Spokesman for German Science* (Berkeley: University of California Press, 1986).

———, "Introductory Essay," in Frängsmyr et al., *Quantifying Spirit*, 1–23.

———, "The Measure of Enlightenment," in Frängsmyr et al., *Quantifying Spirit*, 207–242.

Heilbron, John, and Robert Seidel, *Lawrence and His Laboratory* (Berkeley: University of California Press, 1989).

Henderson, James P., "Induction, Deduction and the Role of Mathematics: The Whewell Group vs. the Ricardian Economists," *Research in the History of Economic Thought and Methodology*, 7 (1990), 1–36.

Henry, Ernest, *Les formes des Enquêtes administratives en matière de travaux d'intérêt public* (Paris and Nancy: Berger-Levrault et Cie., 1891).

Hill, Austin Bradford, *Principles of Medical Statistics* (London: The Lancet, 1937 and many subsequent editions).

———, "The Clinical Trial-II" (1952) in Hill, *Statistical*, 29–43.

———, "The Philosophy of the Clinical Trial" (1953), in Hill, *Statistical*, 3–14.

———, "Smoking and Carcinoma of the Lung," in Hill, *Statistical*, 384–413.

———, "Observation and Experiment," in Hill, *Statistical*, 369–383.

———, "Aims and Ethics," in Hill, *Controlled Clinical Trials*, 3–7.

———, ed., *Controlled Clinical Trials* (Oxford: Blackwell Scientific Publication, 1960).

———, *Statistical Methods in Clinical and Preventive Medicine* (Edinburgh, U.K.: E. & S. Livingston Ltd., 1962).

Hilts, Victor, "*Aliis exterendum*, or the Origins of the Statistical Society of London," *Isis*, 69 (1978), 21–43.

Himmelfarb, Gertrude, *Poverty and Compassion: The Moral Imagination of the Late Victorians* (New York: Alfred A. Knopf, 1991).

Hines, Lawrence G., "Precursors to Benefit-Cost Analysis in Early United States Public Investment Projects," *Land Economics*, 49 (1973), 310–317.

Hoffmann, Robert G., *New Clinical Laboratory Standardization Methods* (New York: Exposition Press, 1974).

Hoffmann, Stanley, "Paradoxes of the French Political Community," in Hoffmann et al., *In Search of France* (Cambridge, Mass.: Harvard University Press, 1963), 1–117.

Hofstadter, Douglas R., *Gödel, Escher, Bach: An Eternal Golden Braid* (New York: Vintage Books, 1980).

Hofstadter, Richard, *Anti-Intellectualism in American Life* (New York: Alfred A. Knopf, 1963).

Hollander, Samuel, "William Whewell and John Stuart Mill on the Methodology of Political Economy," *Studies in the History and Philosophy of Science*, 14 (1983), 127–168.

Hollinger, David, "Free Enterprise and Free Inquiry: The Emergence of Laissez-Faire Communitarianism in the Ideology of Science in the United States," *New Literary History*, 21 (1990), 897–919.

Holt, Robert R., *Methods in Clinical Psychology*, vol. 2: *Prediction and Research* (New York: Plenum Press, 1978).

Holton, Gerald, "Fermi's Group and the Recapture of Italy's Place in Physics," in Holton, *The Scientific Imagination: Case Studies* (Cambridge, U.K.: Cambridge University Press, 1978).

———, "On Doing One's Damndest: The Evolution of Trust in Scientific Findings," forthcoming.

Hopwood, Anthony, *An Accounting System and Managerial Behaviour* (Lexington, Mass.: Lexington Books, 1973).

Horkheimer, Max, and Theodor W. Adorno, *Dialektik der Aufklärung: Philosophische Fragmente* (Frankfurt: S. Fischer Verlag, 1948).

Hornstein, Gail A., "Quantifying Psychological Phenomena: Debates, Dilemmas, and Implications," in Jill G. Morawski, ed., *The Rise of Experimentation in American Psychology* (New Haven, Conn.: Yale University Press, 1988).

Hoskin, Keith W., and Richard R. Macve, "Accounting and the Examination: A Genealogy of Disciplinary Power," *Accounting, Organizations, and Society*, 11 (1986), 105–136.

Hoslin, C., *Les limites de l'intérêt public dans l'établissement des chemins de fer* (Marseille: Imprimerie Saint-Joseph, 1878).

Hudson, Liam, *The Cult of the Fact* (London: Jonathan Cape Ltd., 1972).

Hufbauer, Karl, *The Formation of the German Chemical Community (1720–1905)* (Berkeley: University of California Press, 1982).

Hundley, Norris, Jr., *The Great Thirst: Californians and Water* (Berkeley: University of California Press, 1992).

Hunt, Bruce J., *The Maxwellians* (Ithaca, N.Y.: Cornell University Press, 1991).

Hunter, J. S., "The National System of Scientific Measurement," *Science*, 210 (November 21, 1980), 869–874.

Ijiri, Yuji, and Robert K. Jaedicke, "Reliability and Objectivity of Accounting Measurements," *Accounting Review*, 41 (1986), 474–483.

Ingrao, Bruna, and Giorgio Israel, *The Invisible Hand: Equilibrium in the History of Science*, Ian McGilvray, trans. (Cambridge, Mass.: MIT Press, 1987).

Jacobs, Stuart, "Scientific Community: Formulations and Critique of a Sociological Motif," *British Journal of Sociology*, 38 (1987), 266–276.

Jaffé, William, ed., *Correspondence of Léon Walras and Related Papers*, 3 vols. (Amsterdam: North Holland, 1965).

Jasanoff, Sheila, "The Misrule of Law at OSHA," in Dorothy Nelkin, ed., *The Language of Risk* (Beverly Hills: Sage, 1985), 155–178.

———, *Risk Management and Political Culture* (New York: Russell Sage Foundation, 1986).

———, "The Problem of Rationality in American Health and Safety Regulation," in Smith and Wynne, *Expert Evidence*, 151–183.

———, *The Fifth Branch: Science Advisers as Policymakers* (Cambridge, Mass.: Harvard University Press, 1990).

———, "Science, Politics, and the Renegotiation of Expertise at EPA," *Osiris*, 7 (1991), 194–217.

Jellicoe, Charles, "On the Rate of Premiums to be charged for Assurances on the Lives of Military Officers serving in Bengal," *Assurance Magazine*, 1 (1850–51), no. 3, 166–178.

———, "On the Rates of Mortality Prevailing . . . in the Eagle Insurance Company," *Assurance Magazine*, 4 (1853–54), 199–215.

Jenkin, Fleeming, "Trade Unions: How Far Legitimate" (1868), in *Papers*, vol. 2, 1–75.

———, "The Graphic Representation of the Laws of Supply and Demand" (1870), in *Papers*, vol. 2, 76–106.

———, "On the Principles which Regulate the Incidence of Taxes" (1871–72), in *Papers*, vol. 2, 107–121.

Jenkin, Fleeming, *Papers: Literary, Scientific, &c.*, ed. Sidney Colvin and J. A. Ewing, 2 vols. (London: Longman, Green, and Co., 1887).

Johnson, H. Thomas, "Management Accounting in an Early Multidivisional Organization: General Motors in the 1920's," *Business History Review*, 52 (1978), 490–517.

———, "Toward a New Understanding of Nineteenth-Century Cost Accounting," *Accounting Review*, 56 (1981), 510–518.

Johnson, H. T., and R. S. Kaplan, *Relevance Lost: The Rise and Fall of Management Accounting* (Boston: Harvard Business School Press, 1987).

Johnson, Hildegard Binder, *Order upon the Land: The US Rectangular Land Survey and the Upper Mississippi Country* (New York: Oxford University Press, 1976).

Jones, Edgar, *Accountancy and the British Economy, 1840–1980* (London: B. T. Batsford, 1981).

Jouffroy, Louis-Maurice, *La Ligne de Paris à la Frontière d'Allemagne (1825–1852): Une étape de la construction des grandes lignes de chemins de fer en France*, 3 vols. (Paris: S. Barreau & Cie., 1932), vol. 1.

Jouvenel, Bertrand de, *The Art of Conjecture*, Nikita Lant, trans. (New York: Basic Books, 1967).

Jullien, Ad., "Du prix des transports sur les chemins de fer," *Annales des Ponts et Chaussées* [2], 8 (1844), 1–68.

Jungnickel, Christa, and Russell McCormmach, *Intellectual Mastery of Nature: Theoretical Physics from Ohm to Einstein*, 2 vols. (Chicago: University of Chicago Press, 1986).

Kang, Zheng, "Lieu de savoir social: La Société de Statistique de Paris au XIXe siècle (1860–1910)," Thèse de Doctorat en Histoire, Ecole des Hautes Etudes en Sciences Sociales, 1989.

Keller, Evelyn Fox, *Reflections on Gender and Science* (New Haven, Conn.: Yale University Press, 1985).

———, "The Paradox of Scientific Subjectivity," *Annals of Scholarship*, 9 (1992), 135–153.

Keller, Morton, *Affairs of State: Public Life in Late Nineteenth Century America* (Cambridge, Mass.: Harvard University Press, 1977).

———, *Regulating a New Economy: Public Policy and Economic Change in America, 1900–1933* (Cambridge, Mass.: Harvard University Press, 1990).

Kevles, Daniel J., "Testing the Army's Intelligence: Psychologists and the Military in World War I," *Journal of American History*, 55 (1968–69), 565–581.

Keyfitz, Nathan, "The Social and Political Context of Population Forecasting," in Alonso and Starr, *Politics of Numbers*, 235–258.

Kindleberger, Charles P., "Technical Education and the French Entrepreneur," in E. C. Carter et al., eds., *Enterprise and Entrepreneurs in Nineteenth- and Twentieth-Century France* (Baltimore: Johns Hopkins University Press, 1976).

Klein, Judy, *Time and the Science of Means: The Statistical Analysis of Changing Phenomena, 1830–1940* (Cambridge, U.K.: Cambridge University Press, forthcoming).

Kleinmuntz, Benjamin, "The Scientific Study of Clinical Judgment in Psychology and Medicine" (1984), in Hal R. Arkes and Kenneth R. Hammond, eds., *Judgement and Decision Making* (Cambridge, U.K.: Cambridge University Press, 1986).

Knoll, Elizabeth, "The Communities of Scientists and Journal Peer Review," *Journal of the American Medical Association*, 263, no. 10, March 9, 1990, 1330–1332.

Knorr-Cetina, Karin D., *The Manufacture of Knowledge* (Oxford: Pergamon, 1981).

Kranakis, Eda, "The Affair of the Invalides Bridge," *Jaarboek voor de Geschiedenis van Bedrijf in Techniek*, 4 (1987), 106–130.

———, "Social Determinants of Engineering Practice: A Comparative View of France and America in the Nineteenth Century," *Social Studies of Science*, 19 (1989), 5–70.

Krüger, Lorenz, Lorraine Daston, and Michael Heidelberger, eds., *The Probabilistic Revolution*, vol. 1: *Ideas in History* (Cambridge, Mass.: MIT Press, 1987).

Krüger, Lorenz, Gerd Gigerenzer, and Mary Morgan, eds., *The Probabilistic Revolution*, vol. 2: *Ideas in the Sciences* (Cambridge, Mass.: MIT Press, 1987).

Krutilla, John, and Otto Eckstein, *Multiple-Purpose River Development* (Baltimore: Johns Hopkins University Press, 1958).

Kuhn, Thomas, *The Structure of Scientific Revolutions*, 2d ed. (Chicago: University of Chicago Press, 1970).

Kuhn, Tillo E., *Public Enterprise Economics and Transport Problems* (Berkeley: University of California Press, 1962).

Kuisel, Richard F., *Ernest Mercier: French Technocrat* (Berkeley: University of California Press, 1967).

———, *Capitalism and the State in Modern France* (Cambridge, U.K.: Cambridge University Press, 1981).

Kula, Witold, *Measures and Men*, Richard Szreter, trans. (Princeton, N.J.: Princeton University Press, 1986).

Labiche, Eug., and Ed. Martin, *Les Vivacités du Capitaine Tic* (Paris: Calmann-Levy, n.d.; first performed 1861).

Labry, Félix de, "A Quelles conditions les travaux publics sont-ils rémunerateurs," *Journal des économistes*, 10 (November 1875), 301–307.

———, "Note sur le profit des travaux," *Annales des Ponts et Chaussées* [5], 19 (1880), 76–85.

———, "L'outillage national et la dette de l'état: Replique à M. Doussot," *Annales des Ponts et Chaussées* [5], 20 (1880), 131–144.

La Gournerie, Jules de, "Essai sur le principe des tarifs dans l'exploitation des chemins de fer" (1879), in La Gournerie, *Etudes économiques.*

———, *Etudes économiques sur l'exploitation des chemins de fer* (Paris: Gauthier-Villars, 1880).

Lalanne, Léon, "Sur les tables graphiques et sur la géometrie anamorphique appliquée à diverses questions qui se rattachent à l'art de l'ingénieur," *Annales des Ponts et Chaussées* [2], 11 (1846).

Lamoreaux, Naomi, "Information Problems and Banks' Specialization in Short-

Term Commercial Lending: New England in the Nineteenth Century," in Temin, *Inside*, 161–195.

Lance, William, "Paper upon Marine Insurance," *Assurance Magazine*, 2 (1851–52), 362–376.

Landes, David, *Revolution in Time* (Cambridge, Mass.: Harvard University Press, 1983).

Latour, Bruno, *Science in Action* (Cambridge, Mass.: Harvard University Press, 1987).

———, *The Pasteurization of France*, Alan Sheridan and John Law, trans. (Cambridge, Mass.: Harvard University Press, 1988).

———, *We Have Never Been Modern*, Catherine Porter, trans. (Cambridge, Mass.: Harvard University Press, 1993).

Laurent, Hermann, *Petit traité d'économie politique, rédigé conformément aux préceptes de l'école de Lausanne* (Paris: Charles Schmid, 1902).

———, *Statistique mathématique* (Paris: Octave Doin, 1908).

Lave, Jean, "The Values of Quantification," in Law, *Power*, 88–111.

Lavoie, Don, "The Accounting of Interpretations and the Interpretation of Accounts," *Accounting, Organizations, and Society*, 12 (1987), 579–604.

Lavoisier, Antoine, *De la richesse territoriale de France*, Jean-Claude Perrot, ed. (Paris: Editions du C.T.H.S., 1988).

Lavollée, Hubert, "Les Chemins de fer et le budget," *Revue des Deux Mondes*, 55, February 15, 1883, 857–885.

Law, John, ed., *Power, Action, and Belief: A New Sociology of Knowledge?* (London: Routledge, 1986).

Lawrence, Christopher, "Incommunicable Knowledge: Science Technology and the Clinical Art in Britain, 1850–1914," *Journal of Contemporary History*, 20 (1985), 503–520.

Lécuyer, Bernard-Pierre, "L'hygiène en France avant Pasteur," in Claire Salomon-Bayet, ed., *Pasteur et la révolution pastorienne* (Paris: Payot, 1986), 65–142.

———, "The Statistician's Role in Society: The Institutional Establishment of Statistics in France," *Minerva*, 25 (1987), 35–55.

Legendre, Pierre, *Histoire de l'administration de 1750 jusqu'à nos jours* (Paris: Presses Universitaires de France, 1968).

Legoyt, A., "Les congrès de statistique et particulièrement le congrès de statistique de Berlin," *Journal de la Société de Statistique de Paris*, 4 (1863), 271–285.

———, untitled remarks, *Journal de la Société de Statistique de Paris*, 8 (1867), 284.

Leibnitz, Heinz-Maier, "The Measurement of Quality and Reputation in the World of Learning," *Minerva*, 27 (1989), 483–504.

Léon, A., review of Michel Chevalier, *Travaux publics de la France, Annales des Ponts et Chaussées*, 16 (1838), 201–246.

Leonard, Robert, "War as a 'Simple Economic Problem': The Rise of an Economics of Defense," in Craufurd D. Goodwin, ed., *Economics and National Security: A History of Their Interactions* (Durham, N.C.: Duke University Press, 1991), 261–283.

Leopold, Luna B., and Thomas Maddock, Jr., *The Flood Control Controversy: Big Dams, Little Dams, and Land Management* (New York: Ronald Press, 1954).

Le Play, Frédéric, "Vues générales sur la statistique," *Journal de la Société de Statistique de Paris*, 26 (1885), 6–11.

Leuchtenberg, William, *Flood Control Politics: The Connecticut River Valley Problem, 1927–1950* (Cambridge, Mass.: Harvard University Press, 1953).

Levasseur, E., "Le département du travail et les bureaux de statistique aux Etats-Unis," *Journal de la Société de Statistique de Paris*, 35 (1894), 21–29.

Lewis, Gene D., *Charles Ellet, Jr.: The Engineer as Individualist* (Urbana: University of Illinois Press, 1968).

Lewis, Jane, "The Place of Social Investigation, Social Theory, and Social Work in the Approach to Late Victorian and Edwardian Social Problems: The Case of Beatrice Webb and Helen Bosanquet," in Bulmer et al., *Social Survey*, 148–170.

Lexis, Wilhelm, "Zur mathematisch-ökonomischen Literatur," *Jahrbücher für Nationalökonomie und Statistik*, N.F. 3 (1881), 427–434.

Liebenau, Jonathan, *Medical Science and Medical Industry* (London: Macmillan, 1987).

Liesse, Andre, *La statistique: Ses difficultés, ses procédés, ses résultats*, 5th ed. (Paris: Félix Alcan, 1927).

Lindqvist, Svante, "Labs in the Woods: The Quantification of Technology in the Late Enlightenment," in Frängsmyr et al., *Quantifying Spirit*, 291–314.

Loft, Anne, "Towards a Critical Understanding of Accounting: The Case of Cost Accounting in the U.K., 1914–1925," *Accounting, Organizations and Society*, 12 (1987), 235–265.

Loisne, Henri Menche de, "De l'influence des rampes sur le prix de revient des transports en transit," *Annales des Ponts et Chaussées* [5], 17 (1879), 283–298.

Loua, Toussaint, comment, *Journal de la Société de Statistique de Paris*, 10 (1869), 65–67.

———, "A nos lecteurs," *Journal de la Société de Statistique de Paris*, 15 (1874), 57–59.

Louvois, Marquis de, "Au rédacteur," *Journal des débats politiques et littéraires*, January 14, 1842, 1.

Lowi, Theodore J., *The End of Liberalism: Ideology, Policy, and the Crisis of Public Authority*, 2d ed. (New York: Norton, 1979).

———, "The State in Political Science: How We Become What We Study," *American Political Science Review*, 86 (1992), 1–7.

Löwy, Ilana, "Tissue Groups and Cadaver Kidney Sharing: Sociocultural Aspects of a Medical Controversy," *International Journal of Technology Assessment in Health Care*, 2 (1986), 195–218.

Luethy, Herbert, *France against Herself*, Eric Mosbacher, trans. (New York: Praeger, 1955).

Lundgreen, Peter, "Measures for Objectivity in the Public Interest," part 2 of his *Standardization—Testing—Regulation*, *Report Wissenschaftsforschung*, 29 (Bielefeld, Germany: Kleine Verlag, 1986).

Lundgreen, Peter, "Engineering Education in Europe and the U.S.A., 1750–1930: The Rise to Dominance of School Culture and the Engineering Profession," *Annals of Science*, 47 (1990), 37–75.

Maass, Arthur, *Muddy Waters: The Army Engineers and the Nation's Rivers* (Cambridge, Mass.: Harvard University Press, 1951).

Maass, Arthur, and Raymond L. Anderson, . . . *And the Desert Shall Rejoice: Conflict, Growth, and Justice in Arid Environments* (Cambridge, Mass.: MIT Press, 1978).

McCandlish, J. M., "Fire Insurance," *Encyclopaedia Britannica*, 9th ed., 1881, vol. 13.

McCloskey, Donald N., *The Rhetoric of Economics* (Madison: University of Wisconsin Press, 1985).

———, "Economics Science: A Search through the Hyperspace of Assumptions?" *Methodus*, June 3, 1991, 6–16.

McCraw, Thomas K., ed., *Regulation in Perspective: Historical Essays* (Cambridge, Mass.: Harvard University Press, 1981).

McKean, Roland N., *Efficiency in Government through Systems Analysis, with Emphasis on Water Resource Development: A RAND Corporation Study* (New York: John Wiley & Sons, 1958).

MacKenzie, Donald, "Negotiating Arithmetic, Constructing Proof: The Sociology of Mathematics and Information Technology," *Social Studies of Science*, 23 (1993), 37–65.

MacLeod, Roy, ed., *Government and Expertise: Specialists, Administrators and Professionals, 1860–1919* (Cambridge, U.K.: Cambridge University Press, 1988).

McPhee, John, *The Control of Nature* (New York: Farrar, Straus & Giroux, 1989).

Maier, Paul, Jerome Sacks, and Sandy Zabell, "What Happened in Hazelwood: Statistics, Employment Discrimination, and the 80% Rule," in DeGroot et al., *Statistics*, 1–40.

Mainland, Donald, *The Treatment of Clinical and Laboratory Data* (Edinburgh, U.K.: Oliver and Boyd, 1938).

Makeham, William Matthew, "On the Law of Mortality and the Construction of Annuity Tables," *Assurance Magazine*, 8 (1858–60), 301–330.

Marcuse, Herbert, *Reason and Revolution: Hegel and the Rise of Social Theory* (New York: Oxford University Press, 1941).

Margolis, Julius, "Secondary Benefits, External Economies, and the Justification of Public Investment," *Review of Economics and Statistics*, 39 (1957), 284–291.

———, "The Economic Evaluation of Federal Water Resource Development," *American Economic Review*, 49 (1959), 96–111.

Marks, Harry M., "Ideas as Reforms: Therapeutic Experiments and Medical Practice, 1900–1980" (Ph.D. dissertation, MIT, 1987).

———, "Notes from the Underground: The Social Organization of Therapeutic Research," in Russell C. Maulitz and Diana E. Long, eds., *Grand Rounds: One Hundred Years of Internal Medicine* (Philadelphia: University of Pennsylvania Press, 1988), 297–336.

Martinez-Alier, Juan, *Ecological Economics* (New York: Basil Blackwell, 1987).

Matthews, J. Rosser, *Mathematics and the Quest for Medical Certainty* (Princeton, N.J.: Princeton University Press, 1995).

Mauskopf, Seymour, and Michael R. McVaugh, *The Elusive Science: Origins of Experimental Psychical Research* (Baltimore: Johns Hopkins University Press, 1980).

Meehl, Paul E., *Clinical versus Statistical Prediction* (Minneapolis: University of Minnesota Press, 1954).

Megill, Allan, "Introduction: Four Senses of Objectivity," in Megill, *Rethinking Objectivity*, 301–320.

———, ed., *Rethinking Objectivity*, special issue of *Annals of Scholarship*, 8 (1991), parts 3–4, and 9 (1992), parts 1–2.

Mehrtens, Herbert, *Moderne—Sprache—Mathematik: Eine Geschichte des Streits um die Grundlagen der Disziplin und des Subjects formaler Systeme* (Frankfurt am Main: Suhrkamp Verlag, 1990).

Mellet and Henry, MM., *L'arbitraire administratif des ponts et chaussées dévoilé aux chambres* (Paris: Giraudet et Jouaust, 1835).

Ménard, Claude, *La formation d'une rationalité économique: A. A. Cournot* (Paris: Flammarion, 1978).

———, "La machine et le coeur: Essai sur les analogies dans le raisonnement économique," in André Lichnérowicz, ed., *Analogie et connaissance* (Paris: Librairie Maloine, 1981).

Mendoza, Eric, "Physics, Chemistry, and the Theory of Errors," *Archives internationales d'histoire des sciences*, 41 (1991), 282–306.

Merchant, Carolyn, *Ecological Revolutions: Nature, Gender, and Science in New England* (Chapel Hill: University of North Carolina Press, 1989).

Meynaud, Jean, "A propos des spéculations sur l'avenir. Esquisse bibliographique," *Revue française de la science politique*, 13 (1963), 666–688.

———, *Technocracy*, Paul Barnes, trans. (New York: Free Press, 1969).

Michel, Louis-Jules, "Etude sur le trafic probable des chemins de fer d'intérêt local," *Annales des Ponts et Chaussées* [4] 1868, 145–179.

Miles, A. A., "Biological Standards and the Measurement of Therapeutic Activity," *British Medical Bulletin*, 7 (1951), no. 4, (special number on "Measurement in Medicine"), 283–291.

Miller, Leslie A., "The Battle That Squanders Billions," *Saturday Evening Post*, 221, May 14, 1949, 30–31.

Miller, Peter, and Ted O'Leary, "Accounting and the Construction of the Governable Person," *Accounting, Organizations, and Society*, 12 (1987), 235–265.

Miller, Peter, and Nikolas Rose, "Governing Economic Life," *Economy and Society*, 19 (1991), 1–31.

Minard, Charles-Joseph, "Tableau comparatif de l'estimation et de la dépense de quelques canaux anglais," *Annales des Ponts et Chaussées*, 1832 (offprint).

———, *Second mémoire sur l'importance du parcours partiel sur les chemins de fer* (Paris: Imprimerie de Fain et Thunot, 1843).

Mirowski, Philip, *More Heat than Light: Economics as Social Physics. Physics as Nature's Economics* (New York: Cambridge University Press, 1989).

Mirowski, Philip, "Looking for Those Natural Numbers: Dimensions Constants and the Idea of Natural Measurement," *Science in Context*, 5 (1992), 165–188.

Mirowski, Philip, and Steven Sklivas, "Why Econometricians Don't Replicate (Although They Do Reproduce)," *Review of Political Economy*, 3 (1991), 146–162.

Moore, Jamie W., and Dorothy P. Moore, *The Army Corps of Engineers and the Evolution of Federal Flood Plain Management Policy* (Boulder: Institute of Behavioral Science, University of Colorado, 1989).

Moyer, Albert E., *Simon Newcomb: A Scientist's Voice in American Culture* (Berkeley: University of California Press, 1992).

Nathan, Richard P., *Social Science in Government: Uses and Misuses* (New York: Basic Books, 1988).

Navier, C.L.M.H., "De l'exécution des travaux, et particulièrement des concessions," *Annales des Ponts et Chaussées*, 3 (1832), 1–31.

———, "Note sur la comparaison des avantages respectifs de diverses lignes de chemins de fer," *Annales des Ponts et Chaussées*, 9 (1835), 129–179.

Nelson, John S., Allan Megill, and Donald McCloskey, eds., *The Rhetoric of the Human Sciences* (Madison: University of Wisconsin Press, 1987).

Nelson, William E., *The Roots of American Bureaucracy, 1830–1900* (Cambridge, Mass.: Harvard University Press, 1987).

Newcomb, Simon, *Principles of Political Economy* (New York, 1885).

Noël, Octave, *La question des tarifs des chemins de fer* (Paris: Guillaumin, 1884).

Nordling, Wilhelm, "Note sur le prix de revient des transports par chemin de fer," *Annales des Ponts et Chaussées* [6] 11 (1886), 292–303.

Novick, Peter, *That Noble Dream: The 'Objectivity Question' and the American Historical Profession* (Cambridge, U.K.: Cambridge University Press, 1988).

Nyhart, Lynn K., "Writing Zoologically: The *Zeitschrift für wissenschaftliche Zoologie* and the Zoological Community in Late Nineteenth-Century Germany," in Peter Dear, ed., *The Literary Structure of Scientific Argument: Historical Studies* (Philadelphia: University of Pennsylvania Press, 1991).

Oakeshott, Michael, "Rationalism in Politics" (1947), in *Rationalism in Politics and Other Essays* (Indianapolis: Liberty Press, 1991), 1–36.

Olesko, Kathryn M., *Physics as a Calling: Discipline and Practice in the Königsberg Seminar for Physics* (Ithaca, N.Y.: Cornell University Press, 1991).

O'Malley, Michael, *Keeping Watch: A History of American Time* (New York: Viking, 1990).

Orlans, Harold, "Academic Social Scientists and the Presidency: From Wilson to Nixon," *Minerva*, 24 (1986), 172–204.

Osborne, Thomas R., *A Grande Ecole for the Grand Corps: The Recruitment and Training of the French Administrative Elite in the Nineteenth Century* (Boulder: Social Science Monographs, 1983).

Owens, Larry, "Patents, the 'Frontiers' of American Invention, and the Monopoly Committee of 1939: Anatomy of a Discourse," *Technology and Culture*, 32 (1992), 1076–1093.

Palmer, Robert R., *The Age of the Democratic Revolution*, 2 vols. (Princeton, N.J.: Princeton University Press, 1959–1964).

Parker, James E., "Testing Comparability and Objectivity of Exit Value Accounting," *Accounting Review*, 50 (1975), 512–524.

Parker, R. H., *The Development of the Accountancy Profession in Britain in the Early Twentieth Century* (London: Academy of Accounting Historians, 1986).

Parkhurst, David F., "Statistical Hypothesis Tests and Statistical Power in Pure and Applied Science," in George M. von Furstenberg, ed., *Acting under Uncertainty: Multidisciplinary Conceptions* (Boston: Kluwer, 1990).

Patriarca, Silvana, *Numbers and the Nation: The Statistical Representation of Italy, 1820–1871* (Cambridge, U.K.: Cambridge University Press, forthcoming).

Pauly, Philip, *Controlling Life: Jacques Loeb and the Engineering Ideal in Biology* (New York: Oxford University Press, 1987).

Pearson, Karl (published under pseudonym Loki), *The New Werther* (London: C. Kegan Paul and Co., 1880).

Pearson, Karl, "The Ethic of Freethought," in *The Ethic of Freethought and Other Essays* (London: T. F. Unwin, 1888).

———, *The Grammar of Science* (1892; New York: Meridian, reprint of 3d [1911] ed., 1957).

Perkin, Harold, *The Rise of Professional Society: England since 1880* (London: Routledge, 1989).

Peterson, Elmer T., *Big Dam Foolishness: The Problem of Modern Flood Control and Water Storage* (New York: Devin-Adair Co., 1954).

Peterson, William, "Politics and the Measurement of Ethnicity," in Alonso and Starr, *Politics of Numbers*, 187–233.

Picard, Alfred, "Enquêtes relatives aux travaux publics," in *Les chemins de fer français: Etude historique sur la constitution et le régime du réseau*, 6 vols. (Paris: J. Rothschild, 1884).

———, *Les chemins de fer: Aperçu historique, résultats généraux de l'ouverture des chemins de fer . . .* (Paris: H. Dunod et E. Pinat, 1918).

Pickering, Andrew, ed., *Science as Practice and Culture* (Chicago: University of Chicago Press, 1992).

Picon, Antoine, "Les ingénieurs et la mathématisation: L'exemple du génie et de la construction," *Revue d'histoire des sciences*, 42 (1989): 155–172.

———, *L'invention de l'ingénieur moderne: L'Ecole des Ponts et Chaussées, 1747–1851* (Paris: Presses de l'Ecole Nationale des Ponts et Chaussées, 1992).

———, "Les années d'enlisement: L'Ecole Polytechnique de 1870 à l'entre-deux-guerres," in Belhoste et al., *Formation*, 143–179.

Pinch, Trevor, *Confronting Nature: The Sociology of Solar-Neutrino Detection* (Dordrecht, Holland: D. Reidel, 1986).

Pingle, Gautam, "The Early Development of Cost-Benefit Analysis," *Journal of Agricultural Economics*, 29 (1978), 63–71.

Pinkney, David H., *Decisive Years in France, 1840–1848* (Princeton, N.J.: Princeton University Press, 1986).

Poirrier, A., *Tarifs des chemins de fer: Rapport . . . présenté à la Chambre de Commerce de Paris* (Havre: Imprimerie Brennier & Cie., 1882).

Polanyi, Michael, *Personal Knowledge: Towards a Post-Critical Philosophy* (Chicago: University of Chicago Press, 1958).

Popper, Karl, *The Open Society and Its Enemies*, 2 vols., 4th ed. (London: Routledge and Kegan Paul, 1962).

Porter, Henry W., "On Some Points Connected with the Education of an Actuary," *Assurance Magazine*, 4 (1853–54), 108–118.

Porter, Theodore M., "The Promotion of Mining and the Advancement of Science: The Chemical Revolution of Mineralogy," *Annals of Science*, 38 (1981), 543–570.

———, *The Rise of Statistical Thinking, 1820–1900* (Princeton, N.J.: Princeton University Press, 1986).

———, "Objectivity and Authority: How French Engineers Reduced Public Utility to Numbers," *Poetics Today*, 12 (1991), 245–265.

———, "Quantification and the Accounting Ideal in Science," *Social Studies of Science*, 22, (1992), 633–652.

———, "Objectivity as Standardization: The Rhetoric of Impersonality in Measurement, Statistics, and Cost-Benefit Analysis," in Megill, *Rethinking Objectivity, Annals of Scholarship*, 9 (1992), 19–59.

———, "Statistics and the Politics of Objectivity," *Revue de Synthèse*, 114 (1993), 87–101.

———, "Information, Power, and the View from Nowhere," in Bud-Frierman, *Information Acumen*, 217–230.

———, "The Death of the Object: Fin-de-siècle Philosophy of Physics," in Ross, *Modernist Impulses*, 128–151.

———, "Rigor and Practicality: Rival Ideals of Quantification in Nineteenth-Century Economics," in Philip Mirowski, ed., *Natural Images in Economic Thought: Markets Read in Tooth and Claw* (New York: Cambridge University Press, 1994), 128–170.

———, "Precision and Trust: Early Victorian Insurance and the Politics of Calculation," in Wise, *Values of Precision*, 173–197.

———, "Information Cultures," *Accounting, Organizations, and Society*, forthcoming.

Power, Michael, "After Calculation? Reflections on *Critique of Economic Reason* by André Gorz," *Accounting, Organizations, and Society*, 17 (1992), 477–499.

Prest, A. R., and R. Turvey, "Cost-Benefit Analysis: A Survey," *Economic Journal*, 75 (1965), 683–735.

Price, Don K., *The Scientific Estate* (Cambridge, Mass.: Harvard University Press, 1965).

Proctor, Robert N., *Value-Free Science?: Purity and Power in Modern Knowledge* (Cambridge, Mass.: Harvard University Press, 1991).

Quirk, Paul J., "Food and Drug Administration," in James Q. Wilson, ed., *The Politics of Regulation* (New York: Basic Books, 1980), 191–235.

Ratcliffe, Barrie M., "Bureaucracy and Early French Railroads: The Myth and the Reality," *Journal of European Economic History*, 18 (1989), 331–370.

Reader, W. J., *Professional Men: The Rise of the Professional Classes in Nineteenth-Century England* (London: Weidenfeld and Nicolson, 1966).

Regan, Mark M., and E. L. Greenshields, "Benefit-Cost Analysis of Resource Development Programs," *Journal of Farm Economics*, 33 (1951), 866–878.

Reisner, Marc, *Cadillac Desert: The American West and Its Disappearing Water* (New York: Viking Penguin, 1986).

Resnick, Daniel, "History of Educational Testing," in Alexandra K. Wigdor and Wendell R. Garner, eds., *Ability Testing: Uses, consequences, and Controversies*, 2 vols. (Washington, D.C.: National Academy Press, 1982), vol. 2, 173–194.

Reuss, Martin, *Water Resources, People and Issues: Interview with Arthur Maass* (Fort Belvoir, Va.: Office of History, U.S. Army Corps of Engineers, 1989).

———, "Coping with Uncertainty: Social Scientists, Engineers, and Federal Water Resource Planning," *Natural Resources Journal*, 32 (1992), 101–135.

Reuss, Martin, and Paul K. Walker, *Financing Water Resources Development: A Brief History* (Fort Belvoir, Va.: Historical Division, Office of the Chief of Engineers, 1983).

Revel, Jacques, "Knowledge of the Territory," *Science in Context*, 4 (1991), 133–162.

Reynaud, "Tracé des routes et des chemins de fer," *Annales des Ponts et Chaussées* [2], 2 (1841), 76–113.

Ribeill, Georges, *La révolution ferroviaire* (Paris: Belin, 1993).

Richards, Joan, *Mathematical Visions: The Pursuit of Geometry in Victorian England* (Boston: Academic Press, 1988).

———, "Rigor and Clarity: Foundations of Mathematics in France and England, 1800–1840," *Science in Context*, 4 (1991), 297–319.

Ricour, Théophile, "Notice sur la répartition du trafic des chemins de fer français et sur le prix de revient des transports," *Annales des Ponts et Chaussées* [6], 13 (1887), 143–194.

———, "Le prix de revient sur les chemins de fer," *Annales des Ponts et Chaussées* [6], 15 (1888), 534–564.

Rider, Robin, "Measures of Ideas, Rule of Language: Mathematics and Language in the 18th Century," in Frängsmyr et al., *Quantifying Spirit*, 113–140.

Ringer, Fritz, *The Decline of the German Mandarins, 1890–1933* (Cambridge, Mass.: Harvard University Press, 1969).

Roberts, Lissa, "A Word and the World: The Significance of Naming the Calorimeter," *Isis*, 82 (1991), 198–222.

Rorty, Richard, *Objectivity, Relativism, and Truth* (Cambridge, Mass.: Cambridge University Press, 1991).

Rose, Nikolas, *Governing the Soul* (London: Routledge, 1990).

Ross, Dorothy, *The Origins of American Social Science* (Cambridge, U.K.: Cambridge University Press, 1991).

———, ed., *Modernist Impulses in the Human Sciences* (Baltimore: Johns Hopkins University Press, 1994).

Rothermel, Holly, "Images of the Sun: Warren De la Rue, George Biddell Airy and Celestial Photography," *British Journal for the History of Science*, 26 (1993), 137–169.

Rudwick, Martin J. S., *The Great Devonian Controversy* (Chicago: University of Chicago Press, 1985).

S., M. "La mesure de l'utilité des chemins de fer," *Journal des économistes*, 7 (1879), 231–243.

Sagoff, Mark, *The Economy of the Earth: Philosophy, Law, and the Environment* (Cambridge, U.K.: Cambridge University Press, 1988).

Salomon-Bayet, Claire, ed., *Pasteur et la révolution pastorienne* (Paris: Payot, 1986).

Samelson, Franz, "Was Mental Testing (a) Racist Inspired, (b) Objective Science, (c) a Technology for Democracy, (d) the Origin of Multiple-Choice Exams, (e) None of the Above? (Mark the Right Answer)," in Sokal, *Psychological Testing*, 113–127.

Schabas, Margaret, *A World Ruled by Number: William Stanley Jevons and the Rise of Mathematical Economics* (Princeton, N.J.: Princeton University Press, 1989).

Schaffer, Simon, "Glass Works: Newton's Prisms and the Uses of Experiments," in Gooding et al., *Uses of Experiment*, 67–104.

———, "Astronomers Mark Time: Discipline and the Personal Equation," *Science in Context*, 2 (1988), 115–145.

———, "Late Victorian Metrology and Its Instrumentation: A Manufactory of Ohms," in Robert Bud and Susan E. Cozzens, eds., *Invisible Connections: Instruments, Institutions, and Science* (Bellingham, Wash.: SPIE Optical Engineering Press, 1992), 23–56.

Schiesl, Martin J., *The Politics of Efficiency: Municipal Administration and Reform in America* (Berkeley: University of California Press, 1977).

Schneider, Ivo, "Forms of Professional Activity in Mathematics before the Nineteenth Century," in Herbert Mehrtens, H. Bos, and I. Schneider, eds., *Social History of Nineteenth-Century Mathematics* (Boston: Birkhäuser, 1981), 89–110.

———, "Maß und Messen bei den Praktikern der Mathematik vom 16. bis 19. Jahrhundert," in Harald Witthöft et al., eds., *Die historische Metrologie in den Wissenschaften* (St. Katharinen, Switz.: Scripta Mercaturae Verlag, 1986).

Schuster, John A., and Richard R. Yeo, eds., *The Politics and Rhetoric of Scientific Method* (Dordrecht, Holland: Reidel, 1986).

Select Committee on Assurance Associations (SCAA), *Report*, British Parliamentary Papers, 1853, vol. 21.

Self, Peter, *Econocrats and the Policy Process: The Politics and Philosophy of Cost-Benefit Analysis* (London: Macmillan, 1975).

Sellers, Christopher, "The Public Health Service's Office of Industrial Medicine," *Bulletin of the History of Medicine*, 65 (1991), 42–73.

Servos, John, "Mathematics and the Physical Sciences in America," *Isis*, 77 (1986), 611–629.

Sewell, William, *Work and Revolution in France: The Language of Labor from the Old Regime to 1848* (New York: Cambridge University Press, 1980).

Shabman, Leonard A., "Water Resources Management: Policy Economics for an Era of Transitions," *Southern Journal of Agricultural Economics*, July 1984, 53–65.

Shallat, Todd, "Engineering Policy: The U.S. Army Corps of Engineers and the Historical Foundation of Power," *The Public Historian*, 11 (1989), 7–27.

Shapin, Steven, *A Social History of Truth: Gentility, Credibility, and Scientific Knowledge in Seventeenth-Century England* (Chicago: University of Chicago Press, 1994).

Shapin, Steven, and Simon Schaffer, *Leviathan and the Air-Pump: Hobbes, Boyle, and the Experimental Life* (Princeton, N.J.: Princeton University Press, 1985).

Sharp, Walter Rice, *The French Civil Service: Bureaucracy in Transition* (New York: Macmillan, 1931).

Shinn, Terry, *Savoir scientifique et pouvoir social: L'Ecole Polytechnique (1794–1914)* (Paris: Presses de la Fondation Nationale des Sciences Politiques, 1980).

Simon, Marion J., *The Panama Affair* (New York: Charles Scribner's Sons, 1971).

Sklar, Kathryn Kish, "*Hull House Maps and Papers*: Social Science as Women's Work in the 1890s," in Bulmer et al., *Social Survey*, 111–147.

Skrowonek, Stephen, *Building a New American State: The Expansion of National Administrative Capacities, 1877–1920* (Cambridge, U.K.: Cambridge University Press, 1982).

Sloan, Alfred P., Jr., *My Years with General Motors* (Garden City, N.Y.: Doubleday, 1964).

Smith, Adam, *The Wealth of Nations*, 2 vols. (1776; New York: Dutton, 1971).

Smith, Cecil O., Jr., "The Longest Run: Public Engineers and Planning in France," *American Historical Review*, 95 (1990), 657–692.

Smith, Crosbie, and M. Norton Wise, *Industry and Empire: A Biographical Study of Lord Kelvin* (Cambridge, U.K.: Cambridge University Press, 1989).

Smith, Roger, and Brian Wynne, "Introduction," to Smith and Wynne, *Expert Evidence*, 1–22.

———, eds., *Expert Evidence: Interpreting Science in the Law* (London: Routledge, 1989).

Smith, V. Kerry, ed., *Environmental Policy under Reagan's Executive Order: The Role of Benefit-Cost Analysis* (Chapel Hill: University of North Carolina Press, 1984).

Sokal, Michael M., ed., *Psychological Testing and American Society, 1890–1930* (New Brunswick, N.J.: Rutgers University Press, 1987).

Starr, Paul, *The Social Transformation of American Medicine* (New York: Basic Books, 1982).

———, "The Sociology of Official Statistics," in Alonso and Starr, *Politics of Numbers*, 7–57.

Statistical Society of Paris, Excerpt from statutes, *Journal de la Société de Statistique de Paris*, 1 (1860), 7–9.

Stechl, Peter, "Biological Standardization of Drugs before 1928" (Ph.D. dissertation, University of Wisconsin, 1969).

Stigler, George, *Memoirs of an Unregulated Economist* (New York: Basic Books, 1988).

Stigler, Stephen M., *The History of Statistics: The Measurement of Uncertainty before 1900* (Cambridge, Mass.: Harvard University Press, 1986).

Stine, Jeffrey K., "Environmental Politics in the American South: The Fight over the Tennessee-Tombigbee Waterway," *Environmental History Review*, 15 (1991), 1–24.

Suleiman, Ezra N., *Politics, Power, and Bureaucracy in France: The Administrative Elite* (Princeton, N.J.: Princeton University Press, 1974).

———, *Elites in French Society: The Politics of Survival* (Princeton, N.J.: Princeton University Press, 1978).

———, "From Right to Left: Bureaucracy and Politics in France," in Suleiman, ed., *Bureaucrats and Policy Making: A Comparative Overview* (New York: Holmes and Meier, 1985), 107–135.

Supple, Barry, *Royal Exchange Assurance: A History of British Assurance: 1720–1970* (Cambridge, U.K.: Cambridge University Press, 1970).

Sutherland, Gillian, *Ability, Merit and Measurement: Mental Testing and English Education, 1880–1940* (Oxford: Clarendon Press, 1984).

Sutherland, Ian, "The Statistical Requirements and Methods," in Hill, *Controlled Clinical Trials*, 47–51.

Swijtink, Zeno, "The Objectification of Observation: Measurement and Statistical Methods in the Nineteenth Century," in Krüger et al., *Probabilistic Revolution*, vol. 1, 261–285.

Tarbé de Saint-Hardouin, *Quelques mots sur M. Dupuit* (Paris: Dunod, 1868).

Tarbé des Vauxclairs, M. le chevalier, *Dictionnaire des travaux publics, civils, militaires et maritimes* (Paris: Carillan-Goeury, 1835).

Tavernier, René, "Note sur l'exploitation des grandes compagnies et la nécessité de réformes décentralisatrices," *Annales des Ponts et Chaussées* [6] 15 (1888), 637–683.

———, "Note sur les principes de tarification et d'exploitation du trafic voyageurs," *Annales des Ponts et Chaussées* [6], 18 (1889), 559–654.

Teisserenc, Edmond, "Des principes généraux qui doivent présider au choix des tracés des chemins de fer: Observations sur le rapport présenté par M. Le Comte Daru au nom de la sous-commission supérieure d'enquête," extrait de la *Revue indépendante*, September 10, 1843, 6–8.

Temin, Peter, *Taking Your Medicine: Drug Regulation in the United States* (Cambridge, Mass.: Harvard University Press, 1980)

———, ed., *Inside the Business Enterprise* (Chicago: University of Chicago Press, 1991).

Terrall, Mary, "Maupertuis and Eighteenth-Century Scientific Culture" (Ph.D. dissertation, University of California, Los Angeles, 1987).

———, "Representing the Earth's Shape: The Polemics Surrounding Maupertuis's Expedition to Lapland," *Isis*, 83 (1992), 218–237.

Tézenas du Montcel, A., and C. Gérentet, *Rapport de la commission des tarifs de chemins de fer* (Saint-Etienne, France: Imprimerie Théolier Frères, 1877).

Thévenez, René, *Legislation des chemins de fer et des tramways* (Paris: H. Dunod et E. Pinat, 1909).

Thévenot, Laurent, "La Politique des statistiques: Les Origines des enquêtes de mobilité sociale," *Annales: Economies, sociétés, civilisations*, no. 6 (1990), 1275–1300.

Thoenig, Jean-Claude, *L'ère des technocrates: Le cas des Ponts et Chaussées* (Paris: Editions d'Organisation, 1973).

Thompson, E. P., "Time, Work Discipline, and Industrial Capitalism," *Past and Present*, 38 (December 1967), 56–97.

Thuillier, Guy, *Bureaucratie et bureaucrates en France au XIXe siècle* (Geneva: Librairie Droz, 1980).

Todhunter, Isaac, ed., *William Whewell, D.D., An Account of his Writings*, 2 vols. (London: Macmillan, 1876).

Tompkins, H., "Remarks upon the present state of Information relating to the Laws of Sickness and Mortality . . .," *Assurance Magazine*, 3 (1852–53), 7–15; "Editorial Note," ibid., 15–17.

Traweek, Sharon, *Beamtimes and Lifetimes: The World of High Energy Physicists* (Cambridge, Mass.: Harvard University Press, 1988).

Trebilcock, Clive, *Phoenix Assurance and the Development of British Insurance*, vol. 1: 1782–1870 (Cambridge, U.K.: Cambridge University Press, 1985).

Trevan, J. W., "The Error of Determination of Toxicity," *Proceedings of the Royal Society of London*, B, 101 (July 1927), 483–514.

Tribe, Lawrence, "Trial by Mathematics: Precision and Ritual in the Legal Process," *Harvard Law Review*, 84 (1971), 1329–1393, 1801–1820.

Tudesq, André-Jean, *Les grands notables en France (1840–1849): Etude historique d'une psychologie sociale*, 2 vols. (Paris: Presses Universitaires de France, 1964).

Turhollow, Anthony F., *A History of the Los Angeles District, U.S. Army Corps of Engineers* (Los Angeles: Los Angeles District, Corps of Engineers, 1975).

Vogel, David, "The 'New' Social Regulation in Historical and Comparative Perspective," in McCraw, *Regulation*, 155–185.

Von Mayrhauser, Richard T., "The Manager, the Medic,and the Mediator: The Clash of Professional Styles and the Wartime Origins of Group Mental Testing," in Sokal, *Psychological Testing*, 128–157.

Wagner, John W., "Defining Objectivity in Accounting," *Accounting Review*, 40 (1965), 599–605.

Ward, Stephen H., "Treatise on the Medical Estimate of Life for Life Assurance," *Assurance Magazine*, 8 (1858–60), 248–263, 329–343.

Warner, John Harley, *The Therapeutic Perspective: Medical Practice, Knowledge, and Identity in America, 1820–1885* (Cambridge, Mass.: Harvard University Press, 1986).

Weber, Eugen, *Peasants into Frenchmen: The Modernization of Rural France, 1870–1914* (Stanford, Calif.: Stanford University Press, 1976).

Weber, Max, *Economy and Society*, Guenther Ross and Claus Wittich, eds., 2 vols. (Berkeley: University of California Press, 1978).

Weisbrod, Burton A., *Economics of Public Health: Measuring the Economic Impact of Diseases* (Philadelphia: University of Pennsylvania Press, 1961).

———, "Costs and Benefits of Medical Research: A Case Study of Poliomyelitis," in *Benefit-Cost Analysis: An Aldine Annual, 1971* (Chicago: Aldine-Atherton, 1972), 142–160.

Weiss, John H., *The Making of Technological Man: The Social Origins of French Engineering Education* (Cambridge, Mass.: MIT Press, 1982).

———, "Bridges and Barriers: Narrowing Access and Changing Structure in the French Engineering Profession, 1800–1850," in Geison, *Professions*, 15–65.

Weiss, John H., "Careers and Comrades," unpublished manuscript.

Weisz, George, *The Emergence of Modern Universities in France, 1863–1914* (Princeton, N.J.: Princeton University Press, 1983).

———, "Academic Debate and Therapeutic Reasoning in Mid-19th Century France," in Ilana Löwy et al., eds., *Medicine and Change: Historical and Sociological Studies of Medical Innovation* (Paris and London: John Libbey Eurotext, 1993).

Westbrook, Robert B., *John Dewey and American Democracy* (Ithaca, N.Y.: Cornell University Press, 1991).

Whewell, William, "Mathematical Exposition of some of the leading Doctrines in Mr. Ricardo's 'Principles of Political Economy and Taxation,'" reprinted in Whewell, *Mathematical Exposition of Some Doctrines of Political Economy* (1831; New York: Augustus M. Kelley, 1971).

———, review of Richard Jones, *An Essay on the Distribution of Wealth and on the Sources of Taxation*, *The British Critic*, 10 (1831), 41–61.

White, Gilbert F., "The Limit of Economic Justification for Flood Protection," *Journal of Land and Public Utility Economics*, 12 (1936), 133–148.

White, James Boyd, "Rhetoric and Law: The Arts of Cultural and Communal Life," in Nelson et al., *Rhetoric*, 298–318.

Wiebe, Robert, *The Search for Order* (New York: Hill and Wang, 1967).

Wiener, Martin J., *Reconstructing the Criminal: Culture, Law, and Policy in England, 1830–1914* (Cambridge, U.K.: Cambridge University Press, 1990).

Williams, Alan, "Cost-Benefit Analysis: Bastard Science? And/Or Insidious Poison in the Body Politick," *Journal of Public Economics*, 1 (1972), 199–225.

Williams, L. Pearce, "Science, Education, and Napoleon I," *Isis*, 47 (1956), 369–382.

Wilson, James Q., *Bureaucracy: What Government Agencies Do and Why They Do It* (New York: Basic Books, 1989).

Wise, M. Norton, "Work and Waste: Political Economy and Natural Philosophy in Nineteenth-Century Britain," *History of Science*, 27 (1989), 263–317, 391–449; 28 (1990), 221–261.

———, "Exchange Value: Fleeming Jenkin Measures Energy and Utility," unpublished manuscript.

———, ed., *The Values of Precision* (Princeton, N.J.: Princeton University Press, 1995).

Wise, M. Norton, and Crosbie Smith, "The Practical Imperative: Kelvin Challenges the Maxwellians," in Robert Kargon and Peter Achinstein, eds., *Kelvin's Baltimore Lectures and Modern Theoretical Physics* (Cambridge, Mass.: MIT Press, 1987), 324–348.

Wojdak, Joseph F., "Levels of Objectivity in the Accounting Process," *Accounting Review*, 45 (1970), 88–97.

Wolman, Abel, Louis R. Howson, and R. T. Veatch, *Flood Protection in Kansas River Basin* (Kansas City: Kansas Board of Engineers, May, 1953).

Wood, Gordon S., *The Radicalism of the American Revolution* (New York: Alfred A. Knopf, 1992).

Worster, Donald, *Nature's Economy* (Cambridge, Mass.: Cambridge University Press, 1985).

———, *Rivers of Empire: Water, Aridity, and the Growth of the American West* (New York: Pantheon, 1985).

Wynne, Brian, *Rationality and Ritual: The Windscale Inquiry and Nuclear Decisions in Britain* (Chalfont St. Giles, U.K.: British Society for the History of Science, 1982).

———, "Establishing the Rules of Laws: Constructing Expert Authority," in Smith and Wynne, *Expert Evidence*, 23–55.

Yeo, Richard, "Scientific Method and the Rhetoric of Science in Britain, 1830–1917," in Schuster and Yeo, *Politics and Rhetoric*, 259–297.

Zahar, Elie, "Einstein, Meyerson, and the Role of Mathematics in Physical Discovery," *British Journal for the Philosophy of Science*, 31 (1980), 1–43.

Zeff, Stephen A., "Some Junctures in the Evolution of the Process of Establishing Accounting Principles in the USA: 1917–1972," *Accounting Review*, 59 (1984), 447–468.

———, ed., *Accounting Principles through the Years: The Views of Professional and Academic Leaders, 1938–1954* (New York: Garland, 1982).

Zeldin, Theodore, *France, 1848–1945*, vol. 1: *Ambition, Love and Politics* (Oxford: Clarendon Press, 1973).

———, *France, 1848–1945*, vol. 2: *Intellect, Taste, and Anxiety* (Oxford: Clarendon Press, 1977).

Zenderland, Leila, "The Debate over Diagnosis: Henry Goddard and the Medical Acceptance of Intelligence Testing," in Sokal, *Psychological Testing*, 46–74.

Ziman, John, *Reliable Knowledge: An Exploration of the Grounds for Belief in Science* (Cambridge, U.K.: Cambridge University Press, 1978).

Zinoviev, Alexander, *Homo Sovieticus*, Charles Janson, trans. (Boston: Atlantic Monthly Press, 1985).

Zwerling, Craig, "The Emergence of the Ecole Normale Supérieure as a Centre of Scientific Education in the Nineteenth Century," in Fox and Weisz, *Organization*.

Zylberberg, André, *L'économie mathématique en France, 1870–1914* (Paris: Economica, 1990).

Index

THEODORE PORTER IS ASSOCIATE PROFESSOR OF HISTORY AT THE UNIVERSITY OF CALIFORNIA, LOS ANGELES. HE IS THE AUTHOR OF *THE RISE OF STATISTICAL THINKING, 1820–1900* (PRINCETON, 1986), AND A CO-AUTHOR OF *THE EMPIRE OF CHANCE: HOW PROBABILITY CHANGED SCIENCE AND EVERYDAY LIFE* (CAMBRIDGE, 1989).